The

Molecular World

Elements of the p Block

edited by Charlie Harding,
Rob Janes and David Johnson

This publication forms part of an Open University course, S205 *The Molecular World*. Most of the texts which make up this course are shown opposite. Details of this and other Open University courses can be obtained from the Call Centre, PO Box 724, The Open University, Milton Keynes MK7 6ZS, United Kingdom: tel. +44 (0)1908 653231, e-mail ces-gen@open.ac.uk

Alternatively, you may visit the Open University website at http://www.open.ac.uk where you can learn more about the wide range of courses and packs offered at all levels by The Open University.

The Open University, Walton Hall, Milton Keynes, MK7 6AA

First published 2002

Edited, designed and typeset by The Open University.

Published by the Royal Society of Chemistry, Thomas Graham House, Science Park, Milton Road, Cambridge CB4 0WF, UK.

Printed in the United Kingdom by Bath Press Colourbooks, Glasgow.

ISBN 0 85404 690 9

A catalogue record for this book is available from the British Library.

1.1

s205book 9 i1.1

The Molecular World

This series provides a broad foundation in chemistry, introducing its fundamental ideas, principles and techniques, and also demonstrating the central role of chemistry in science and the importance of a molecular approach in biology and the Earth sciences. Each title is attractively presented and illustrated in full colour.

The Molecular World aims to develop an integrated approach, with major themes and concepts in organic, inorganic and physical chemistry, set in the context of chemistry as a whole. The examples given illustrate both the application of chemistry in the natural world and its importance in industry. Case studies, written by acknowledged experts in the field, are used to show how chemistry impinges on topics of social and scientific interest, such as polymers, batteries, catalysis, liquid crystals and forensic science. Interactive multimedia CD-ROMs are included throughout, covering a range of topics such as molecular structures, reaction sequences, spectra and molecular modelling. Electronic questions facilitating revision/consolidation are also used.

The series has been devised as the course material for the Open University Course S205 *The Molecular World*. Details of this and other Open University courses can be obtained from the Course Information and Advice Centre, PO Box 724, The Open University, Milton Keynes MK7 6ZS, UK; Tel +44 (0)1908 653231; e-mail: ces-gen@open.ac.uk. Alternatively, the website at www.open.ac.uk gives more information about the wide range of courses and packs offered at all levels by The Open University.

Further information about this series is available at www.rsc.org/molecularworld.

Orders and enquiries should be sent to:

Sales and Customer Care Department, Royal Society of Chemistry, Thomas Graham House, Science Park, Milton Road, Cambridge, CB4 0WF, UK

Tel: +44 (0)1223 432360; Fax: +44 (0)1223 426017; e-mail: sales@rsc.org

The titles in *The Molecular World* series are:

THE THIRD DIMENSION
    edited by Lesley Smart and Michael Gagan

METALS AND CHEMICAL CHANGE
    edited by David Johnson

CHEMICAL KINETICS AND MECHANISM
    edited by Michael Mortimer and Peter Taylor

MOLECULAR MODELLING AND BONDING
    edited by Elaine Moore

ALKENES AND AROMATICS
    edited by Peter Taylor and Michael Gagan

SEPARATION, PURIFICATION AND IDENTIFICATION
    edited by Lesley Smart

ELEMENTS OF THE p BLOCK
    edited by Charles Harding, David Johnson and Rob Janes

MECHANISM AND SYNTHESIS
    edited by Peter Taylor

The Molecular World Course Team

Course Team Chair
Lesley Smart

Open University Authors
Eleanor Crabb (Book 8)
Michael Gagan (Book 3 and Book 7)
Charles Harding (Book 9)
Rob Janes (Book 9)
David Johnson (Book 2, Book 4 and Book 9)
Elaine Moore (Book 6)
Michael Mortimer (Book 5)
Lesley Smart (Book 1, Book 3 and Book 8)
Peter Taylor (Book 5, Book 7 and Book 10)
Judy Thomas (*Study File*)
Ruth Williams (skills, assessment questions)
*Other authors whose previous contributions to the earlier
courses S246 and S247 have been invaluable in the
preparation of this course:* Tim Allott, Alan Bassindale, Stuart
Bennett, Keith Bolton, John Coyle, John Emsley, Jim Iley, Ray
Jones, Joan Mason, Peter Morrod, Jane Nelson, Malcolm
Rose, Richard Taylor, Kiki Warr.

Course Manager
Mike Bullivant

Course Team Assistant
Debbie Gingell

Course Editors
Pat Forster
Ian Nuttall
Bina Sharma
Dick Sharp
Peter Twomey
Pamela Wardell

CD-ROM Production
Andrew Bertie
Greg Black
Matthew Brown
Philip Butcher
Chris Denham
Spencer Harben
Peter Mitton
David Palmer

BBC
Rosalind Bain
Stephen Haggard
Melanie Heath
Darren Wycherley
Tim Martin
Jessica Barrington

Course Reader
Cliff Ludman

Course Assessor
Professor Eddie Abel, University of Exeter

Audio and Audiovisual recording
Kirsten Hintner
Andrew Rix

Design
Steve Best
Carl Gibbard
Sarah Hack
Lee Johnson
Mike Levers
Sian Lewis
Jenny Nockles
Jon Owen
John Taylor
Howie Twiner
Liz Yeomans

Library
Judy Thomas

Picture Researchers
Lydia Eaton
Deana Plummer

Technical Assistance
Brandon Cook
Pravin Patel

Consultant Authors
Ronald Dell (*Case Study:* Batteries and Fuel Cells)
Adrian Dobbs (Book 8 and Book 10)
Chris Falshaw (Book 10)
Andrew Galwey (*Case Study:* Acid Rain)
Guy Grant (*Case Study:* Molecular Modelling)
Alan Heaton (*Case Study:* Industrial Organic Chemistry,
    Case Study: Industrial Inorganic Chemistry)
Bob Hill (*Case Study:* Polymers and Gels)
Roger Hill (Book 10)
Anya Hunt (*Case Study:* Forensic Science)
Corrie Imrie (*Case Study:* Liquid Crystals)
Clive McKee (Book 5)
Bob Murray (*Study File*, Book 11)
Andrew Platt (*Case Study:* Forensic Science)
Ray Wallace (*Study File*, Book 11)
Craig Williams (*Case Study:* Zeolites)

CONTENTS

ELEMENTS OF THE p BLOCK

Charlie Harding, Rob Janes and David Johnson

CASE STUDY:
ACID RAIN: SULFUR AND POWER GENERATION

Andrew Galwey

CASE STUDY: INDUSTRIAL INORGANIC CHEMISTRY

Alan Heaton and Rob Janes

Elements of the p Block

edited by Charlie Harding,
Rob Janes and David Johnson

Based on 'The Chemistry of Hydrogen, the Halogens and
the Noble Gases' by David Johnson (1994) and
'The Chemistry of Groups III-VI'
by David Johnson and Lesley Smart (1994)

INTRODUCTION

1

Chemistry is an immensely diverse subject. Nowhere is this more evident than in the properties of the p-Block elements, those in which the outer electronic configuration is s^2p^x. In this Book, we will study the typical elements, focusing on the p-Block elements (Figure 1.1).

IUPAC numbering

Group	1	2	13	14	15	16	17	18

Mendeléev numbering

Group	I	II	III	IV	V	VI	VII	VIII
	s^1	s^2	s^2p^1	s^2p^2	s^2p^3	s^2p^4	s^2p^5	s^2 or s^2p^6

Period								
1								2 He
2	3 Li	4 Be	5 B	6 C	7 N	8 O	9 F	10 Ne
3	11 Na	12 Mg	13 Al	14 Si	15 P	16 S	17 Cl	18 Ar
4	19 K	20 Ca	31 Ga	32 Ge	33 As	34 Se	35 Br	36 Kr
5	37 Rb	38 Sr	49 In	50 Sn	51 Sb	52 Te	53 I	54 Xe
6	55 Cs	56 Ba	81 Tl	82 Pb	83 Bi	84 Po	85 At	86 Rn
7	87 Fr	88 Ra						

Figure 1.1
A periodic arrangement of the typical or main-Group elements, showing the p-Block elements in green. Note that hydrogen has been omitted. Its classification is discussed in Section 5.

Much of the chemistry of the metallic s-Block elements is consistent with an ionic model. With the p-Block elements, this is no longer the case: the proportion of obviously covalent substances is much larger. Before we begin our study of these elements, it is necessary to introduce some essential principles, mainly concerning bonding and thermochemistry, which apply particularly to covalent substances. Following this, we consider the chemistry of the Groups of p-Block elements, beginning with the halogens and noble gases, as these Groups most clearly display covalent properties. These are the two Groups that contain no metallic elements.

1.1 Group numbers and the Periodic Table

Because this Book concentrates upon the typical elements, we shall make particular use of the mini-Periodic Table of Figure 1.1 that contains the typical elements alone. In Figure 1.1, the Groups are numbered in two different ways. In the Mendeléev numbering scheme, which is given in Roman numerals, the Group numbers increase smoothly across each row from I to VIII. These numbers are useful because, for each element, they are almost always equal to the number of outer electrons in the atom, and to the highest oxidation state (see Section 2.1) that is reached in the compounds that the element forms. Use is made of these relationships in this Book.

Most modern textbooks and chemistry research journals, however, use a numbering system recommended by IUPAC (International Union of Pure and Applied Chemistry). In this system, the Groups are numbered in Arabic numerals as Groups 1–18 (Figure 1.2). The p-Block elements of Groups III–VIII are, therefore, numbered as Groups 13–18. Because both numbering schemes are important, we include both alternatives in the headings of the Sections and the figures describing the Groups in this Book. Thus the Group of elements headed by boron is designated Group III/13. Notice that for the typical elements of Figure 1.1, the second digit of a Group number in the IUPAC scheme is equal to the Mendeléev number. This provides an easy way of shifting from one system to another.

IUPAC numbering

Group	1	2	3	4	5	6	7	8	9	10	11	12	13	14	15	16	17	18

Mendeléev numbering

Group	I	II											III	IV	V	VI	VII	VIII
1							1 H											2 He
2	3 Li	4 Be											5 B	6 C	7 N	8 O	9 F	10 Ne
3	11 Na	12 Mg											13 Al	14 Si	15 P	16 S	17 Cl	18 Ar
4	19 K	20 Ca	21 Sc	22 Ti	23 V	24 Cr	25 Mn	26 Fe	27 Co	28 Ni	29 Cu	30 Zn	31 Ga	32 Ge	33 As	34 Se	35 Br	36 Kr
5	37 Rb	38 Sr	39 Y	40 Zr	41 Nb	42 Mo	43 Tc	44 Ru	45 Rh	46 Pd	47 Ag	48 Cd	49 In	50 Sn	51 Sb	52 Te	53 I	54 Xe
6	55 Cs	56 Ba	71 Lu	72 Hf	73 Ta	74 W	75 Re	76 Os	77 Ir	78 Pt	79 Au	80 Hg	81 Tl	82 Pb	83 Bi	84 Po	85 At	86 Rn
7	87 Fr	88 Ra	103 Lr	104 Rf	105 Db	106 Sg	107 Bh	108 Hs	109 Mt	110	111	112						

Period (vertical label on left)

Figure 1.2 The arrangement of the typical elements and transition elements in a Periodic Table that generates the IUPAC numbering. The Mendeléev numbering is also shown.

OXIDATION AND REDUCTION

Oxidation may be defined as a loss of electrons, and reduction as a gain of electrons. In a reaction such as

$$Mg(s) + Cl_2(g) = MgCl_2(s) \tag{2.1}$$

the ionic model implies that magnesium dichloride is composed of the ions Mg^{2+} and Cl^-. Consequently, Reaction 2.1 is a redox reaction: magnesium has been oxidized because a magnesium atom loses two electrons; chlorine has been reduced because a chlorine atom gains an electron. However, this argument assumes that $MgCl_2$ conforms to an ionic model. While this may be justified for compounds such as $MgCl_2$, a difficulty arises when we turn to covalent compounds. Consider the reaction between sulfur and chlorine:

$$S(s) + Cl_2(g) = SCl_2(l) \tag{2.2}$$

Reactions 2.1 and 2.2 appear similar, but at room temperature, SCl_2 is a red liquid containing discrete SCl_2 molecules, whose atoms are held together by covalent bonds (Structure **2.1**); there cannot be complete transfer of two electrons per sulfur atom from sulfur to chlorine. Consequently, our definitions of oxidation and reduction prevent the classification of *both* Reactions 2.1 and 2.2 as redox reactions. To obtain a broader definition of oxidation and reduction, we must abandon the implication that complete electron transfer takes place.

2.1 Oxidation numbers or oxidation states

The solution that has been devised for this problem is to *imagine* that the shared electron pairs in covalent bonds between different elements are completely transferred to *one* of the two bound atoms. The transfer is assumed to occur to the more electronegative of the two.

- Consider Structure **2.1**. Which is the more electronegative type of atom?

- Chlorine is more electronegative than sulfur; it lies to the right of sulfur in Period 3 (Figure 1.1).

If the shared electron pair in each bond in Structure **2.1** were transferred to chlorine, each chlorine would have eight outer electrons, one more than in the free atom. It would therefore become Cl^-, and carry a charge of -1.

- What number of outer electrons would the sulfur be left with, and what charge would it carry?

- Sulfur would be left with four outer electrons, two less than in the free atom, and would have a charge of $+2$.

2.1

These imaginary charges are called **oxidation numbers** or **oxidation states**. The oxidation number of an atom in its elemental form, e.g. graphite, aluminium metal and chlorine gas, is taken to be zero.

> Oxidation is now defined as an increase in oxidation number, and reduction as a decrease in oxidation number.

⬤ Why is Reaction 2.2 a redox reaction? What has been oxidized and reduced in it?

⬤ Sulfur has been oxidized because its oxidation number has increased from zero to +2; chlorine has been reduced because its oxidation number has decreased from zero to −1.

To deduce the oxidation numbers of the elements in a compound or ion, it is unnecessary to write down a Lewis structure, as described above. Instead a set of rules allows the rapid assignment of oxidation numbers in most of the substances that you will normally encounter (Box 2.1).

BOX 2.1 Useful rules for the assignment of oxidation numbers

1   The oxidation number of a monatomic ion is equal to the charge on the ion. Thus, the oxidation numbers of iron and chlorine in Fe^{2+}(aq) and Cl^-(aq) are +2 and −1, respectively.

2   The oxidation number of an atom in its elemental form is zero. Thus, the values for aluminium and bromine atoms in Al(s) and Br_2(l) are zero.

3   The oxidation number of fluorine in compounds is always −1.

4   The oxidation number of oxygen in compounds is −2, except when it is bound to fluorine or, as in compounds such as peroxides, to other oxygen atoms. Thus, in CO_2, the oxygen atom has an oxidation number of −2.

5   The oxidation number of hydrogen in compounds is taken to be +1, except in metallic hydrides such as NaH, where it is −1. Thus, in HCl, the oxidation number of hydrogen is +1.

6   The oxidation number of chlorine, bromine or iodine in compounds is −1, except in the compounds that they form with oxygen and with other halogen atoms. Thus, in HCl the oxidation number of chlorine is −1.

7   The sum of the oxidation numbers of the atoms in a compound or ion is equal to the charge on that compound or ion. Thus, HCl has no charge, and the oxidation numbers of H and Cl are +1 and −1, respectively.

Remember that only in ionic substances does the oxidation number of an element equal the charge that the element carries. In NaCl, which is regarded as an ionic compound, we do indeed take the charge carried by sodium to be equal to its oxidation number of +1.

But in the covalent compound SCl_2, you have seen that the sulfur oxidation number of +2 was obtained by an imaginary distortion of the distribution of electrons, so that the shared electron pairs in the covalent bonds are transferred completely to the more electronegative chlorine. The charge carried by sulfur in the real SCl_2 molecule must therefore be less than the oxidation number of +2. To emphasize this, oxidation numbers are often written as Roman numerals in parentheses after the element. Thus, SCl_2 is written as sulfur(II) chloride, and SO_2 as sulfur(IV) oxide. Oxidation numbers can be used to balance redox equations, which we consider next.

2.2 Balancing redox equations

Much of the chemistry of the p-Block elements involves change in oxidation number. For example, in both of the reactions given above (Reactions 2.1 and 2.2), one element, Mg or S, is oxidized and another, Cl, is reduced. Balancing equations of this kind is often a trivial exercise which can be done by inspection. Consider the reaction of antimony with iodine, in which antimony is oxidized and iodine is reduced. The equation (2.3) is balanced simply by ensuring that the ratio Sb : I is 1 : 3 while avoiding the inconvenience of including fractions:

$$2Sb + 3I_2 = 2SbI_3 \tag{2.3}$$

For more complicated reactions, we can use oxidation numbers to obtain a properly balanced equation, as shown below.

A convenient method of preparing a small amount of chlorine gas in the laboratory is the reaction of potassium permanganate, $KMnO_4$, with concentrated hydrochloric acid. As the chloride ion, $Cl^-(aq)$, is oxidized to chlorine gas, the permanganate ion, $MnO_4^-(aq)$, is reduced to give $Mn^{2+}(aq)$ ions. A preliminary equation for this reaction, in which only the element being oxidized and the element being reduced are balanced, is

$$MnO_4^-(aq) + Cl^-(aq) = Mn^{2+}(aq) + \tfrac{1}{2}Cl_2(g) \tag{2.4}$$

The *first step* is to use the rules in Box 2.1 to calculate the oxidation numbers of the elements which are oxidized or reduced, in this case manganese and chlorine:

Rule 1 tells us that in $Mn^{2+}(aq)$ and $Cl^-(aq)$, the oxidation numbers are +2 and −1, respectively.

Rule 2 tells us that in $Cl_2(g)$, the oxidation number is zero.

What is the oxidation number of manganese in $MnO_4^-(aq)$?

+7. By rule 4, we assign an oxidation number of −2 to each of four oxygen atoms. The oxidation number of manganese must be +7, because when 7 is added to −8, it gives −1, which is the charge on $MnO_4^-(aq)$ (rule 7).

We can now write Equation 2.4 with the oxidation numbers in place:

$$MnO_4^-(aq) + Cl^-(aq) = Mn^{2+}(aq) + \tfrac{1}{2}Cl_2(g) \tag{2.4}$$
$$+7 \qquad\quad -1 \qquad\quad +2 \qquad\quad 0$$

*The symbol ⌨ indicates that this reaction can be viewed in 'The p-Block elements in action' on one of the CD-ROMs accompanying this Book.

Notice that as the equation stands, the oxidation number of manganese falls from +7 to +2, and the oxidation number of chlorine rises from −1 to zero. Notice too, that the total change in oxidation number is −4, because the value for manganese falls by five and that for chlorine rises by one. The following rule now applies:

> In a balanced equation, the total change in oxidation number is zero.

So, in the *second step* we adjust the ratio $MnO_4^- : Cl^-$ to make this so.

● What number of $Cl^-(aq)$ ions must be written to make this true?

○ Five. If each of five chlorine ions increases its oxidation state by one, giving +5 in all, this balances the change of −5 for manganese.

The equation now becomes

$$MnO_4^-(aq) + 5Cl^-(aq) = Mn^{2+}(aq) + \tfrac{5}{2}Cl_2(g) \qquad (2.5)$$

In the *third step*, we balance the equation in the elements that are not oxidized or reduced. Since the reaction takes place in aqueous acid solution, we do this by including the necessary $H^+(aq)$ or H_2O. There are four oxygen atoms on the left side (in MnO_4^-), and so four molecules of H_2O must be added to the right-hand side.

$$MnO_4^-(aq) + 5Cl^-(aq) = Mn^{2+}(aq) + \tfrac{5}{2}Cl_2(g) + 4H_2O(l) \qquad (2.6)$$

We now add the appropriate number of hydrogen ions, $H^+(aq)$, to complete the balancing.

● How many $H^+(aq)$ are required?

○ $8H^+(aq)$ must be added to the left-hand side to balance the hydrogen atoms in four water molecules on the right.

$$MnO_4^-(aq) + 5Cl^-(aq) + 8H^+(aq) = Mn^{2+}(aq) + \tfrac{5}{2}Cl_2(g) + 4H_2O(l) \qquad (2.7)$$

For convenience, we double this equation to eliminate fractions:

$$2MnO_4^-(aq) + 10Cl^-(aq) + 16H^+(aq) = 2Mn^{2+}(aq) + 5Cl_2(g) + 8H_2O(l) \qquad (2.8)$$

The equation is now balanced with respect to all elements and oxidation numbers. If we have done this correctly, it should also be balanced with respect to charge. This is the *fourth and final step*.

● Find the total charge on each side of the equation. Is the equation balanced with respect to charge?

○ On the left the charge is: −2 − 10 + 16 = +4. This balances the charge of +4 on the right.

2.3 Summary of Section 2

1   Oxidation states or oxidation numbers may be regarded as charges, usually fictitious, which are assigned to elements in compounds by assuming that shared electron pairs in covalent bonds are completely transferred to one of the two bound atoms. Transfer is assumed to occur to the more electronegative of the two.

2   Oxidation is an increase in oxidation number; reduction is a decrease.

3   Redox reactions may be represented by balanced equations. Oxidation numbers are assigned to those substances involved in the oxidation/reduction. The total change in oxidation number is set to zero by adjusting the numbers of reactants and products. In aqueous solution, H_2O, $H^+(aq)$ or $OH^-(aq)$ ions are included as necessary to balance the number of hydrogen and oxygen atoms.

QUESTION 2.1 ✴

The ten substances listed below contain sulfur. Work out the oxidation number of sulfur in each one. Then present your results in the form of a table, showing the substances in order of increasing sulfur oxidation number.

(i)   solid sulfur, $S(s)$

(ii) sulfide ion, $S^{2-}(aq)$

(iii) sulfur dioxide, $SO_2(g)$

(iv) sulfur trioxide, $SO_3(s)$

(v)   hydrogen sulfide, $H_2S(g)$

(vi) sulfur dichloride, $SCl_2(l)$

(vii) sulfuric acid, $H_2SO_4(l)$

(viii) sulfite ion, $SO_3^{2-}(aq)$

(ix) sulfur hexafluoride, $SF_6(g)$

(x)   hydrogen sulfate ion, $HSO_4^-(aq)$

QUESTION 2.2 ✴

By working out the relevant oxidation numbers, specify which of the following are redox reactions and state which element is oxidized, and which is reduced.

(i) $C(s) + 2Cl_2(g) = CCl_4(l)$

(ii) $N_2(g) + 3H_2(g) = 2NH_3(g)$

(iii) $H^+(aq) + OH^-(aq) = H_2O(l)$

(iv) $2H_2S(g) + SO_2(g) = 3S(s) + 2H_2O(l)$

QUESTION 2.3

During the manufacture of bromine, bromine gas is concentrated by mixing it with sulfur dioxide, and passing the mixture into water, when a concentrated solution of bromide is produced. This can later be re-oxidized with chlorine. An unbalanced equation for the concentrating process is:

$$SO_2(g) + \tfrac{1}{2}Br_2(g) = Br^-(aq) + SO_4^{2-}(aq)$$

By assigning oxidation numbers, identify the elements that are oxidized and reduced. Then balance the equation by the addition of $H_2O(l)$ and $H^+(aq)$.

QUESTION 2.4

Bromine can also be made by adding acid to a mixture of solutions of sodium bromide (NaBr) and sodium bromate ($NaBrO_3$). Starting with the unbalanced equation,

$$BrO_3^-(aq) + Br^-(aq) = Br_2(aq)$$

give a balanced equation by using oxidation numbers, and adding $H_2O(l)$ and $H^+(aq)$.

STUDY NOTE

A set of interactive self-assessment questions is provided on one of the CD-ROMs accompanying this Book. The questions are scored, and you can come back to the questions as many or as few times as you wish in order to improve your score on some or all of them. This is a good way of reinforcing the knowledge you have gained while studying this Book.

DEFINING ACIDS AND BASES

3

The simplest definition of acids and bases is that of Arrhenius: acids dissociate to give $H^+(aq)$ in an aqueous solution; bases dissociate to give $OH^-(aq)$. In this section we consider the strengths of acids and bases, and two examples of attempts to broaden the definitions of an acid and a base.

3.1 Strengths of acids and bases

When the gas HCl dissolves in water, its solution is called hydrochloric acid. According to the Arrhenius theory, it is an acid because the HCl molecules break down to give aqueous H^+ ions:

$$HCl(aq) = H^+(aq) + Cl^-(aq) \tag{3.1}$$

In this reaction the equilibrium lies far to the right, and so hydrochloric acid is called a **strong acid**. Acid strength may be expressed by the equilibrium constant which describes an equation such as 3.1. For acids, the equilibrium constant, which is also referred to as the dissociation constant, is denoted in general by K_a. So for Reaction 3.1,

$$K_a = \frac{[H^+(aq)][Cl^-(aq)]}{[HCl(aq)]} \tag{3.2}$$

For HCl, K_a has a value of about 10^7 mol litre^{-1}, and as $[H^+(aq)] = [Cl^-(aq)]$, an aqueous solution of HCl (1 mol litre^{-1}) contains more than 10^3 as many H^+ ions as HCl molecules. By contrast, when acetic (ethanoic) acid, CH_3COOH, dissolves in water, the equilibrium lies far to the left:

$$CH_3COOH(aq) = CH_3COO^-(aq) + H^+(aq) \tag{3.3}$$

So acetic acid is only slightly dissociated, and is described as a **weak acid** ($K_a = 1.8 \times 10^{-5}$ mol litre^{-1}).

Similarly we recognize strong and weak bases by the extent to which they dissociate. Thus NaOH dissolves in water with virtually total dissociation to give $Na^+(aq)$ and $OH^-(aq)$ ions, and so is referred to as a strong base.

3.2 The Brønsted–Lowry theory

By the Arrhenius definition, an acid such as HCl dissociates to give H^+ ions:

$$HCl(aq) = H^+(aq) + Cl^-(aq) \tag{3.1}$$

This definition does not give much prominence to the solvent: water only appears in the equation as the symbol (aq). Yet the solvent is important; HCl also dissolves in benzene, for example, yet the solution is non-conducting and fails to give characteristic acid tests. This is because it contains solvated HCl molecules rather than ions. By contrast, in aqueous solution, the hydrogen ion is present as the hydrated proton, which we represent as H_3O^+. The realization that the aqueous proton is bound to a solvent water molecule enables us to put this neglect of the solvent to rights.

- Rewrite Equation 3.1 so $H_3O^+(aq)$ rather than $H^+(aq)$ appears on the right-hand side.

- $HCl(aq) + H_2O(l) = H_3O^+(aq) + Cl^-(aq)$ (3.4)

This equation recognizes the solvent by including a solvent molecule on the left-hand side. What happens is that a proton, H^+, is transferred from HCl to H_2O (Figure 3.1). This transfer of protons is the basis of the **Brønsted–Lowry theory**:

An acid is a substance from which a proton can be removed; a base is a substance that can accept a proton from an acid.

Thus, in Equation 3.4, HCl is the acid: removal of a proton leaves the ion $Cl^-(aq)$.

- What acts as a base in Equation 3.4?

- Water; the H_2O molecule accepts a proton and forms H_3O^+.

For the weak acid, acetic acid, by the Brønsted-Lowry theory, we write

$$CH_3COOH(aq) + H_2O(l) = H_3O^+(aq) + CH_3COO^-(aq)$$ (3.5)

Here equilibrium lies well to the left: vinegar, a 6% solution of acetic acid, contains many more CH_3COOH molecules than aqueous hydrogen ions and acetate ions. This means that the right-to-left reaction is dominant. If sodium acetate, a source of CH_3COO^- ions, is dissolved in hydrochloric acid, where $H_3O^+(aq)$ ions are plentiful, the reverse reaction occurs, and $CH_3COOH(aq)$ is formed.

- In this reverse reaction, what is the acid, and what is the base?

- $H_3O^+(aq)$ is the acid; it loses a proton. $CH_3COO^-(aq)$ is the base; it accepts a proton.

Thus, Equations 3.4 and 3.5 have the form:

$$acid(1) + base(2) = acid(2) + base(1)$$ (3.6)

In such a system, the pairs acid(1)/base(1) and acid(2)/base(2) are called a **conjugate acid and base**. Thus, HCl and Cl^- are a conjugate acid and base pair; so are H_3O^+ and H_2O. An acid is transformed into its conjugate base by losing a proton; a base is transformed into its conjugate acid by acquiring a proton (Figure 3.2, overleaf). Now look again at Equation 3.4; HCl is a strong acid, and equilibrium lies to the right.

- Is $Cl^-(aq)$ a strong or weak base?

- Equilibrium lies to the right, so Cl^- shows very little tendency to take up a proton, and so it is a weak base.

This illustrates the general point that the conjugate bases of strong acids are weak; similarly, conjugate acids of strong bases are weak (Figure 3.3, overleaf).

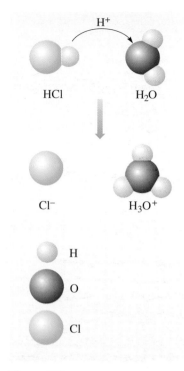

Figure 3.1
According to the Brønsted–Lowry theory, the key feature of the dissolution of HCl in water is the transfer of a proton from a molecule of HCl to a water molecule.

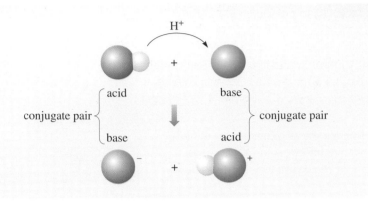

Figure 3.2
According to the Brønsted–Lowry theory, an acid–base reaction produces the conjugate base of the reacting acid and the conjugate acid of the reacting base.

ACID		BASE
$HClO_4$		ClO_4^-
HCl	100% ionized in H_2O	Cl^-
HNO_3	no HX(aq)	NO_3^-
H_2SO_4		HSO_4^-
H_3O^+		H_2O
HF		F^-
HNO_2	equilibrium mixtures of	NO_2^-
CH_3COOH	HX and X^- in aqueous solution	CH_3COO^-
HOCl		OCl^-
NH_4^+		NH_3
H_2O		OH^-
NH_3	100% reacted in H_2O	NH_2^-
H_2	no B^-(aq)	H^-
OH^-		O^{2-}

acid strength increases → (upward)

base strength increases → (downward)

Figure 3.3
The relative strengths of conjugate acid–base pairs in water. The strongest acids ($HClO_4$, HCl, HNO_3, H_2SO_4) appear at the top of the left-hand column; the strongest bases (O^{2-}, H^-, NH_2^-, OH^-) are at the bottom of the right-hand column. An acid will transfer a proton to a base *below* it in the table.

The Arrhenius theory only recognizes the solvent when forced to do so. For example, ammonia is a base: it produces aqueous OH^-. To explain this, the Arrhenius theory invokes a reaction with a water molecule:

$$NH_3(aq) + H_2O(l) = NH_4^+(aq) + OH^-(aq) \qquad (3.7)$$

On the Brønsted–Lowry theory, this is a standard acid–base reaction.

● What are the conjugate acid–base pairs?

● H_2O is an acid; OH^- is its conjugate base. The other acid–base pair is NH_4^+/NH_3.

Notice that the solvent, H_2O, can act as either an acid or a base: it is an acid in Equation 3.7, but a base in Equations 3.4 and 3.5. This dual possibility is most apparent in the familiar self-ionization reaction of water. In Arrhenius notation this is written

$$H_2O(l) = H^+(aq) + OH^-(aq) \qquad (3.8)$$

In Brønsted–Lowry notation it becomes

$$H_2O(l) + H_2O(l) = H_3O^+(aq) + OH^-(aq) \qquad (3.9)$$

One of the two water molecules on the left acts as an acid; the other acts as a base. During the familiar acid–base neutralization reaction, Reaction 3.9 occurs in reverse (Figure 3.4).

The Brønsted–Lowry theory is stated in terms of proton transfer, rather than through ions in aqueous solution, and so it can be applied in non-aqueous systems. Thus, if gaseous ammonia and hydrogen chloride are mixed, a white smoke consisting of the ionic salt, ammonium chloride, is formed

$$NH_3(g) + HCl(g) = NH_4^+Cl^-(s) \qquad (3.10)$$

This is a Brønsted–Lowry acid–base reaction in which two gases combine to form a solid.

● What are the conjugate acid–base pairs?

● HCl is one acid with Cl^- as its conjugate base; the other acid–base pair is NH_4^+/NH_3.

Acetic acid is a weak acid in water (Equation 3.3); however, if it is dissolved in liquid ammonia, it is completely dissociated

$$CH_3COOH(amm) + NH_3(l) = NH_4^+(amm) + CH_3COO^-(amm) \qquad (3.11)$$

● Why is acetic acid more dissociated in liquid ammonia than in water?

● Ammonia is a stronger base than water (Figure 3.3).

So here, virtually every acetic acid molecule is able to transfer a proton to an ammonia molecule acting as the base.

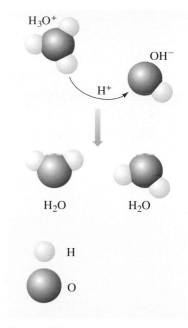

Figure 3.4
On the Brønsted–Lowry theory, the Arrhenius H^+/OH^- neutralization reaction involves proton transfer from H_3O^+ to OH^-, with the production of two water molecules.

3.3 The pH scale and acid strength

Pure water consists mainly of undissociated molecules, H_2O, although a small fraction of the molecules are dissociated, according to Equation 3.8

$$H_2O(l) = H^+(aq) + OH^-(aq) \qquad (3.8)$$

At 25 °C the concentration of $H^+(aq)$ ions, which equals that of $OH^-(aq)$ ions, is 1×10^{-7} mol litre^{-1}. The equilibrium constant for Equation 3.8 is given by:

$$K = \frac{[H^+(aq)][OH^-(aq)]}{[H_2O(l)]} \qquad (3.12)$$

In dilute aqueous solutions, the concentration of water is effectively constant. As K is constant, it follows that, at equilibrium, the product $[H^+(aq)][OH^-(aq)]$ will also be constant. From the special case of pure water, we know that this constant is 1×10^{-14} mol^2 litre^{-2}. So in dilute aqueous solutions at 25 °C, $[H^+(aq)][OH^-(aq)]$ will always have the value 1×10^{-14} mol^2 litre^{-2}.

Consider a dilute solution of hydrochloric acid of concentration 0.1 mol litre^{-1}. Note that the most concentrated hydrochloric acid is more than 100 times this value. In this dilute solution, $[H^+(aq)] = 0.1$ mol litre^{-1}, one million times the value in pure water.

Now consider a dilute solution of sodium hydroxide of the same concentration. In this solution, the hydroxide ion concentration is $[OH^-(aq)] = 1 \times 10^{-1}$ mol litre^{-1}.

● What is the concentration of hydrogen ions in this solution?

● According to the equilibrium constant, Equation 3.12, the product
$[H^+(aq)][OH^-(aq)] = 1 \times 10^{-14}$ mol^2 litre^{-2}.
So $[H^+(aq)] = 1 \times 10^{-14}$ mol^2 litre$^{-2}/1 \times 10^{-1}$ mol litre$^{-1} = 1 \times 10^{-13}$ mol litre^{-1}.

The range of hydrogen ion concentration between the dilute solutions of hydrochloric acid and sodium hydroxide encompasses a factor of 1×10^{12}. Because these concentrations cover such a vast range, even for dilute solutions, a scale has been devised to make the numbers more manageable. Notice that the index in the concentration terms changes from -1 for the dilute hydrochloric acid to -13 for the dilute sodium hydroxide solution. This is the basis of the scale, known as the **pH** scale, in which pH is defined as

$$pH = -\log[H^+(aq)/\text{mol litre}^{-1}] \tag{3.13}$$

For a solution of sodium hydroxide of concentration 0.1 mol litre^{-1}, the pH is therefore

$$pH = -\log(1 \times 10^{-13} \text{ mol litre}^{-1}/\text{mol litre}^{-1})$$
$$pH = -\log(1 \times 10^{-13})$$
$$pH = 13$$

● What is the pH of a solution of hydrochloric acid in which $[H^+] = 1$ mol litre^{-1}?

● Here the pH is given by

$$pH = -\log(1)$$
$$pH = 0$$

More concentrated solutions will have a negative pH.

The effect of the negative sign in the definition of pH is twofold. All except relatively concentrated solutions of acid have a positive pH. The value of the pH increases as the acidity of the solution decreases. Values of pH of some common chemical, natural and household solutions are given in Table 3.1.

● From the data in Table 3.1, which is more acidic: vinegar or gastric juice; blood or seawater?

● pH is defined so that low pH corresponds to high acidity. So gastric juice is more acidic than vinegar, and blood is more acidic than seawater.

Strengths of acids, as measured by K_a (Equation 3.2), span several orders of magnitude. For the strong acid, hydrochloric acid, K_a is about 1×10^7 mol litre^{-1}, whereas for the weak acid, acetic acid, K_a is 1.8×10^{-5} mol litre^{-1}. Again, a logarithmic scale is a convenient way of expressing these large differences, and so acid strength is often described by a quantity **pK_a**, which is defined by Equation 3.14.

$$pK_a = -\log(K_a/\text{mol litre}^{-1}) \tag{3.14}$$

As in the definition of pH, the negative sign indicates that strong acids have low values of pK_a, and vice versa.

● What is the value of pK_a for acetic acid?

● According to Equation 3.14:

$$pK_a = -\log(1.8 \times 10^{-5} \text{ mol litre}^{-1}/\text{mol litre}^{-1})$$

$$pK_a = 4.74$$

Table 3.1 pH values for some common solutions

Solution	pH
hydrochloric acid (0.1 mol litre^{-1})	1.0
gastric juice (human)	1.0–2.5
lemon juice	about 2.1
acetic acid (0.1 mol litre^{-1})*	2.9
orange juice	about 3.0
tomato juice	about 4.1
urine	6.0
rainwater (unpolluted)	5.2–6.5
saliva (human)	6.8
milk	about 6.9
pure water (25 °C)	7.0
blood (human)	7.4
seawater	7.9–8.3
ammonia (0.1 mol litre^{-1})†	11.1
sodium hydroxide (0.1 mol litre^{-1})	13.0

*Approximately the concentration of household vinegar.
†Approximately the concentration of household ammonia solutions.

3.4 Summary of Section 3

1 According to the Brønsted–Lowry theory, an acid is a substance from which a proton can be removed; a base is a substance that can accept a proton from an acid.

2 Conjugate acids and bases are related as shown in Figure 3.2:

conjugate acid = H$^+$ + conjugate base

Acid–base reactions then take the form:

acid(1) + base(2) = acid(2) + base(1)   (3.6)

and occur if acid(1) is stronger than acid(2).

3 For convenience, hydrogen ion concentration is often expressed on a logarithmic scale using the quantity pH.

$$pH = - \log([H^+(aq)]/\text{mol litre}^{-1})$$

Acid strength is similarly expressed on a logarithmic scale, using pK_a.

$$pK_a = - \log(K_a/\text{mol litre}^{-1})$$

QUESTION 3.1

(a) What are the conjugate bases of the acids HI, HClO$_4$ and H$_2$O? (b) What are the conjugate acids of the bases Br$^-$, NO$_3^-$ and NH$_3$? (c) When the ion HSO$_4^-$ acts as a Brønsted–Lowry acid, what is its conjugate base? When it acts as a base, what is its conjugate acid?

QUESTION 3.2

Using Figure 3.3, predict whether a reaction will occur or not when the following acid–base combinations are put into water. (a) Perchloric acid, $HClO_4$, and a solution of sodium fluoride containing aqueous fluoride ions, $F^-(aq)$. (b) Acetic acid, CH_3COOH, and a solution of sodium nitrate containing aqueous nitrate ions, $NO_3^-(aq)$. (c) Water to which sodium hydride, $Na^+ H^-$, is added. Where a reaction occurs, write it out in the form defined by Equation 3.6, and identify the conjugate acid–base pairs.

QUESTION 3.3 ✳

If a dilute solution of acetic acid has a concentration of hydrogen ions of 4×10^{-3} mol litre^{-1}, what is the pH of this solution?

SOME ASPECTS OF CHEMICAL BONDING

4

In this Section, we introduce some new theoretical ideas, which will prove useful in the descriptive chemistry of the non-metals.

4.1 Lewis acids and bases

The Arrhenius definition of an acid and a base attributed acidity to the presence of $H^+(aq)$, and alkalinity to $OH^-(aq)$. Brønsted–Lowry theory generalizes the acid–base concept by focusing on proton transfer, rather than on particular aqueous ions. Here, we discuss an attempt to generalize it further by focusing on the changes in *electronic structure* that occur when acid–base reactions take place, ideas introduced by G. N. Lewis.

When ammonia, a base, accepts a proton from an acid, the process can be written:

$$\text{structure} \tag{4.1}$$

Here the base, ammonia, donates a non-bonding pair of electrons to the proton, and the product is the ammonium ion in which each atom has a noble gas configuration. Lewis suggested that acids should be thought of as non-bonding-pair acceptors, and bases as non-bonding-pair donors; hence the names **Lewis acid** and **Lewis base**. This idea breaks the link between the acid–base concept and hydrogen ion chemistry. For example, BF_3 and NH_3 are colourless gases with trigonal planar and pyramidal molecules (Structures **4.1** and **4.2**, respectively).

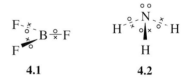

4.1   **4.2**

- ○ By how many electrons is the boron atom in BF_3 short of a noble gas configuration?

- ○ Two electrons give it the structure of neon.

The boron atom acquires these two electrons if BF_3 and NH_3 are mixed:

$$\text{structure} \tag{4.2}$$

A solid compound, H_3NBF_3, is formed as a white smoke (🖳). In this compound the coordination around both the boron and the nitrogen atoms is tetrahedral.

- ○ Identify the Lewis acid and Lewis base in this reaction.

- ○ Ammonia donates the non-bonding pair — it is the Lewis base; BF_3 is the Lewis acid — it accepts the non-bonding pair.

The nitrogen–boron bond in the solid compound is a *dative bond* because both shared electrons come from the nitrogen atom of the ammonia. The bond can therefore be written $H_3N{\rightarrow}BF_3$. An alternative, but precisely equivalent way of writing it is $H_3\overset{+}{N}{-}\overset{-}{B}F_3$. The latter provides a more accurate description. Consider Equation 4.1 in which NH_3 forms a dative bond with H^+. The first representation gives Structure **4.3**, which suggests that one of the H atoms is different from the other three. However, in the new representation (Structure **4.4**), the four N—H bonds are equivalent, consistent with the fact that they are of equal length.

A feature of this idea is that metal complexes are composed of Lewis acids and Lewis bases. Thus, the formation of $[Cu(H_2O)_4]^{2+}$ may be thought of as a combination of the Cu^{2+} ion with four ligand water molecules, which donate their non-bonding pairs to the central copper ion: Cu^{2+} is a Lewis acid; H_2O is a Lewis base.

4.2 The strengths of chemical bonds

The strength of the chemical bond in a diatomic molecule can be expressed by the bond dissociation enthalpy, D. This is equal to the standard enthalpy change, $\Delta H_m^{\ominus}$, for a reaction in which the bond is broken. For the dissociation of hydrogen chloride:

$$HCl(g) = H(g) + Cl(g) \tag{4.3}$$

$$\begin{aligned}\Delta H_m^{\ominus} &= \Delta H_f^{\ominus}(H, g) + \Delta H_f^{\ominus}(Cl, g) - \Delta H_f^{\ominus}(HCl, g)\\ &= [218.0 + 121.7 - (-92.3)]\,\text{kJ mol}^{-1}\\ &= 432.0\,\text{kJ mol}^{-1}\end{aligned}$$

This value tells us that the H—Cl bond is very strong: nearly as strong as the H—H bond, the strongest two-electron bond between identical atoms, for which $D(H{-}H)$ = 436 kJ mol^{-1}. When we turn to polyatomic molecules, such as those in the gaseous water molecule, things are less straightforward. There are two O—H bonds in each water molecule, and if they are broken in turn, they have different bond dissociation enthalpies:

$$H_2O(g) = H(g) + OH(g); \quad D(HO{-}H) = 499\,\text{kJ mol}^{-1} \tag{4.5}$$

$$OH(g) = H(g) + O(g); \quad D(O{-}H) = 428\,\text{kJ mol}^{-1} \tag{4.6}$$

So how can we choose a measure of the O—H bond strength in $H_2O(g)$? If we take the sum of Reactions 4.5 and 4.6, we get:

$$H_2O(g) = 2H(g) + O(g); \quad \Delta H_m^{\ominus} = 927\,\text{kJ mol}^{-1} \tag{4.7}$$

In this reaction, gaseous water is broken into free atoms. The standard enthalpy change is therefore the standard enthalpy of atomization of gaseous water — that is, of the free water molecule. Since two identical O—H bonds are broken in this reaction, we can assign half of the enthalpy change to each one. This gives us a new measure of the strengths of bonds in a molecule, which we call the **bond enthalpy term**, B, to distinguish it from a bond dissociation enthalpy, D.

⬤ What is the value of $B(O{-}H)$ in $H_2O(g)$?

⬤ It is $\frac{1}{2}\Delta H_m^{\ominus}$ for Reaction 4.7: $B(O{-}H)$ = 463.5 kJ mol^{-1}.

Bond enthalpy terms are quantities, which when summed over all the bonds in a *free molecule*[*] give the atomization energy. An essential part of this idea is the assumption that the bond enthalpy term for a particular type of bond does not

[*]Note the stress on dealing with free (which usually means gaseous) molecules. This eliminates intermolecular forces, and ensures that the enthalpy of atomization depends only on the energies of the bonds in the molecules. Thus, in this Section, we work with gaseous H_2O and H_2O_2 and their thermodynamic properties at 25 °C, even though both substances are liquids at this temperature.

change from molecule to molecule. For example, the hydrogen peroxide molecule is shown in Figure 4.1. For the atomization process:

$$H_2O_2(g) = 2H(g) + 2O(g) \qquad (4.8)$$

$$\begin{aligned}\Delta H_m^{\ominus} &= 2\Delta H_f^{\ominus}(H,\ g) + 2\Delta H_f^{\ominus}(O,\ g) - \Delta H_f^{\ominus}(H_2O_2,\ g)\\ &= [2 \times 218.0 + 2 \times 249.2 - (-136.3)]\ kJ\ mol^{-1}\\ &= 1\,071\ kJ\ mol^{-1}\end{aligned}$$

As the molecule contains one O—O and two O—H bonds, this quantity is $[B(O-O) + 2B(O-H)]$. Assuming that $B(O-H)$ here is the same as in H_2O, $B(O-O) = 144\ kJ\ mol^{-1}$. Some bond enthalpy terms are recorded in Table 4.1. The assumption that they do not change from one type of molecule to another is only approximately correct, but it is sufficiently valid to make it useful. Bond enthalpy terms can, for example, be used to estimate the enthalpies of formation of non-existent compounds.

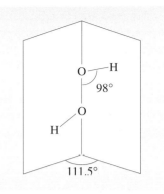

Figure 4.1
The most stable conformation of the hydrogen peroxide molecule. There is fairly free rotation about the O—O bond, but in the conformation of lowest energy the two O—H bonds are at an angle of 111.5°.

Table 4.2 Some values of bond enthalpy terms at 298.15 K

Bond	$B/kJ\ mol^{-1}$	Bond	$B/kJ\ mol^{-1}$
H—H	436	C—H	413
C—C	347	Si—H	318
C=C	612	O—H	464
C≡C	838	S—H	364
Si—Si	226	H—F	568
N≡N	945	H—Cl	432
O—O	144	C—O	358
O=O	498	Si—O	466
S—S	266	C—F	467
F—F	158	Si—F	597
Cl—Cl	243	O—F	191

Oxygen and sulfur both occur in Group VI/16. Elemental sulfur is a yellow solid consisting of S_8 puckered rings within which there are S—S single bonds (Figure 4.2a); the S_8 molecules are held together in the solid by the usual weak intermolecular forces. By contrast, elemental oxygen is a diatomic gas containing O=O molecules. Is it possible that oxygen could also exist as O_8 rings like sulfur (Figure 4.2b)?

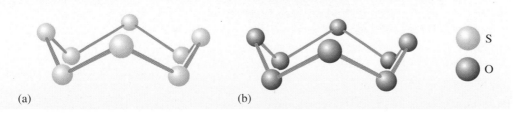

(a)   (b)   S   O

Figure 4.2 Oxygen and sulfur occur in the same Group of the Periodic Table. The S_8 ring, (a), occurs in solid sulfur, and in sulfur vapour; (b) the O_8 ring (see text).

To find out, we use Figure 4.3 in which, at the bottom, the unknown $O_8(g)$, is formed from the familiar oxygen gas. Since $O_2(g)$ is the standard reference state of oxygen, the enthalpy change for this step is $\Delta H_f^{\ominus}(O_8, g)$. In the remainder of the cycle, this reaction is performed in two steps. In step 1, four O_2 molecules are broken down into eight oxygen atoms. For this, $\Delta H_m^{\ominus}$ is $8\,\Delta H_f^{\ominus}(O, g)$, or, what is the same thing, $4D(O{=}O)$. So $\Delta H_m^{\ominus} = 8 \times 249.2$ kJ mol^{-1} or $1\,994$ kJ mol^{-1}. In step 2, these 8 atoms form an O_8 ring in which there are 8 O—O bonds. The enthalpy change for this step is $-8B(O{-}O)$, where the minus sign indicates that we are dealing with the *reverse* of the atomization of $O_8(g)$.

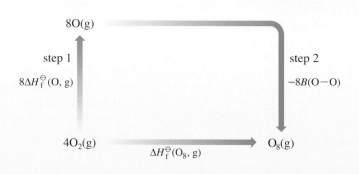

step 1

$8\Delta H_f^{\ominus}(O, g)$

step 2

$-8B(O{-}O)$

$8O(g)$

$4O_2(g)$

$\Delta H_f^{\ominus}(O_8, g)$

$O_8(g)$

Figure 4.3
A thermodynamic cycle that can be used to estimate an approximate value of the standard enthalpy of formation of $O_8(g)$.

⬤ Calculate $\Delta H_f^{\ominus}(O_8, g)$ using the value of $B(O{-}O)$ derived above. P$_{\mathsf J}$ 33 DATA Book

⬤ $\Delta H_f^{\ominus}(O_8, g)$ = $1\,994$ kJ mol$^{-1} - 8B(O{-}O)$

$= (1\,994 - 1\,152)$ kJ mol^{-1}

$= 842$ kJ mol^{-1}

The positive value tells us that $O_8(g)$ must be thermodynamically very unstable to the reverse reaction, the decomposition into diatomic oxygen:

$$O_8(g) = 4O_2(g); \quad \Delta H_m^{\ominus} = -842 \text{ kJ mol}^{-1} \qquad (4.10)$$

This is why oxygen consists of O_2 molecules rather than O_8 rings. The large negative value arises because the O—O single bond is very weak; $B(O{-}O)$ is much less than half $B(O{=}O)$.

4.3 Bond enthalpy terms and electronegativities

It was through bond enthalpy terms that Linus Pauling arrived at the famous **Pauling electronegativity scale**. In the previous Section, you saw that $B(H{-}Cl) = 432$ kJ mol^{-1}. The homonuclear molecules, H_2 and Cl_2, formed by these two atoms have bond enthalpy terms of 436 and 243 kJ mol^{-1}, respectively. The average is 340 kJ mol^{-1}, yet $B(H{-}Cl)$ exceeds this by 92 kJ mol^{-1}. This shows that the bond in the HCl molecule is unusually strong.

Pauling suggested that if the electron pair in the H—Cl bond were shared equally between the hydrogen and chlorine atoms, then $B(H{-}Cl)$ would indeed be the average of $B(H{-}H)$ and $B(Cl{-}Cl)$. The additional 92 kJ mol^{-1} is caused by an *unequal* sharing of the electron pair. The charges on the atoms result in an additional attraction which makes the molecule more stable, and so the binding energy is greater. The shift of electron density from hydrogen to chlorine is the result of a difference in electronegativity.

Pauling therefore argued that, for *single* bonds between two different atoms, A and B, the quantity $B(A-B) - \frac{1}{2}[(B(A-A) + B(B-B)]$ is a measure of the electronegativity difference. He called this quantity the **ionic resonance energy**, and found that he got the best fit if he assumed that it was proportional to the *square* of the electronegativity difference $(\chi_A - \chi_B)$, where χ_A and χ_B are the electronegativities of A and B:

$$B(A-B) - \frac{1}{2}[B(A-A) + B(B-B)] = C(\chi_A - \chi_B)^2 \qquad (4.11)$$

Here C is a constant. The scale of electronegativities is tied to a fixed point by assigning a value of 96.5 kJ mol^{-1} [*] to C, and χ_H is fixed at 2.1. Thus, rearranging Equation 4.11:

$$C(\chi_A - \chi_B)^2 = B(A-B) - \frac{1}{2}[B(A-A) + B(B-B)] \qquad (4.12)$$

⬤ Calculate the electronegativity difference between hydrogen and chlorine from the data on HCl.

⬤ We calculated $B(H-Cl) - \frac{1}{2}[B(H-H) + B(Cl-Cl)]$ to be 92 kJ mol^{-1}, so,

$$\left(\chi_H - \chi_{Cl}\right)^2 = \frac{92}{96.5}$$

$$\left(\chi_H - \chi_{Cl}\right) = \sqrt{\frac{92}{96.5}} = \pm 1.0 \text{ (to one decimal place)}$$

We must now choose either +1.0 or −1.0, and here we use chemical intuition. The chemical properties of HCl, such as its ionization in water, suggest that chlorine is more electronegative than hydrogen. Thus, −1.0 is appropriate and, as $\chi_H = 2.1$, $\chi_{Cl} = 3.1$.

Some Pauling electronegativities are shown in Figure 4.4.

Equations 4.11 and 4.12 contain the *arithmetic* mean of $B(A-A)$ and $B(B-B)$. Subsequently, Pauling found that his electronegativity values were more internally consistent if he used a modified equation which used the *geometric* mean, the square root of the product of $B(A-A)$ and $B(B-B)$. Equation 4.12 now becomes:

$$C(\chi_A - \chi_B)^2 = B(A-B) - \sqrt{[B(A-A) \times B(B-B)]} \qquad (4.13)$$

This is one of several other attempts to define electronegativity, but here we shall use the Pauling scale described by Equation 4.12.

				H 2.1
B 2.0	C 2.5	N 3.0	O 3.5	F 4.0
	Si 1.8	P 2.1	S 2.5	Cl 3.0
	Ge 1.8	As 2.0	Se 2.4	Br 2.8
	Sn 1.8	Sb 1.9	Te 2.1	I 2.5

Figure 4.4
Pauling electronegativities of some non-metallic elements.

4.4 Dipole moments

In a free atom, the electron density distribution is spherical, and centred on the nucleus, with the number of electrons equal to the nuclear charge. So diagrams often represent an atom by a point at its nucleus, the atom bearing a net charge of zero. In a homonuclear diatomic molecule, bonding electrons are shared equally between the

[*]This particular energy was chosen because it is equivalent to 1 electronvolt.

two identical atoms, as in H_2.* This represents one extreme type of chemical bonding, covalent bonding, in which the charge distribution between the atoms is equal. At the other extreme lie ionic compounds such as NaCl, in which electron transfer occurs to generate an array of ions such as Na^+ and Cl^-. The distinction between ionic and covalent bonding is not sharp. For example, in the HCl molecule, hydrogen has acquired a *fractional* positive charge of $+ne$, where e is the electronic charge, and n is simply a number less than one, and chlorine has acquired an equal negative charge. These two separate and equal but opposite charges, like those in Figure 4.5a, constitute a **dipole**. Molecules such as HCl, which can be represented as a dipole, are said to be **polar molecules**. The extent of the charge separation is described by the **dipole moment**, μ, of a polar molecule which is defined by

$$\mu = qr \tag{4.14}$$

where q is the charge at the positive end, and r is the distance between the charges. The SI unit of charge is the coulomb (C), the charge on a proton being 1.602×10^{-19} C. Thus, the SI unit of dipole moment is C m. For HCl, the value is 3.603×10^{-30} C m. However, it is more convenient to use a unit called the debye † (D), where $1\,D = 3.336 \times 10^{-30}$ C m.

● What is μ(HCl) in debyes?

● $\mu = \dfrac{3.603 \times 10^{-30}}{3.336 \times 10^{-30}}\,D = 1.08\,D.$

The dipole moment of a molecule is represented by the arrow-like symbol, $\longmapsto$, and we write it so that the + end of the arrow is located at the positive end of the molecule and the arrow points to the negative end of the dipole. This leads to the representation of the HCl dipole moment shown in Figure 4.5b. Dipole moments can be determined for polyatomic as well as for diatomic molecules. It is then often useful to think of how the dipole moments of *individual bonds* in the molecule might contribute to the overall dipole moment, even though these cannot be measured. Let us consider the example of water, which contains two O—H bonds.

● Would you expect the dipole moment arrow for an O—H bond to point towards the hydrogen or towards the oxygen?

● It should point to the oxygen (O◄—+H) as oxygen is more electronegative than hydrogen; the oxygen has a greater share of the bonding electrons, and should bear a fractional negative charge: $O^{\delta-}$—$H^{\delta+}$.

Now water is a bent molecule, Figure 4.6, and so the individual bond dipoles combine to produce the overall dipole of the water molecule. If water were a linear molecule, the individual dipoles would be equal and opposite and would cancel out.

A simple experimental observation (Figure 4.7) shows that water *does* have a dipole moment. This proves that the molecule is bent. Two other important molecular shapes where individual bond dipoles cancel are shown in Figure 4.8.

---

*Molecular bonding is discussed in *Molecular Modelling and Bonding*.[1] See the references in Further Reading (p. 241) for details of other titles in *The Molecular World* series that are relevant.

†Named after P. J.W. Debye. Debye (1884–1966) was a physicist from The Netherlands who worked in universities in The Netherlands, Switzerland, Germany and the USA. He studied molecular structure using dipole moments, X-ray diffraction and spectroscopy, work for which he received the Nobel Prize for Chemistry in 1936.

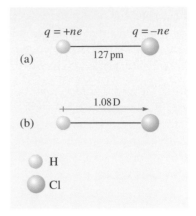

Figure 4.5
(a) A representation of the hydrogen chloride molecule showing a partial positive charge on the hydrogen atom and a partial negative charge on the chlorine atom. (b) The H—Cl molecule with its dipole moment; the arrow points towards the negative end of the molecule.

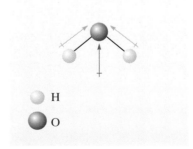

Figure 4.6
In the water molecule, the individual bond dipoles (light brown) combine and contribute to the overall dipole moment (darker brown).

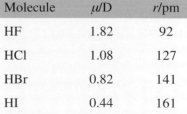

Figure 4.7
Water molecules have a dipole moment. If a plastic ball-point pen is rubbed vigorously with a woollen cloth, electrons are removed and the pen becomes positively charged. If the pen is then held next to a slow smooth stream of water from a tap, the water is deflected towards it. The negative ends of the water molecule dipoles become preferentially orientated towards, and attracted by, the positively charged pen, so the stream is deflected towards it.

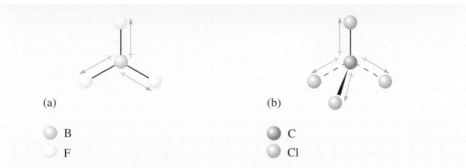

(a) ● B ○ F
(b) ● C ● Cl

Figure 4.8
Two other molecules where bond dipoles cancel to give a net dipole moment of zero: the symmetrical trigonal planar case (e.g. BF_3) and the symmetrical tetrahedral case (e.g. CCl_4).

So far, we have treated dipole moments as a consequence of electronegativity differences between atoms. Table 4.2 shows the values of the dipole moments and internuclear distances for the hydrogen halides: the dipole moments decrease down the halogen Group.

● Is this the trend expected from the electronegativity differences?

● Yes; like the electronegativity differences, the dipole moments decrease from HF to HI, as expected.

But although electronegativity differences make a very important contribution to dipole moments, that contribution is not the only one. In the HCl molecule (Structure **4.5**), we attribute the dipole moment to the shift of the electrons in the electron-pair bond towards chlorine. But when the hydrogen and chlorine atoms interact, the other outer, but non-bonding, electrons around chlorine will also be affected. They too will shift in some way along the H—Cl axis.

Table 4.2 Dipole moments and internuclear distances in the hydrogen halide molecules

Molecule	μ/D	r/pm
HF	1.82	92
HCl	1.08	127
HBr	0.82	141
HI	0.44	161

4.5

Indeed, according to valence-shell electron-pair repulsion (VSEPR) theory, these non-bonding electrons develop a strongly directional character and influence molecular shape. For example, VSEPR theory predicts that the molecules of ammonia and nitrogen trifluoride are pyramidal, because one of four tetrahedrally disposed repulsion axes is occupied by a non-bonding pair of electrons on nitrogen (Figure 4.9). Now the electronegativity difference between nitrogen and hydrogen is very similar to that between nitrogen and fluorine (Figure 4.4). This, by itself, suggests that the dipole moments of NH_3 and NF_3 should be similar, but opposed in direction. In fact, the values are 1.47 D and 0.23 D, respectively. We can explain this difference by arguing that in both cases, the presence of the non-bonding pair leads to a shift of electron density in the upward direction in Figure 4.9, and makes its own contribution to the dipole moment as shown in Figure 4.10.

⬤ How then does Figure 4.10 account for the different dipole moments of NH_3 and NF_3?

⬤ In NH_3, the bond polarities are $N^{\delta-}$—$H^{\delta+}$, and the bond and non-bonding pair contributions reinforce each other. In NF_3, the bond polarities are $N^{\delta+}$—$F^{\delta-}$ and the two kinds of contribution are opposed.

This example, therefore, shows that non-bonding pairs as well as electronegativity differences can make important contributions to dipole moments.

4.5 Intermolecular forces

If a gas is gradually cooled, the average speed of the molecules decreases. Eventually, there comes a temperature when the gas turns into a liquid or solid, which occupies a much smaller volume than the gas did. How do we explain this?

The condensation of a gas and the freezing of a liquid show that there are attractive forces between molecules. When, in a gas, a collision brings two molecules together, the attractive force is not strong enough to quench the relative motion of the molecules and to prevent them from escaping each other's clutches. But if the relative motion in the gas is made less vigorous by cooling, the intermolecular forces eventually become dominant, and a liquid or solid is formed.

⬤ If the gas is cooled at 1 atmosphere pressure, and at some temperature, T, a liquid is formed, what is T called?

⬤ It is the boiling temperature of the substance.

Boiling temperatures are therefore a crude measure of the intermolecular forces: if these forces are strong, a high temperature is needed to free the molecules from each other's influence.

In molecular substances, distances between the molecules depend on whether they are in the solid, liquid or gaseous state. For example, in solid chlorine (Figure 4.11) we can recognize Cl_2 molecules as pairs of atoms separated by 197 pm. Atoms within one Cl_2 molecule are separated from their nearest neighbouring atom in an adjacent Cl_2 molecule by 332 pm. The forces that hold the molecules together in the solid and operate at distances such as the 332 pm in Figure 4.11 are very different

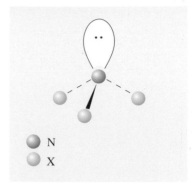

Figure 4.9
In ammonia and the nitrogen trihalides, there is a non-bonding pair of electrons, leading to a pyramidal molecular shape.

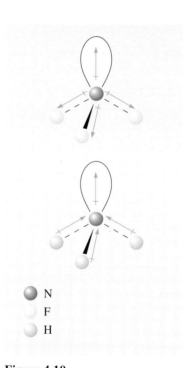

Figure 4.10
In NF_3 the dipole moment developed by the non-bonding pair opposes that developed in the N—F bonds: in NH_3, it reinforces it. The dipole moment of NF_3 is therefore much less than that of NH_3.

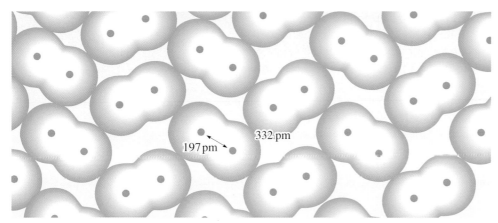

Figure 4.11 The structure of a layer in solid chlorine: the atoms are grouped in pairs around the shortest internuclear distance of 197 pm. The shortest distance between the atoms of different pairs is 332 pm.

from forces of the type that hold the atoms in Cl_2 together. Intermolecular forces of this kind are called van der Waals forces, and are of more than one type. For example, molecules with permanent dipole moments or bond dipoles will tend to arrange themselves so that the positive end of one dipole is close to the negative end of another (Figure 4.12). This gives rise to a dipole–dipole interaction.

○ Could such an interaction account for the forces that hold solid chlorine together?

○ Cl_2 has no permanent dipole moment and so these forces cannot be what holds chlorine molecules together in a solid.

Nevertheless, a type of dipole-based interaction is still believed to be the source of the binding in solid chlorine. To see how, consider a helium atom, which has the ground state electronic configuration $1s^2$. This atom is spherical, *but only on a time-averaged basis*. The two electrons are in motion about the nucleus, and at any instant they will be in positions such as those shown in Figure 4.13a. *At this instant*, the atom has an instantaneous dipole. If another helium atom is close by, the instantaneous dipole in the first atom will polarize the second, as shown in Figure 4.13b. A dipole is induced in the second atom, which creates an attractive force between the two atoms. As the charge cloud of one atom fluctuates as its electrons move about the nucleus, the charge cloud of the other fluctuates in sympathy to preserve arrangements of the type shown in Figure 4.13b, and permanent binding can result. This is called an **induced dipole–induced dipole interaction**. This kind of intermolecular force is the most important van der Waals force, and is usually much stronger than any permanent dipole–dipole interaction.

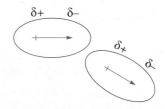

Figure 4.12
In a fluid containing molecules with a permanent dipole moment, arrangements of the type shown here will have lower energies and be more common than those where the arrangement of the dipoles tends to juxtapose like charges. This arrangement gives rise to a dipole–dipole interaction, which is of a bonding nature.

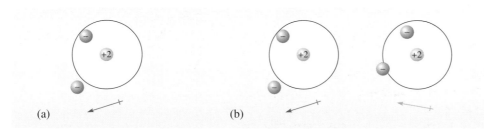

(a) (b)

Figure 4.13 (a) A helium atom, although spherical on a time-averaged basis, may at a particular instant have a net dipole moment. (b) This dipole moment (dark brown arrow) can then induce a dipole moment (pale brown arrow) in an adjacent atom or molecule in a direction that leads to a binding force between the two atoms.

The strength of the induced dipole–induced dipole interaction between molecules increases with their polarizabilities. The **polarizability** is a measure of how easily the electron distribution in a molecule can be distorted by an external charge or electric field: when the polarizability is larger, the induced dipole will also be larger. In a small atom such as helium, the ionization energies are high and so the electrons are tightly held by the nucleus. They are not easily polarized and do not form strong intermolecular forces. In keeping with this, helium has the lowest boiling temperature (4.2 K) of any liquid. But suppose a molecule contains large atoms, like those of iodine, where there are many electrons, which are some distance from the nuclei and have relatively low ionization energies. Then the electron distribution will be responsive to neighbouring charges because it is less strongly constrained by the nuclei.

● Will the polarizability be smaller or larger as a result?

● It will be larger: the electron cloud will be more easily distorted by an external charge.

Thus, substances that consist of large molecules containing many electrons are likely to have strong intermolecular forces and are not expected to be volatile. And since the number of electrons in a molecule tends to increase with its mass, we see a reason for the general increase of melting and boiling temperatures with molecular mass.

Molecular fluorides provide a particularly good example of the influence of polarizability on boiling temperature. Consider a molecule like CF_4 (Structure **4.6**), whose exterior is composed of fluorine atoms. Fluorine is a small, highly electronegative atom with an ionization energy that is exceeded only by those of the noble gases helium and neon. Its electrons will be tightly constrained, and so the CF_4 molecule will have low polarizability. Table 4.3 shows that CF_4 is much more volatile than Cl_2, despite the fact that it contains more electrons and has a higher molar mass than Cl_2. One well-known application of the low polarizability of fluorides is in the use of the polymer Teflon® or poly(tetrafluoroethene), PTFE (Figure 4.14). In addition to its familiar use as a non-stick coating for frying pans, PTFE forms a layer in breathable fabric (Gore-Tex®) (Figure 4.15); it also has many other specialized uses, such as the fabrication of artificial veins.

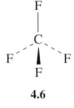

4.6

Table 4.3 Properties of the gases CF_4 and Cl_2

	CF_4	Cl_2
relative molar mass	88	71
number of electrons	42	34
boiling temperature/K	144	239

Figure 4.14
Teflon®, or poly(tetrafluoroethene), PTFE, is a polymer with the molecular formula $-(CF_2)_n-$. The helical chains present a sheath of fluorine atoms to neighbouring molecules. The low polarizability of the sheath means that the interactions with adjacent molecules will be very weak: hence the non-stick properties.

Figure 4.15
The layered structure of breathable waterproof fabric (100 × magnification). The layer of microporous PTFE (coloured yellow), is sandwiched between layers of other polymers such as nylon (coloured pink). The pores in the PTFE layer are less than 10^{-6} m in diameter.

4.6 Summary of Section 4

1   The Lewis theory regards bases as non-bonding-pair donors and acids as non-bonding-pair acceptors. According to this definition, the formation of a complex from a metal ion and ligands can be envisaged as an acid–base reaction.

2   The standard enthalpy change for the conversion of a gaseous covalent molecule into gaseous atoms is the sum of the bond enthalpy terms for the bonds in that molecule. Bond enthalpy terms do not vary greatly from one molecule to another. *

3   The bond enthalpy term of a bond A—B nearly always exceeds the arithmetic or geometric mean of the bond enthalpy terms for the bonds A—A and B—B. According to Pauling, this excess is $C(\chi_A - \chi_B)^2$, where C is 96.5 kJ mol^{-1}, and χ_A and χ_B are the electronegativities of A and B.

4   A heteronuclear diatomic molecule constitutes a dipole: two equal but opposite charges, $+q$ and $-q$, separated by the internuclear distance, r. The dipole moment, μ, is qr.

5   Both bond dipoles, which are influenced by electronegativity differences, and non-bonding pairs, contribute to the overall dipole moment of a polyatomic molecule.

6   The main intermolecular or van der Waals force is the induced dipole–induced dipole interaction. It is largest between molecules of high polarizability: those with electron clouds that are easily distorted by an external charge, and have low ionization energies.

*This generalization is true if the valency of the atoms does not change.

QUESTION 4.1 ✳

The statement below is a description of the reaction

$$[Cu(H_2O)_4]^{2+} + 4Cl^- = [CuCl_4]^{2-} + 4H_2O \qquad (4.17)$$

which occurs in concentrated chloride solution:

The complex 1 is a combination of Lewis acid 2 with Lewis base 3. In the reaction, Lewis base 3 is replaced by Lewis base 4 to give a new complex 5. This is a combination of Lewis acid 2 and Lewis base 4.

Identify the species 1–5.

Handwritten notes:
1 – $[Cu(H_2O)_4]^{2+}$
2 – Cu^{2+}
3 – H_2O
4 – Cl^-
5 – $[CuCl_4]^{2-}$

QUESTION 4.2

(a) When solid $AlCl_3$ is heated, the vapour contains Al_2Cl_6 molecules, as represented by the structural formula **4.7**.

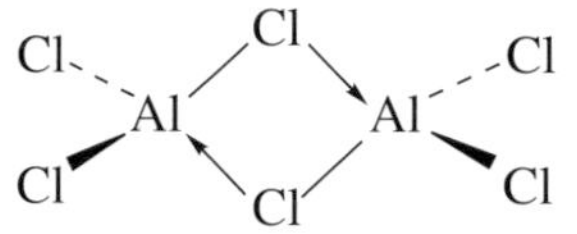

4.7

Write an alternative structural formula. Experiment shows that the bond lengths in the Al—Cl—Al bridges are identical; which of the possible structural formulae best fits this observation?

(b) The nitrate ion, NO_3^-, can be represented as a resonance hybrid of three structural formulae in Structure **4.8**.

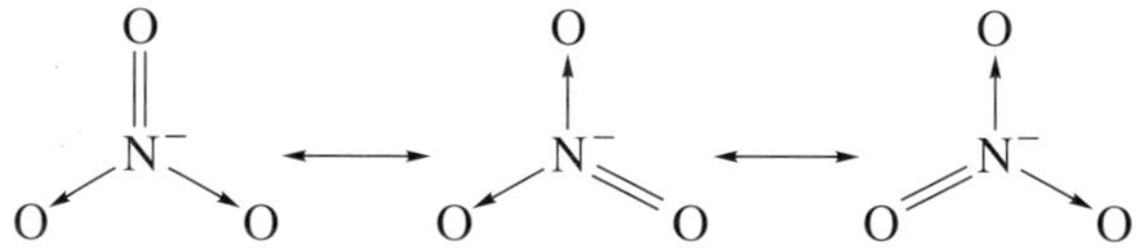

4.8

Write an alternative form of this representation.

QUESTION 4.3 ✳

Hydrazine, $H_2N—NH_2$, is a volatile liquid composed of molecules containing one N—N and four N—H bonds. Use the data in Table 4.4 to estimate bond enthalpy terms for these two kinds of bond. Which datum in Table 4.4 is not required?

Table 4.4 Thermodynamic data for the nitrogen hydrides at 298.15 K

Substance	State	$\Delta H_f^{\ominus}$ /kJ mol^{-1}
N	g	472.7
H	g	218.0
NH_3	g	−46.1
N_2H_4	l	50.6
N_2H_4	g	95.4

QUESTION 4.4

$B(H—H) = 436\,kJ\,mol^{-1}$. Use this value, along with the values of $B(N—H)$ and $B(N—N)$ that you calculated in Question 4.3, to obtain the difference in electronegativity between nitrogen and hydrogen from Equation 4.12. How well does your difference agree with that obtained from Figure 4.4?

QUESTION 4.5

Explain the following observations: (a) The molecules CO_2 and PF_5 have zero dipole moment; the molecules SO_2 and BrF_5 do not. (b) The dipole moment of the H_2O molecule is 1.85 D; that of the F_2O molecule is only 0.25 D.

QUESTION 4.6

The following substances are molecular in the liquid and gaseous states. In each case the number of electrons in the molecule is given in parentheses. By assessing the likely polarizabilities, predict the order of the boiling temperatures.

(a) Ar (18) and Kr (36)

(b) SiF_4 (50) and SCl_2 (50)

(c) SnH_4 (54) and BCl_3 (56)

(d) GeH_4 (36), AsH_3 (36) and H_2Se (36)

(e) SF_6 (70) and Br_2 (70).

THE CHEMISTRY OF HYDROGEN

5

Hydrogen is the most abundant of all the elements; 90–95% of the atoms or atomic nuclei in the universe are believed to be those of hydrogen. One sign of this abundance is the energy that we receive from the Sun. This is supplied by a sequence of nuclear fusion reactions in which hydrogen ($^1_1$H) is converted into helium ($^4_2$He). In the Sun, some 600 million tonnes of hydrogen are consumed every second in this way.

The situation is rather different on Earth, where hydrogen is only the ninth most abundant element by mass in the Earth's crust, oceans and atmosphere, and the fourth most common type of atom. Nevertheless, it seems especially plentiful because it is available in highly accessible forms such as water and the hydrocarbons present in natural gas and crude oil.

5.1 Preparation and properties of hydrogen

At present, most industrial hydrogen gas is made by the reaction of steam with natural gas (Case Study: *Industrial Inorganic Chemistry*), whose chief constituent is methane:

$$CH_4(g) + H_2O(g) \xrightarrow[750\,°C]{Ni\ catalyst} CO(g) + 3H_2(g) \tag{5.1}$$

The carbon monoxide is then used to generate further hydrogen in a second two-stage reaction with steam at a lower temperature:

$$CO(g) + H_2O(g) \xrightarrow[Cu\ catalyst\,/\,200\,°C]{Fe\ catalyst/400\,°C} CO_2(g) + H_2(g) \tag{5.2}$$

Hydrogen is also a byproduct of some industrial electrolytic industries, such as the production of chlorine and sodium hydroxide in the chlor-alkali industry (Case Study: *Industrial Inorganic Chemistry*).

In the laboratory, hydrogen is conveniently made by the reaction of metals, such as zinc or magnesium, with acids (Figure 5.1):

$$Zn(s) + 2H^+(aq) = Zn^{2+}(aq) + H_2(g) \tag{5.3}$$

At room temperature and pressure, hydrogen is a colourless, odourless gas consisting of diatomic molecules,[*] H_2, and has the lowest density of all known substances (Figure 5.1). It burns in air (Figure 5.2). Hydrogen can be condensed to a solid or liquid only by the most extreme cooling: solid hydrogen melts at 14 K and liquid hydrogen boils at only 20.4 K. Whether solid, liquid or gaseous, it contains the diatomic molecule, H_2.

---

[*]It is therefore often referred to as *dihydrogen*.

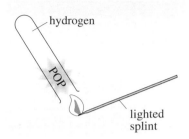

Figure 5.2
A simple laboratory test for
hydrogen; the gas burns in air
with a squeaky pop.

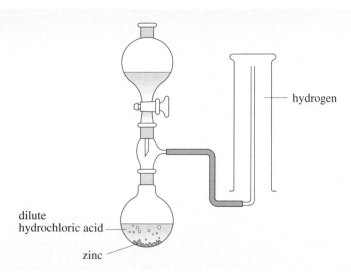

Figure 5.1 A cheap way of making hydrogen in the laboratory. It works because hydrogen
has a density only one-fifteenth that of air. The hydrogen collects at the top of the gas jar and
gradually fills it by displacing the air downwards.

5.2 Industrial uses of hydrogen

The major industrial use of hydrogen gas is in the synthesis of ammonia by the
Haber process, which is usually carried out at about 500 °C and 200 atmospheres
pressure, in the presence of an iron catalyst:

$$N_2(g) + 3H_2(g) \xrightarrow[\text{500 °C; 200 atm}]{\text{Fe catalyst}} 2NH_3(g) \tag{5.4}$$

Large amounts of hydrogen are also consumed in the manufacture of certain organic
compounds. In these reactions, the H—H bond is broken, and the hydrogen atoms
become attached to the carbon atoms of a C=C bond:

$$\text{C=C} + H_2(g) \longrightarrow \begin{array}{c} | \ | \\ -C-C- \\ | \ | \\ H \ \ H \end{array} \tag{5.5}$$

Processes of this kind, known as **hydrogenation**, occur, for example, in the
manufacture of margarine from vegetable oils. An important related reaction is the
manufacture of methanol from carbon monoxide, which is performed at about
250 °C and 30 atmospheres pressure in the presence of a copper catalyst:

$$\overset{-}{C}\equiv\overset{+}{O}(g) + 2H_2(g) \xrightarrow[\text{250 °C; 30 atm}]{\text{Cu catalyst}} CH_3-OH(g) \tag{5.6}$$

How is this reaction related to Reaction 5.5, and how does it differ?

In both reactions, hydrogen atoms from H_2 are added to atoms at either end of a
multiple bond to produce a single bond; in Reaction 5.6, however, a C≡O triple
bond is destroyed and in Reaction 5.5, a C=C double bond.

5.3 Hydrogen as a fuel

The major sources of energy in the world derive from the burning of the fossil fuels coal, oil and gas, in spite of their deleterious effects on the environment. Emission of CO_2 is predicted to cause global warming, and oxides of sulfur are a major source of acid rain (Case Study: *Acid Rain: sulfur and power generation*). As pressure is brought to bear on governments to reduce dependence on fossil fuels, other energy sources are being investigated. One attractive alternative is hydrogen, which might one day become the staple fuel of a so-called hydrogen economy. In such an economy, fuel combustion would be relatively pollution-free because the only substantial product would be water:

$$H_2(g) + \tfrac{1}{2}O_2(g) = H_2O(l) \tag{5.7}$$

Two formidable obstacles bar the way to the hydrogen economy. First, an economical method must be found of making hydrogen from water by using energy sources such as solar or nuclear energy. At present, electrolysis seems the most favoured method. The second obstacle is very damaging to the prospect of using hydrogen as a fuel for transportation: hydrogen is a gas at all but the very lowest temperatures (Section 5.1). Unlike our familiar liquid fuels such as petrol, it would occupy too large a space in an automobile, unless we can devise effective and economical ways of storing it in a much smaller volume.

The most obvious response to this problem is to reduce the volume of the hydrogen by some physical change, for example by compressing the gas in cylinders, or by supplying vehicles with a vacuum-insulated fuel tank, which can be loaded with, and store, liquid hydrogen.

The alternative is to store the hydrogen as a solid or liquid chemical compound until it is needed, and then to release it when it is required. Metal hydrides are one possibility. For example, the lanthanum–nickel alloy, $LaNi_5$, absorbs nearly 3 moles of H_2 at only $2\tfrac{1}{2}$ atmospheres pressure and 20 °C to give a hydride with a stoichiometry approaching $LaNi_5H_6$. It then releases this hydrogen at 1 atmosphere pressure. This fully-charged alloy, however, contains less than 2% hydrogen by weight. Recently, ECD Ovonics have developed a magnesium alloy in which this figure rises to nearly 7%. At 300 °C, and under a hydrogen pressure of 1.7 atmospheres, the alloy can be charged with about 90% of its hydrogen capacity in 2 minutes. The hydrogen release then occurs at lower pressures and can be controlled by varying the temperature. Even after 650 charge/discharge cycles, the alloy shows no deterioration in performance. There are plans for a vehicle with a range of over 300 miles (Figure 5.3).

A very different type of system is being explored by an Anglo–Swiss research project launched in 1991. It uses a version of Reaction 5.5:

$$\tag{5.8}$$

toluene   methylcyclohexane

Toluene, on the left, is a liquid that boils at 111 °C, and we draw it here as one of its two resonance structures, both of which include a ring containing three C=C bonds.

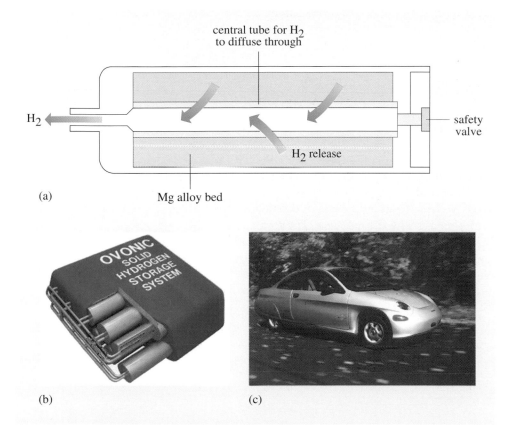

central tube for H$_2$
to diffuse through

H$_2$

safety
valve

H$_2$ release

(a)

Mg alloy bed

(b)

(c)

Figure 5.3
Ovonic metal hydride storage
systems. (a) A cross-section of a
metal hydride storage unit. The
arrows indicate the flow of hydrogen
in discharge mode. In the charge
mode the flow of hydrogen is
reversed. (b) A collection of such
units designed to provide hydrogen
for the internal combustion engine
of a vehicle with a range of over
300 miles. (c) The vehicle in
question.

● What happens to these bonds in the reaction?

● They undergo the same process as the C=C bonds in Reaction 5.5, and are
converted to C—C single bonds.

The product is methylcyclohexane, a liquid with a boiling temperature of 100 °C.
This change takes place readily at 200 °C, at which temperature the equilibrium
lies to the right. The Anglo–Swiss project is developing a bus with two fuel tanks.
One of these is filled with methylcyclohexane that has been made in this way.
When the bus is running, methylcyclohexane is vaporized and heated to 300 °C;
at this higher temperature the equilibrium in Equation 5.8 lies to the left. In the
presence of a catalyst made from platinum metals, the reaction proceeds and
methylcyclohexane decomposes to toluene and hydrogen. The hydrogen is burnt
in the internal combustion engine of the bus, and the toluene is cooled, condensed
and passed to the second tank. The bus is refuelled by filling the first tank with
methylcyclohexane again and draining the toluene out of the second. It can then be
reconverted in a chemical plant to fresh methylcyclohexane fuel. More recently,
attempts have been made to find less expensive catalysts operating at less extreme
conditions to convert cycloalkanes into cycloalkenes and hydrogen.

Another new development in the use of hydrogen as a fuel is in the fuel cell car
developed by Daimler–Chrysler, the NECAR. This prototype electric car uses
hydrogen as a fuel for a fuel cell* in which the energy of hydrogen oxidation to
water is converted to electricity. The car runs on a mixture of methanol and water.
Using a relatively low-temperature catalyst, these substances react to produce
hydrogen and carbon dioxide, in the gas phase:

$$CH_3OH(g) + H_2O(g) = CO_2(g) + 3H_2(g)$$

* *Metals and Chemical Change* includes a
Case Study on *Batteries and fuel cells*.[3]

As the conversion is clean, the only product of this process is the CO_2, but this is a greenhouse gas.

5.4 Atomic and ionic properties of hydrogen

Important features of hydrogen chemistry have now emerged. The element occurs in three oxidation numbers, +1, zero and −1. The oxidation number +1 occurs in non-metallic hydrides such as water, oxidation number zero in elemental hydrogen, $H_2(g)$, and oxidation number −1 in metallic hydrides such as $LaNi_5H_6$ or NaH. In Figure 2.1, hydrogen has not been given a specific position in the Periodic Table, but *chemically* there is a good case for grouping it with the halogens. The formation of a diatomic molecule akin to F_2 and Cl_2, and of metal hydrides which often share the same structure as halides (see Figure 5.4) supports the classification of hydrogen with the halogens. There are also, however, important differences between the hydrogen and halogen molecules, and between metal hydrides and halides. Let us look at the question more closely.

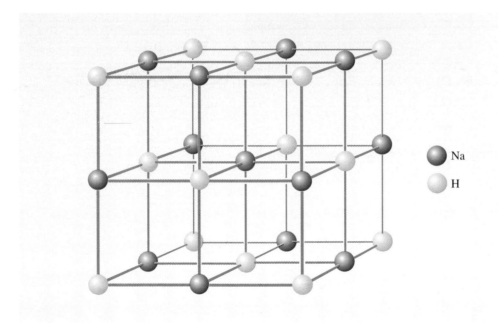

Na

H

Figure 5.4
A resemblance between hydrogen and the halogens. Hydrogen forms alkali metal hydrides like NaH, which has the same structure as the sodium halides.

Whether portrayed as a Lewis structure (Structure **5.1**) or by molecular orbital theory, the bond holding together the two atoms in the H_2 molecule turns out to have a bond order of 1.

H ⦂ H

5.1

The hydrogen atom contains only a single electron moving around a single proton. This suggests that the atom should be small, and the covalent radius of only 37 pm, shown in Table 5.1, supports this. Table 5.1 compares some atomic and molecular properties of hydrogen with those of fluorine, the smallest of the halogen atoms. Let us look at some of the data.

Table 5.1 A comparison of some properties of hydrogen and fluorine

	H	F
electronic configuration	$1s^1$	$1s^22s^22p^5$
ionization energy/kJ mol^{-1}	1 314	1 681
electron affinity/kJ mol^{-1}	73	328
Pauling electronegativity	2.1	4.0
covalent radius/pm	37	71
$D(X–X)$/kJ mol^{-1}	436	158
X–X distance in X_2/pm	74	143
ionic radius of X^-/pm	126	119
melting temperature/K	14	55
boiling temperature/K	20	85
relative molar mass of X_2	2	38

● The melting and boiling temperatures of hydrogen are lower than those of fluorine. Why is this?

● The attractive forces between molecules tend to increase with molecular mass, and the mass of H_2 is less than that of F_2. One would therefore expect F_2 to be less volatile than H_2. Intermolecular forces were discussed in Section 4.5.

Another important difference is provided by the electron affinities. The value for hydrogen is much less than that of fluorine. The small electron affinity of hydrogen becomes especially important when set alongside the much higher bond dissociation energy of the H_2 molecule.

Such differences explain why metal hydrides are much more easily decomposed into their elements than metal fluorides, and indeed all other halides: why, for example, sodium fluoride melts at about 1 000 °C, and boils at 1 700 °C without decomposition, whereas sodium hydride breaks down into sodium and hydrogen at only 400 °C.

To see why, consider the following process in the cases of hydrogen and fluorine:

$$\tfrac{1}{2}X_2(g) + e^-(g) = X^-(g) \tag{5.9}$$

It is the sum of the changes:

$$\tfrac{1}{2}X_2(g) = X(g) \tag{5.10}$$
$$X(g) + e^-(g) = X^-(g) \tag{5.11}$$

● Use the data in Table 5.1 to calculate $\Delta H_m^{\ominus}$ for Reaction 5.9 for the cases X = H and X = F.

● For Reaction 5.10, $\Delta H_m^{\ominus} = \tfrac{1}{2}D(X-X)$; for Reaction 5.11, $\Delta H_m^{\ominus} = -E(X)$. Thus, for Reaction 5.9, $\Delta H_m^{\ominus} = \tfrac{1}{2}D(X-X) - E(X)$. This is $\tfrac{1}{2}(436) - 73 = +145$ kJ mol^{-1} for hydrogen, and $\tfrac{1}{2}(158) - 328 = -249$ kJ mol^{-1} for fluorine.

Thus, because of the high bond dissociation energy of H_2 and the small electron affinity of the hydrogen atom, Reaction 5.9 is about 400 kJ mol^{-1} less favourable for hydrogen than for fluorine.

Now Figure 5.5 shows a Born–Haber cycle for the formation of a sodium salt, NaX. Suppose we compare sodium hydride and sodium fluoride. Because it is concerned only with sodium, step 1 of Figure 5.5 will be the same in both cases. We have seen that step 2 is about $400 \, kJ \, mol^{-1}$ more negative in the fluoride case. Step 3 is the lattice energy of NaX(s). Notice from Table 5.1 that the ionic radii of F^- and H^- are rather similar.

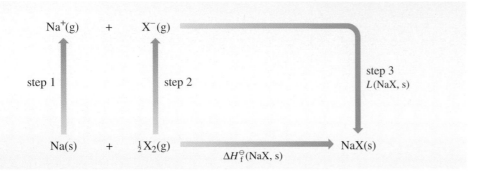

Figure 5.5
A Born–Haber cycle for the formation of sodium fluoride or hydride. The standard enthalpy change for step 2 is that of Reaction 5.9; the electron used up with $\frac{1}{2}X_2(g)$ is not shown in the Figure, but it is obtained from sodium when it is converted to $Na^+(g)$ in step 1.

● What does this tell us about the lattice energies of NaF and NaH?

● They will be reasonably similar as well.

Now $\Delta H_f^{\ominus}$ (NaX, s) is the sum of steps 1, 2 and 3. It follows that the difference between $\Delta H_f^{\ominus}$ (NaF, s) and $\Delta H_f^{\ominus}$ (NaH, s) will be dictated by the large difference in step 2.

● Which is less stable with respect to its constituent elements, NaF(s) or NaH(s)?

● NaH(s) by a long way; it has the less negative value of step 2, so its $\Delta H_f^{\ominus}$ value will be less negative as well.

It follows that if we reverse the formation reaction

$$NaX(s) = Na(s) + \tfrac{1}{2}X_2(g) \tag{5.12}$$

$\Delta H_m^{\ominus}$ will be much more positive for NaF than for NaH. In this type of reaction, stability is determined by the decomposition temperature, which is approximately equal to $\Delta H_m^{\ominus} / \Delta S_m^{\ominus}$, where $\Delta S_m^{\ominus}$ is the standard entropy change. * As $\Delta S_m^{\ominus}$ is similar in the two cases, the decomposition temperature will be much larger for NaF than for NaH.

This comparison is appropriate not just for fluorine. The electron affinity of hydrogen is much less than that of any halogen atom, and the dissociation energy of H_2 is much larger than that of any of the halogen molecules. This is the main reason why ionic hydrides tend to decompose into the metal and hydrogen much more easily than do halides.

In this Section we have tackled a problem about hydrogen in oxidation numbers zero and −1. Hydrogen, however, occurs also in oxidation number +1, and we now turn to the most obvious example — the aqueous ion, H^+(aq), which is the source of acidity in water.

---

* See *Metals and Chemical Change* for more information.[3]

5.5 The aqueous hydrogen ion

When the aqueous hydrogen ion is written as $H^+(aq)$, it indicates only that when a free proton, H^+, is transferred to water, it becomes bound, in some way, to the water molecules. But what is the nature of this binding? Ascertaining the structure of liquids is very difficult, but it is believed that the proton becomes bound primarily to just one water molecule in the ion, H_3O^+ (Section 3.2):

$$H^+ + H_2O(l) - H_3O^+(aq) \qquad (5.13)$$

The ion H_3O^+ can also be found in solids. Thus, if concentrated hydrochloric or perchloric acids are cooled, the hydrates $HCl.H_2O$ and $HClO_4.H_2O$ crystallize out. Their structures show that they are better formulated as $H_3O^+Cl^-$ and $H_3O^+ClO_4^-$, because the grouping of the hydrogen and oxygen atoms (Figure 5.6) is consistent with the presence of H_3O^+.

● Which neutral molecule does H_3O^+ resemble?

● Ammonia, NH_3. The nitrogen atom and the O^+ ion have the same number of electrons. They can combine with three hydrogen atoms to give NH_3 and H_3O^+, respectively. In both cases, VSEPR theory predicts a pyramidal shape.

A Lewis structure for H_3O^+ is shown as Structure **5.2**; the ion O^+ has five outer electrons, and gains a noble gas structure by forming three electron-pair bonds.

If we write the aqueous proton as $H_3O^+(aq)$, this in turn raises the question of how the H_3O^+ group is bound to the surrounding water molecules. The most favoured answer is the one proposed by the Chemistry Nobel Prize winner, Manfred Eigen: each hydrogen in H_3O^+ links the ion to the oxygen atom of a water molecule through a hydrogen bond (Figure 5.7). The resulting unit has the formula $[H_9O_4]^+$ or $[H_3O(H_2O)_3]^+$. Like H_3O^+, it has been found in solid acid hydrates, and the geometry shown in Figure 5.7 is taken from a neutron diffraction study of one of them.

We shall consider hydrogen-bonding in more detail in Section 5.8, but the type shown in Figure 5.7 provides an explanation of the great speed with which the aqueous proton apparently moves through aqueous solutions during, for example, electrolysis; its mobility is much greater than that of any other ion. That the solvent is an important factor is shown by the fact that the mobility of the proton in acetone, CH_3COCH_3, is not exceptionally high. The explanation is called the **Grotthuss mechanism** and is shown in Figure 5.8: a hydrogen in H_3O^+ breaks away as H^+ along the line of the hydrogen bond, and becomes covalently bound to the neighbouring water molecule to the right, forming another H_3O^+ ion.

This ion, in its turn, can pass a proton to a water molecule to its right, and so on. Thus, there can be a *net* transfer of a proton between two points in an aqueous solution without one particular proton having to cover the entire distance.

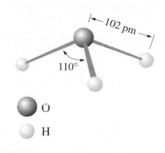

Figure 5.6
The structure of the ion H_3O^+ in crystals. The ion is a flattened pyramid.

5.2

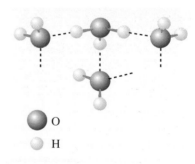

Figure 5.7
The most widely accepted proposal for the arrangement of water molecules around the ion $H_3O^+(aq)$. Each O—H bond of the H_3O^+ ion points towards and binds the oxygen atom of an adjacent water molecule through a hydrogen bond. The hydrogen bonds are shown as dashed purple lines. The whole unit has the formula $[H_9O_4]^+$.

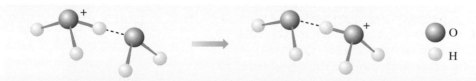

Figure 5.8 The Grotthuss mechanism, which explains why substances that yield H^+ in aqueous solution greatly enhance the conductivity of water.

5.6 Summary of Sections 5.1–5.5

1   Hydrogen is the most abundant element in the universe. On Earth, it occurs in the very accessible forms of water, natural gas and crude oil.

2   Industrial hydrogen is made by the reaction of natural gas with steam. In the laboratory, hydrogen can be made by the reaction of zinc or magnesium with acids. It is the least dense of all gases, and is highly flammable.

3   Major industrial uses of hydrogen include ammonia synthesis, and addition reactions across C=C bonds. The storage of hydrogen fuel is difficult because of its volatility. Possible solutions include storage as metal hydrides (e.g. $LaNi_5H_6$) or as methylcyclohexane, which dissociates to hydrogen and to toluene at 300 °C.

4   The bond in the H_2 molecule is short and very strong, much stronger than in any halogen molecule, and the electron affinity of the hydrogen atom is much smaller than that of any halogen atom. In a Born–Haber cycle, these two effects are the chief cause of the lower stability of hydrides with respect to their constituent elements.

5   A more precise and probably accurate presentation of the aqueous proton, $H^+(aq)$, is $H_3O^+(aq)$, the H_3O^+ ion being a flattened pyramid. In water, this is probably hydrogen-bonded to three water molecules, giving an $[H_9O_4]^+$ unit.

6   Conductive protons can be transmitted very quickly through solutions containing $H_3O^+(aq)$ by the Grotthuss chain mechanism in which O—H covalent and O—H hydrogen bonds are interconverted.

QUESTION 5.1

By selecting compounds from the following list, write balanced equations to describe *three* methods of preparing hydrogen gas:

water, methane, pure liquid H_2SO_4, and magnesium

You may assume access to a battery and electrodes, and to suitable catalysts.

QUESTION 5.2

The electron affinity of chlorine is 349 kJ mol^{-1} and $D(Cl—Cl) = 243$ kJ mol^{-1}. Calculate the value of $\Delta H_m^\ominus$ for the reaction

$$\tfrac{1}{2}Cl_2(g) + e^-(g) = Cl^-(g) \tag{5.14}$$

and compare it with the value of +145 kJ mol^{-1} for hydrogen. What relevance does the difference in *the two values of* $\Delta H_m^\ominus$ have to the fact that $MgH_2(s)$ decomposes into its elements on mild heating, but $MgCl_2(s)$ melts at over 700 °C without decomposition?

5.7 The hydrides of the typical elements

Hydrogen combines with most of the chemical elements. Here we concentrate on binary hydrides of the typical elements, ignoring those of the transition elements, lanthanides and actinides, which often have metallic properties and so resemble alloys. Binary hydrides are compounds of hydrogen and one other element. A useful classification of the highest hydrides of the typical elements is shown in Figure 5.9. It divides them into three classes: salt-like, macromolecular and molecular.

IUPAC numbering

| Group | 1 | 2 | 13 | 14 | 15 | 16 | 17 | 18 |

Mendeléev numbering

| Group | I | II | III | IV | V | VI | VII | VIII |

Period								
1							H_2	He
2	LiH	BeH_2	B_2H_6	CH_4	NH_3	H_2O	HF	Ne
3	NaH	MgH_2	AlH_3	SiH_4	PH_3	H_2S	HCl	Ar
4	KH	CaH_2	GaH_3	GeH_4	AsH_3	H_2Se	HBr	Kr
5	RbH	SrH_2	In	SnH_4	SbH_3	H_2Te	HI	Xe
6	CsH	BaH_2	Tl	PbH_4	BiH_3	H_2Po	HAt	Rn
7	FrH	RaH_2						

☐ salt-like

☐ macromolecular

☐ molecular

Figure 5.9 A classification of the highest hydrides of the typical elements, based mainly on structure. (The hydrides of indium and thallium are not well characterized, being extremely unstable to decomposition to the metal and hydrogen.)

5.7.1 Salt-like hydrides

The salt-like or ionic hydrides are those of the Group I and Group II elements, with the exception of beryllium. They have structures consistent with the presence of ions: each atom is surrounded by atoms of a different type.

● In what sense is this true of the Group I hydrides, MH?

● They have the sodium chloride structure (Figure 5.4).

When, as in the case of LiH, they can be melted without decomposition, salt-like hydrides conduct electricity. On electrolysis, the melts yield hydrogen at the positive electrode. They all react with water, giving hydrogen and an alkaline solution:

$$CaH_2(s) + 2H_2O(l) = Ca^{2+}(aq) + 2OH^-(aq) + 2H_2(g) \qquad (5.15)$$

● Is hydride oxidized or reduced in this reaction?

● It is oxidized: hydrogen in oxidation number −1 is oxidized by water to $H_2(g)$ (oxidation number zero).

In other words, the aqueous hydride ion, $H^-(aq)$, cannot exist in aqueous solution: it reduces water.

5.7.2 Macromolecular hydrides

Macromolecular hydrides have extended structures, which differ from those that would be expected of a collection of ions. Thus, BeH_2 has a chain structure like that of $BeCl_2$. Each beryllium is surrounded tetrahedrally by four hydrogens, but each hydrogen has just two berylliums on one side (Figure 5.10, overleaf).

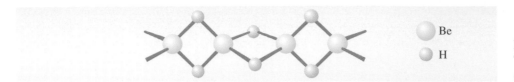

Figure 5.10
The structure of beryllium
dihydride, BeH_2.

Aluminium and beryllium hydrides react with water in the same way as the salt-like hydrides, to give hydrogen and metal hydroxides. However, the hydroxides are insoluble, so alkaline solutions are not produced; for example

$$BeH_2(s) + 2H_2O(l) = Be(OH)_2(s) + H_2(g) \qquad (5.16)$$

5.7.3 Molecular hydrides

Boron and the elements of Groups IV–VII/14–17 form molecular hydrides, those in Figure 5.9 being gases at room temperature. Boron hydrides contain novel types of bonding, and will be discussed in Section 9.1.4.

The hydrides of Groups IV–VII consist of EH_x molecules, in which x is the number of electrons needed to bring the electronic configuration of E up to that of the subsequent noble gas. This is easily explained by simple Lewis theory: the EH_x molecules contain E—H single bonds, and the x bonds add x electrons to the outer electronic configuration of E to give it a noble gas configuration.

⬤ What is the formula and molecular shape of the hydride of arsenic?

⬤ Arsenic in Group V/15 has five outer electrons. Another three will give it the configuration of krypton. Thus, the formula is AsH_3. Three of the five arsenic electrons are involved in the As—H bonds, leaving one non-bonding pair. Thus, there are four tetrahedrally disposed repulsion axes: the molecule is pyramidal, with the fourth axis occupied by the non-bonding pair.

The experimentally determined shape of AsH_3, called arsine, agrees with this prediction (Structure **5.3**). Notice that these principles are consistent with the fact that the noble gases do not form binary hydrides: no electrons, and therefore no E—H bonds, are needed to bring their electronic configurations up to that of a noble gas.

$$H^{-\,-}As^{\,-}{}_{-}H$$
$$H$$

5.3

5.7.4 Trends in structure

In moving across Figure 5.9 from left to right, and especially in Periods 2 and 3, salt-like or ionic hydrides give way first to macromolecular hydrides and then to covalent molecular hydrides. There is a similar trend in the structures of halides. It can be related to electronegativity changes: in Group I, where electronegativities are low, the electronegativity difference in the hydrides is large and ionic bonds are formed. From left to right across a Period, the electronegativity difference diminishes: first macromolecular hydrides and then molecular covalent hydrides appear.

5.8 Hydrogen-bonding

When methane or fluorine gases are chilled, it is van der Waals forces that hold the molecules together in the resulting solids and liquids. They are weak, and a measure of their weakness is the low boiling temperature of liquid methane or liquid fluorine ($-162\,°C$ and $-188\,°C$), respectively.

In the liquid states of some molecular hydrides, however, there is evidence that the van der Waals forces are supplemented by an additional and stronger interaction. Figure 5.11 shows the boiling temperatures of the hydrides of Groups IV–VII/14–17, and of the noble gases.

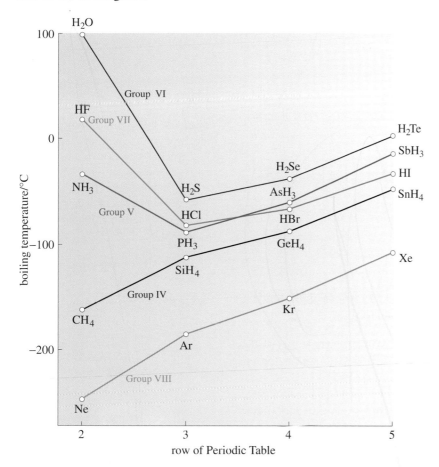

Figure 5.11
The boiling temperatures of the noble gases (Group VIII) and of the hydrides of Groups IV–VII.

Which three hydrides seem to have anomalous values?

H_2O, HF and NH_3: in moving from the bottom to the top of each of Groups IV–VII (right to left in Figure 5.11), the hydride boiling temperatures steadily decrease, except at the tops of Groups V, VI and VII, where there is a sharp *increase*.

This suggests that in liquid H_2O, HF and NH_3 there is a special intermolecular force which is either missing or less marked in the vapour. Further support for such ideas is provided by the structure of the solid compounds. Suppose we compare water, H_2O, with hydrogen sulfide, H_2S, the gas responsible for the smell of rotten eggs. Hydrogen sulfide freezes at −86 °C, and the coordination of each molecule in the solid is shown in Figure 5.12a (overleaf). The H_2S molecules adopt a cubic close-packed arrangement, in which each H_2S molecule is surrounded by twelve others, six in its close-packed plane, three in the plane above, and three in the plane below. This high coordination number maximizes the weak van der Waals bonding between the molecules.

Now compare the H_2S structure with that of ice (Figure 5.12b).

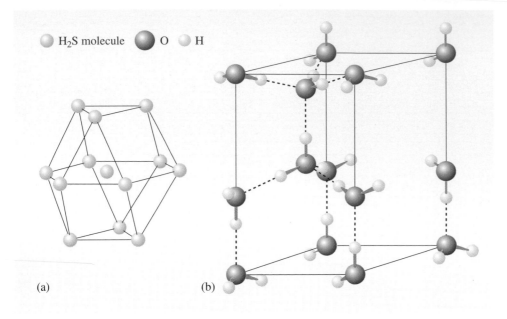

H₂S molecule   O   H

(a)   (b)

Figure 5.12
(a) The coordination of each H_2S molecule in the cubic close-packed structure of the solid; each molecule is surrounded by twelve others.
(b) The structure of ice, H_2O (the hydrogen bonds are shown as dashed purple lines); each molecule has a much lower coordination number than in H_2S.

How many other water molecules is each water molecule surrounded by?

Look at one of the two water molecules which is located *wholly* within the unit cell; each is tetrahedrally surrounded by only four others.

Notice how each oxygen atom is bound not just to two hydrogen atoms by covalent bonds, but also to two other hydrogen atoms which are further away and belong to neighbouring H_2O molecules. As discussed in Section 5.5, the bonds to these more distant hydrogen atoms are **hydrogen bonds** and are shown as purple dashed lines. Bonds of this kind are formed by molecules or ions in which hydrogen is bound to the most electronegative atoms — usually oxygen, nitrogen and fluorine. The bonding can be thought of in electrostatic terms (Figure 5.13). Because of the electronegativity difference between hydrogen and oxygen, the hydrogen in a water molecule carries a fractional positive charge, and can interact electrostatically with the fractional negative charge on the oxygen atom of an adjacent water molecule. When, as in Figures 5.12b and 5.13, a hydrogen atom forms just one hydrogen bond, it lies on or nearly on the line of centres between the two electronegative atoms. In Figure 5.13, the bond lengths are those observed in ice: notice that the covalent O—H bond is only 4 pm longer than the distance of 97 pm found in the free H_2O molecule in steam.

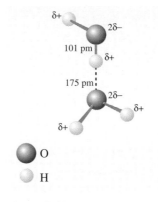

101 pm

175 pm

O

H

Figure 5.13
An electrostatic interpretation of the hydrogen bond between water molecules. Electronegative oxygen carries a fractional negative charge, hydrogen a fractional positive charge.

Which is likely to be the weaker in Figure 5.13: the O—H covalent bond or the O—H hydrogen bond?

The O—H hydrogen bond is longer and therefore likely to be weaker.

Let us try to find out how weak it is. In Figure 5.12b, there is one hydrogen bond on the line of centres between two adjacent oxygen atoms; each water molecule is involved in four hydrogen bonds, one for each of its four neighbours. If we break these four bonds, we release the central water molecule, but we also break $\frac{1}{4}$ of the hydrogen bonds binding each of its four neighbouring water molecules. Since $4 \times \frac{1}{4} = 1$, breaking four hydrogen bonds liberates *two* water molecules, so breaking two hydrogen bonds liberates *one*. The standard enthalpy change for the vaporization of ice, in which one water molecule is liberated, is $50\,kJ\,mol^{-1}$:

$$H_2O(s) = H_2O(g); \quad \Delta H_m^{\ominus} = 50\,kJ\,mol^{-1} \tag{5.18}$$

Thus, the energy of each hydrogen bond is about one half of this, some $25\,kJ\,mol^{-1}$. Typical values for hydrogen bonds lie in the range 10–$30\,kJ\,mol^{-1}$. This is only about 5% of the average energy of the O—H covalent bond ($464\,kJ\,mol^{-1}$: Section 4.2). However, it is some five or six times the energy of the van der Waals bonding which holds the molecules together in solid fluorine or methane; the enthalpy of vaporization of solid methane, for instance, is only $9\,kJ\,mol^{-1}$. Consequently, hydrogen-bonding can offer a significant stabilization, which will prompt a suitable substance to adopt an unexpected structure. Solid HF provides a further example (Figure 5.14).

Figure 5.14 The zigzag chain structure of solid HF; the hydrogen bonds are shown as dashed purple lines. Solid HCl and HBr have a cubic close-packed structure in which each HX molecule is surrounded by twelve others (Figure 5.12a).

HF and H_2O are cases where hydrogen-bonding occurs between identical electronegative atoms of identical molecules. Yet such interaction can occur between different electronegative atoms; moreover, ion–ion or ion–molecule hydrogen bonds are also possible. Consider the ammonium halides, NH_4X, which contain the tetrahedral ammonium ion, NH_4^+. The iodide, NH_4I, has the sodium chloride structure: each NH_4^+ is octahedrally surrounded by six large iodide ions. NH_4Cl and NH_4Br, on the other hand, have the CsCl structure of Figure 5.15: each NH_4^+ is surrounded by eight Cl⁻ or Br⁻ ions at the corners of a cube. It seems that because Cl⁻ and Br⁻ are smaller than I⁻, eight of them, rather than six, can be packed around the ammonium ion.

- If this argument is extended, what should be the minimum number of F⁻ ions packed around NH_4^+ in NH_4F?

- At least eight, because F⁻ is even smaller than Cl⁻.

In fact the number is four, and NH_4F has the structure shown in Figure 5.16 (overleaf). Each ammonium ion forms four N—H—F⁻ hydrogen bonds; this is additional to the usual ionic bonding that exists between NH_4^+ and F⁻. The structure is very similar to that of ice shown in Figure 5.12b. The differences are that the oxygen atoms are replaced by alternate nitrogens and fluorines, and that, instead of each oxygen being tetrahedrally coordinated to two close and two more distant hydrogens, each nitrogen is tetrahedrally coordinated to four close hydrogens in an NH_4^+ ion.

Without the special properties that hydrogen-bonding confers on water, all known forms of life could not exist. In fact, the influence of hydrogen-bonding, which we illustrate here with two examples, is seen at all levels of biological chemistry.

The most common form of DNA is a right-handed double helix. Each strand of the helix consists of a sugar-phosphate polymeric backbone to which are attached the

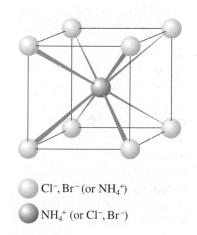

Cl⁻, Br⁻ (or NH_4^+)

NH_4^+ (or Cl⁻, Br⁻)

Figure 5.15
The unit cell of the CsCl structure of ammonium chloride, NH_4Cl, and ammonium bromide, NH_4Br.

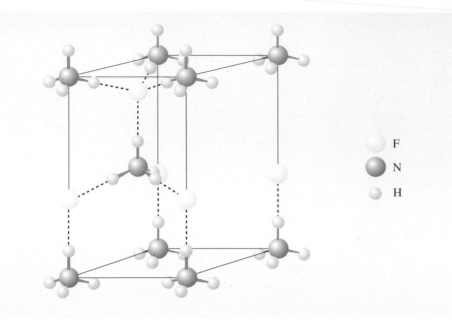

Figure 5.16
The structure of ammonium fluoride; hydrogen bonds are shown in purple. Note the similarity to the structure of ice (Figure 5.12b).

nucleotide bases cytosine, guanine, thymine or adenine (Structures **5.4–5.7**). In this form (Figure 5.17), each horizontal bar of the DNA schematic represents the linkage of a pair of nucleotides, either cytosine (C) to guanine (G) or thymine (T) to adenine (A), by hydrogen-bonding. Nature often selects those atoms and molecules that are best suited to their role in life; in this case the pairing of the nucleotide base pairs illustrated in Figure 5.18 creates the optimum geometry for the formation of hydrogen bonds. In recognition of the work of James Watson and Francis Crick in revealing the double-helix structure of DNA, the structures in Figure 5.18 are known as Watson–Crick base pairs.

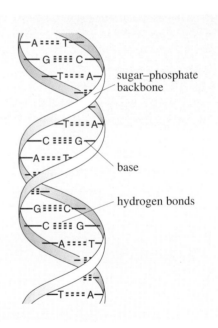

Figure 5.17
The double-helical structure of DNA, showing hydrogen-bonding between the bases on the two strands.

5.4

cytosine (C)

5.5

thymine (T)

5.6

guanine (G)

5.7

adenine (A)

Watson–Crick base pairing

Figure 5.18 The Watson–Crick scheme for the pairing of nucleotide bases.

Our second example is the role of hydrogen-bonding in the structure of proteins in general. Protein structure is often described at three levels. The primary structure is determined by the covalent bonds that join amino acids in a polypeptide chain (Figure 5.19). It is in the secondary structure that hydrogen-bonding comes into play. Here the polypeptide chains of amino acids either coil to allow hydrogen-bonding between members of the same chain, or neighbouring chains join each other through hydrogen bonds to form a sheet. A coiled structure is known as an α–helix (Figure 5.20a, b, overleaf) and the sheet structure as a β-sheet (Figure 5.20c, d). These features are evident in the overall structure of a protein where they are usually depicted as a coiled strip (α–helix) or a series of parallel strips of arrows (β-sheet). An example of a protein containing both features is shown in Figure 5.21 (overleaf).

Figure 5.19
The primary structure of proteins.

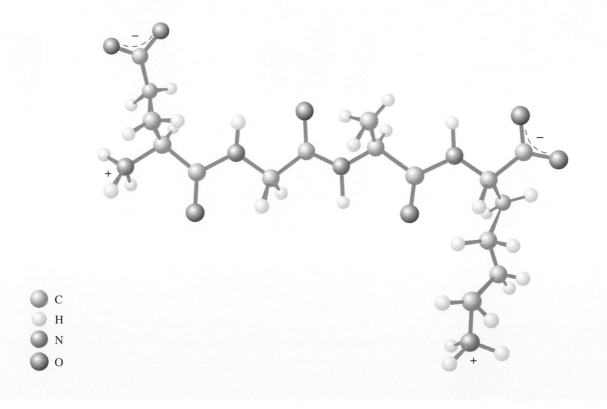

C
H
N
O

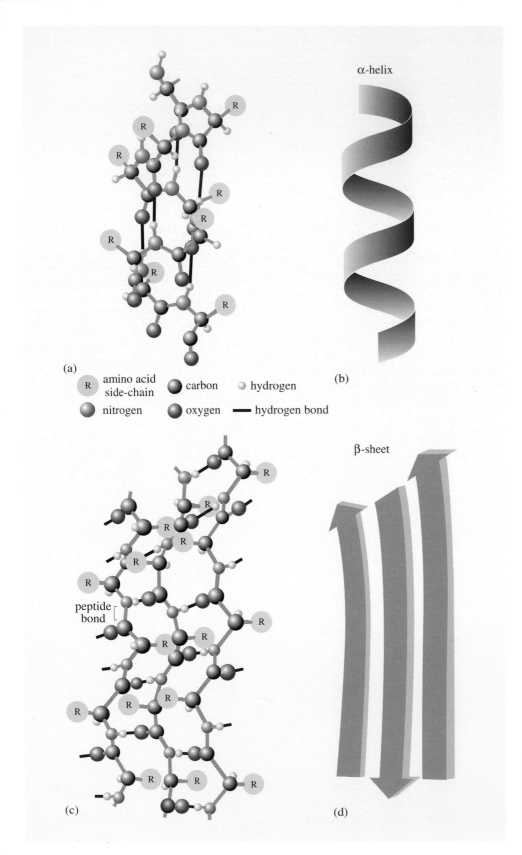

α-helix

(a)

R amino acid side-chain   carbon   hydrogen
nitrogen   oxygen   ▬ hydrogen bond

(b)

β-sheet

peptide bond

(c)   (d)

Figure 5.20 An α-helix (a) and (b), and a β-sheet (c) and (d) of amino acids. In the β-sheet, two or more polypeptide chains run alongside each other and are linked in a regular manner by hydrogen bonds.

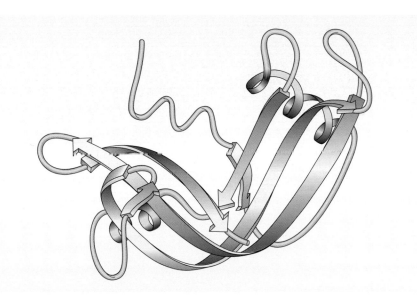

Figure 5.21
The tertiary structure of an enzyme RNAase shown as a 'ribbon' drawing. β-sheets are shown in blue and α-helices in red.

In Figure 5.21 it is also evident that the overall structure of the protein requires interactive forces between various parts of the polypeptide chain. These forces hold the molecule in the globular framework that is known as the tertiary structure. Hydrogen-bonding is also one of several forces acting at this third level of protein structure.

5.9 Summary of Sections 5.7 and 5.8

1   The hydrides of the typical elements can be classified as either salt-like (Group I and Group II except Be), macromolecular (Be and Al) or molecular (boron and Groups IV–VII/14–17.) Salt-like hydrides, which have the characteristics of ionic compounds, and macromolecular hydrides, react with water giving an aqueous or solid hydroxide.

2   The highest molecular hydrides of Groups IV–VIII/14–18 have formulae EH_x, where $x = 8 - N$ and N is the Mendeléev Group number. The shapes are correctly predicted by VSEPR theory.

3   A hydrogen atom can bind two electronegative atoms together by forming a covalent bond to one and a hydrogen bond to the other. The hydrogen bond may be thought of as an electrostatic interaction between partial positive and partial negative charges on hydrogen and the electronegative atom, respectively.

4   Hydrogen-bonding between molecules gives rise to unexpectedly high boiling temperatures (NH_3, H_2O, HF), and solids with unusual structures containing low coordination numbers (ice, HF, NH_4F).

5   Hydrogen-bonding plays many crucial roles in biological chemistry. In the double helix of DNA, the two strands are held together by hydrogen bonding. The function of proteins depends critically on their three-dimensional structure, which is determined at several levels by hydrogen-bonding. The α-helix and β-sheet motifs and the higher-order folded structures of proteins result from the formation of hydrogen bonds.

QUESTION 5.3

When aluminium is heated with tellurium, solid aluminium telluride, Al_2Te_3, is formed. If this is treated with chilled dilute hydrochloric acid, a gaseous hydride of tellurium with a revolting smell is evolved. What is the formula of the hydride, and what is the shape of the molecule?

QUESTION 5.4

In solid sodium hydrogen carbonate, $NaHCO_3$, the hydrogen, carbon and oxygen atoms occur in infinite chains, with the geometry shown in Figure 5.22. Each chain is bound to others running parallel to it, via Na^+ ions in between the chains. Identify the hydrogen bonds in Figure 5.22, and specify their length. Draw a structural formula for the chain in Figure 5.22 which includes both C=O and C—O$^-$ bonds, and which best fits the data in the Figure.

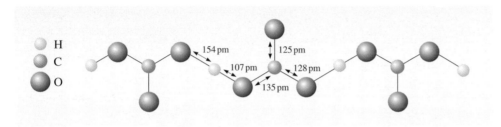

Figure 5.22 Structure of the chains that contain the hydrogen, carbon and oxygen atoms in solid $NaHCO_3$.

GROUP VII/17: HALOGENS AND HALIDES

6

The halogens (Figure 6.1) and the noble gases are the two Groups in the Periodic Table which do not contain a metallic element. Astatine may be a non-metal or semi-metal: its longest-lived isotopes, ^{210}At and ^{211}At, have half lives of only 7–8 hours, so very little is known about it. If, however, we set astatine on one side, the halogens are all non-metals consisting of diatomic molecules, X_2. At room temperature, F_2 and Cl_2 exist as yellow and yellow–green gases, respectively; Br_2 is a deep red liquid; and I_2 is a shiny black solid, which sublimes to give a purple vapour on warming.

These elemental states do not occur naturally. Fluorine, chlorine and bromine are found almost exclusively as solutions or minerals containing halide ions, X^-. Furthermore, most iodine is obtained from iodide solutions, but some occurs as iodates (e.g. $NaIO_3$) in, for example, the famous sodium nitrate deposits in the Atacama desert near the west coast of South America. In industry, the halogens are obtained primarily from halide ions.

	16	17	18
	VI	VII	VIII
	O	F	Ne
	S	Cl	Ar
	Se	Br	Kr
	Te	I	Xe
	Po	At	Rn

Figure 6.1
The place of the halogens in the Periodic Table.

⬤ How is this done in the case of chlorine?

⬤ The chlor-alkali industry uses the electrolysis of sodium chloride solutions.

Once made, chlorine can be used to produce bromine and iodine from seawater, or underground brines like those found in Louisiana. Chlorine is bubbled through the solution to yield the free elements:

$$Cl_2(g) + 2Br^-(aq) = 2Cl^-(aq) + Br_2(aq) \tag{6.1}$$
$$Cl_2(g) + 2I^-(aq) = 2Cl^-(aq) + I_2(aq) \tag{6.2}$$

The bromine and iodine are then carried out of solution as a vapour by using a current of air or steam.

⬤ What *type* of process is Reaction 6.1 and 6.2?

⬤ A redox reaction: the chlorine oxidation number decreases from zero to –1; the bromine or iodine oxidation number increases from –1 to zero.

These reactions show that chlorine is a stronger oxidizing agent than bromine or iodine. Now look at Table 6.1.

⬤ Do the redox potentials agree with this assessment of oxidizing strength?

⬤ Yes; the size of $E^{\ominus}$ is a measure of the strength of the oxidizing agent, and the value for chlorine exceeds that for bromine or iodine.

The preparation of fluorine is much more difficult because, as Table 6.1 shows, fluorine gas is an extremely strong oxidizing agent. It cannot, for example, be obtained from aqueous solution because it oxidizes water. A convenient starting material is the mineral fluorite, CaF_2, from which hydrogen fluoride, HF, can be

Table 6.1 Standard redox potentials for the halogen/halide couples in aqueous solution

Halogen	$\dfrac{E^{\ominus}(\frac{1}{2}X_2 \mid X^-)}{V}$
F	2.89
Cl	1.36
Br	1.10
I	0.53

obtained by treatment with concentrated sulfuric acid:

$$CaF_2(s) + H_2SO_4(l) = CaSO_4(s) + 2HF(g) \tag{6.3}$$

Liquid HF boils at 20 °C. Both it and its aqueous solutions cause deep excruciating burns.

In 1886, Henri Moissan (Figure 6.2) attempted the electrolysis of pure liquid HF. Purification was accomplished by combining the vapour with KF to get potassium hydrogen fluoride, KHF_2:

$$KF(s) + HF(g) = KHF_2(s) \tag{6.4}$$

This compound contains the linear HF_2^- ion, in which the hydrogen lies *midway* between the two fluorines (Structure **6.1**). The bonding in this ion is examined in Question 7.5. A reversal of Reaction 6.4 takes place when KHF_2 is heated, and, if the heating is done carefully, pure HF is obtained:

$$KHF_2(s) = KF(s) + HF(g) \tag{6.5}$$

$$\xleftarrow{\hspace{1em}} 223\ pm \xrightarrow{\hspace{1em}}$$
$$[F\text{——}H\text{——}F]$$

6.1

The HF vapour was condensed and electrolysed in a platinum U-tube with platinum–iridium alloy electrodes at –23 °C. Convinced that fluorine would react violently with silicon, Moissan had long tested likely samples with a crystal of silicon. His crystal had survived many tests unscathed, but on this occasion, when it was exposed to the gas evolved at the positive electrode, it exploded and burst into flames: fluorine had at last been isolated. The French Academy of Science sent a deputation to verify this important discovery. Moissan took no chances: determined that his HF should be as pure as possible, he performed Reaction 6.5 with especial care. But the crucial demonstration was a disaster: no current flowed, and no electrolysis took place. When the embarrassment subsided, Moissan realized that he had been too careful. Pure liquid HF is a very poor conductor, but his first and productive sample had been contaminated by KF from Reaction 6.5. Like K_2O in water, KF in HF creates ions:

$$K_2O(s) + H_2O(l) = 2K^+(aq) + 2OH^-(aq) \tag{6.6}$$

$$KF(s) + HF(l) = K^+(hf) + HF_2^-(hf) \tag{6.7}$$

where (hf) denotes the solvated ion dissolved in liquid HF. When Moissan deliberately added KF to his liquid hydrogen fluoride, he was able to validate his claim. The reactions at the two electrodes are:

positive: $2HF_2^-(hf) = 2HF(l) + F_2(g) + 2e^-$ \hfill (6.8)

negative: $4HF(l) + 2e^- = H_2(g) + 2HF_2^-(hf)$ \hfill (6.9)

total: $2HF(l) = H_2(g) + F_2(g)$ \hfill (6.10)

Today, fluorine is still made by Moissan's method. The chief difference is that, whereas he used a mole ratio KF : HF of about 1 : 13, most modern cells use 1 : 2 (Figure 6.3).

Figure 6.2
The career of Henri Moissan (1852–1907) illustrates one of the less politically correct influences of women upon science: he was very happily married, and Mme Moissan's dowry was large enough to give him the financial independence that he needed to pursue his researches. He won the 1906 Nobel Prize for Chemistry for the isolation of fluorine, but died only 10 weeks after the ceremony. He had once remarked that fluorine would take 10 years off his life.

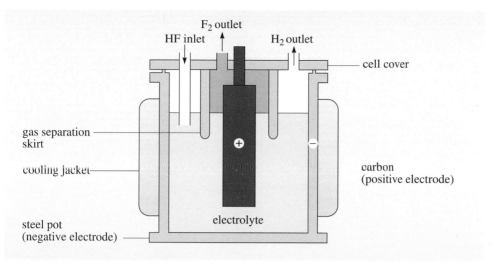

Figure 6.3 A typical, industrial fluorine-generating cell. The electrolysis occurs in a steel pot, which also serves as the negative electrode. The positive electrode is compacted carbon, and a gas separation skirt prevents reaction between the fluorine produced at this electrode and the hydrogen produced at the other. The electrolyte is a blend of one mole of KF per 2 moles of HF, which is liquid above 70 °C. The cell is maintained at 80–100 °C by using a cooling jacket when electrolysis is occurring, and by heating when it is not. The fluorine can be contained within the steel pot because it forms a thin layer of insoluble fluoride, which prevents the bulk of the metal being attacked.

6.1 Industrial uses of the halogens

Chlorine is the most important halogen. About 40 million tonnes are consumed each year, mainly in the synthesis of organochlorine compounds, many of which are used as solvents or converted into plastics. Figure 6.4 shows the distribution of chlorine consumption; the largest *single* use is in the manufacture of chlorinated plastics. For example, it is used to make vinyl chloride (chloroethene) from ethylene (ethene):

$$CH_2{=}CH_2 \;+ Cl_2 \xrightarrow[\text{catalyst}]{FeCl_3} \; ClH_2C-CH_2Cl \qquad (6.11)$$
$$\text{1, 2 - dichloroethane}$$

$$ClH_2C-CH_2Cl \xrightarrow{500\,°C} CH_2{=}CHCl(g) \;+\; HCl(g) \qquad (6.12)$$
$$\text{vinylchloride}$$

Vinyl chloride can then be polymerized to make polyvinyl chloride or PVC, $(CH_2-CHCl-)_n$. Apart from its use in the organic chemicals industry, chlorine is employed mainly as a bleaching agent or disinfectant; these applications draw on its

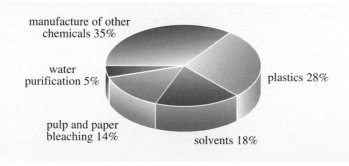

Figure 6.4
The distribution of chlorine consumption.

strength as an oxidizing agent. There is, however, pressure to reduce the use of chlorine gas and organochlorine compounds, because they can lead to the environmental release of chlorinated hydrocarbons, which may cause cancer and other kinds of adverse health effects. The danger compounds include the notorious dioxin (Structure **6.2**), which is a degradation product of organochlorine pesticides and other similar compounds.

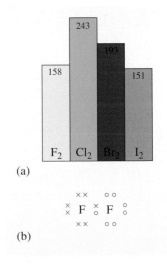

6.2

The pattern for bromine use is similar to that of chlorine, but on a much smaller scale: bromine production is about one per cent that of chlorine. Fluorine production is smaller still and is consumed mainly in the nuclear industry for the manufacture of uranium hexafluoride, UF_6. Natural uranium is 99.3% ^{238}U and only 0.7% ^{235}U. Most nuclear reactors require fuel that contains a higher percentage of ^{235}U (2–5%), and this change is brought about by methods that require the uranium to be in a gaseous form. UF_6 is gaseous above 64 °C, despite the presence of the very heavy uranium atom.

● What explains this remarkable property?

● The sheath of surrounding fluorine atoms, which gives the molecule a low polarizability and hence weakens intermolecular forces (Section 4.5).

6.2 Properties of halogen atoms

Some important properties of the halogens appear in Table 6.2. Beneath the electronic configurations come the covalent radii of the bound atoms and the radii of the negative ions, X^-; both increase down the Group. Then come the first ionization energies and electronegativities, which both decrease down the Group.

● To what may we attribute these changes?

● Descending the Group from fluorine to iodine, the distance of the outer shell of electrons from the nucleus increases.

The same argument suggests that the electron affinities, $E(X)$, should decrease from fluorine to iodine. As you can see, this is so, except in the case of fluorine, where the value is unexpectedly low. A similar anomaly occurs in the bond dissociation energies for the halogen molecules, where the unusually low value for fluorine breaks the otherwise gradual decrease that occurs from the top to the bottom of the Group: fluorine atoms form unexpectedly weak bonds with each other (Figure 6.5a). Notice, however, that this is a special case: fluorine atoms do not usually form weak bonds — quite the reverse, in fact. Two important examples in Table 6.2, the $B(H—X)$ and $B(C—X)$ bond energy terms, have been chosen to illustrate this. This time, there is a gradual decrease down the Group and the fluorine value is very high.

● How will the lengths of the H—X and C—X bonds change from fluorine to iodine?

● They will increase as the covalent radius of the halogen atom increases.

Thus, in both these cases, the smaller the halogen atom, the shorter the bond and the stronger it will be. This is a general tendency in inorganic chemistry: small atoms form short bonds, and short bonds are usually strong. The special case of the weak F—F bond is an exception; an explanation is offered in Figure 6.5b.

(a)

(b)

Figure 6.5
(a) The bond dissociation energies of the gaseous diatomic halogen molecules decrease down the group, except for F_2, which has an unusually low value similar to that of I_2.
(b) The usual explanation: bound fluorine atoms are small, so the F—F distance is short; also each fluorine has six non-bonding electrons. The bond is weakened by repulsion between these non-bonding electrons over the short internuclear distance.

Table 6.2 Some properties of the halogen atoms, molecules and ions

	F	Cl	Br	I
electronic configuration	$[He]2s^22p^5$	$[Ne]3s^23p^5$	$[Ar]3d^{10}4s^24p^5$	$[Kr]4d^{10}5s^25p^5$
covalent radius/pm	71	99	114	133
$r(X^-)$/pm	119	167	182	206
$I(X)$/kJ mol^{-1}	1 681	1 251	1 140	1 008
Pauling electronegativity	4.0	3.0	2.8	2.5
$E(X)$/kJ mol^{-1}	328	349	324	295
$D(X{-}X)$/kJ mol^{-1}	158	243	193	151
$E^\ominus(\frac{1}{2}X_2\vert X^-)$/V	2.89	1.36	1.10	0.53
$B(H{-}X)$/kJ mol^{-1}	568	432	366	298
$B(C{-}X)$/kJ mol^{-1}	467	346	290	228
$\Delta H_f^\ominus(X_2, g)$/kJ mol^{-1}	0	0	30.9	62.4

Table 6.2 also includes the redox potentials, $E^\ominus(\frac{1}{2}X_2\vert X^-)$, which confirm that the oxidizing strength of the elemental halogen decreases from fluorine to iodine. Finally, there is the standard enthalpy of formation of the gaseous molecules, $X_2(g)$. The values are zero for fluorine and chlorine because these substances are already gaseous at 25 °C. However, bromine is a red liquid and iodine a dark solid, so $\Delta H_m^\ominus$ for the reaction

$$X_2(\text{standard state}) = X_2(g) \tag{6.13}$$

has a positive value at 25 °C in these two cases.

⬤ What kind of interaction explains why $\Delta H_m^\ominus$ for Equation 6.13 increases from chlorine to iodine?

◯ The induced dipole–induced dipole interaction (Section 4.5); this increases as the molecules become larger and more polarizable.

6.3 Oxidation number −1: halides

The halogen atoms are one electron short of a noble gas configuration. Halides are substances in which that configuration is attained by chemical combination. If the halide is ionic (e.g. NaCl), this is done by forming the X^- ion; in covalent halides (e.g. SCl_2) it is done by forming a single covalent bond and sharing an electron pair. Every element except helium, neon and argon forms at least one halide. Halides of Groups I and II are ionic. The halides of other typical elements are covered when we deal with their groups in the Periodic Table. The exception is the hydrogen halides, which are covalent, and which we discuss here because they are so prominent in halogen chemistry.

6.3.1 The hydrogen halides

HCl, HBr and HI are all colourless gases at room temperatures; liquid HF has a boiling temperature of 20 °C.

⬤ Why is HF so involatile?

⬤ Because the high electronegativity of fluorine leads to strong hydrogen-bonding in the liquid state (Figure 5.14).

Because of their volatility, HF and HCl can be made by displacing them from fluorides and chlorides with the relatively involatile concentrated sulfuric acid (e.g. Figure 6.6):

$$NaCl(s) + H_2SO_4(l) = NaHSO_4(s) + HCl(g) \qquad (6.14)$$

This is a general way of making volatile acids that are robust enough to resist attack by H_2SO_4. However, it does not work for HBr and HI because they are oxidized by the acid to bromine and iodine, respectively. These two gases can be made by the hydrolysis of phosphorus trihalides; for example

$$PBr_3(l) + 3H_2O(l) = 3HBr(g) + H_3PO_3(aq) \qquad (6.15)$$

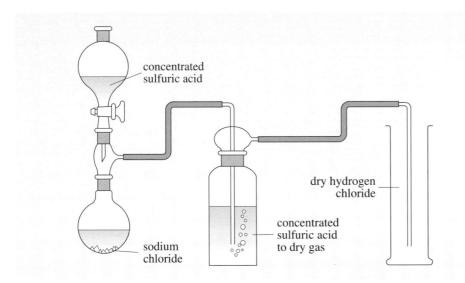

concentrated sulfuric acid

dry hydrogen chloride

concentrated sulfuric acid to dry gas

sodium chloride

Figure 6.6
A simple way of making hydrogen chloride in the laboratory. The gas is dried by bubbling it through concentrated sulfuric acid, which is extremely soluble in water, and mops up water vapour.

Alternative preparative methods for HCl, HBr and HI include the direct combination of hydrogen and halogen, the reaction being facilitated by a platinum catalyst in the case of HBr and HI.

Some physical properties of the compounds are recorded in Table 6.3. As expected, the dipole moment decreases with the difference in electronegativity of H and X from HF to HI. At the same time there is a steep decline in the bond energy, which can also be related to the decreasing electronegativity difference through Pauling's electronegativity equation (Equation 4.12). The declining bond energy is reflected in a decrease in thermal stability. The decomposition of hydrogen bromide into hydrogen and bromine begins above 700 °C, and even at 300 °C, HI contains significant amounts of $H_2(g)$ and $I_2(g)$.

Table 6.3 Properties of the hydrogen halides

	HF	HCl	HBr	HI
boiling temperature/°C	20	−84	−67	−35
$r(H-X)$/pm	92	127	141	161
$D(H-X)$/kJ mol^{-1}	568	432	366	298
dipole moment/D	1.82	1.08	0.82	0.44

All four gases readily dissolve in water to give acid solutions (Figure 6.7). However, HF is a weak acid, whereas HCl, HBr and HI are strong acids. In Brønsted–Lowry form, the dissociation can be represented as:

$$HX(aq) + H_2O(l) = H_3O^+(aq) + X^-(aq) \qquad (6.16)$$

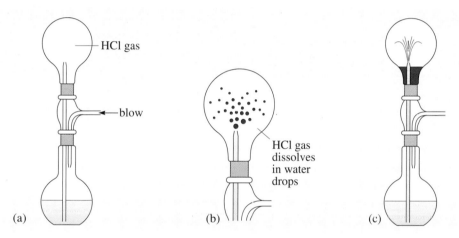

(a)   (b)   (c)

Figure 6.7 The famous 'fountain experiment' illustrates both the very high solubility of hydrogen chloride in water, and the fact that the solution is acid. (a) An inverted flask full of HCl is connected by a tube to a reservoir containing water coloured by blue litmus. By blowing down the auxillary tube, this water is forced up the connecting tube into the upper flask containing HCl. (b) Because HCl is so soluble, the first drops of water dissolve the contents of the flask, leaving a vacuum. (c) Water then fountains in without further assistance, filling the vacuum and the flask and turning red as it does so because of the acidity of the solution and the presence of litmus.

The weakness of HF as an acid has much to do with the fact that because oxygen and fluorine are highly electronegative, the H_3O^+ and F^- ions can form ion pairs held together by hydrogen bonds (Structure **6.3**). This reduces the concentration of free H_3O^+(aq) in dilute aqueous solutions and makes HF appear a weak acid.

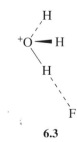

6.3

In the case of HCl, HBr and HI, the equilibrium in Equation 6.16 lies so far to the right that all three acids seem equally strong. However, differences become apparent if something that is less willing to accept the protons from the acids is used as the solvent — that is, something that is a weaker base than water. Liquid acetic acid, CH_3COOH, is suitable.

⬤ Rewrite Equation 6.16 for this new solvent.

⬤ $HX(aca) + CH_3COOH(l) = CH_3CO\overset{+}{O}H_2(aca) + X^-(aca) \qquad (6.17)$
where aca denotes solvation by acetic acid.

This is a more searching test of acidity; ionization is incomplete, and measurements show that equilibrium lies most to the right for HI and least for HCl. Thus, by this test, acid strength runs in the order HI > HBr > HCl > HF. The corrosive nature of HF and its solutions has already been mentioned. They are also remarkable for attacking glass and silica, the final product in the presence of water being fluorosilicic acid, H_2SiF_6:

$$SiO_2(s) + 4HF(aq) = SiF_4(g) + 2H_2O(l) \tag{6.18}$$

$$SiF_4(g) + 2HF(aq) = H_2SiF_6(aq) \tag{6.19}$$

In the laboratory, HF can be handled in containers made from Teflon® (see Figure 4.14).

As in Equation 6.18, many metal halides can be prepared by heating the metals or metal oxides in the hydrogen halide gases; for example

$$2Al(s) + 6HCl(g) = 2AlCl_3(s) + 3H_2(g) \tag{6.20}$$

$$Al_2O_3(s) + 6HF(g) = 2AlF_3(s) + 3H_2O(g) \tag{6.21}$$

6.4 Thermodynamics of the halogen–halide relationship

How easy is it, thermodynamically speaking, to convert the elemental halogens into free halogen atoms or free halide ions? Consider first the conversion into atoms. This occurs in two steps (Figure 6.8), in which a gaseous halogen atom is formed from the elemental halogen in its standard state. The overall energy change is therefore $\Delta H_f^{\ominus}(X, g)$. Then, from Figure 6.8,

$$\Delta H_f^{\ominus}(X, g) = \tfrac{1}{2}\Delta H_f^{\ominus}(X_2, g) + \tfrac{1}{2}D(X{-}X) \tag{6.22}$$

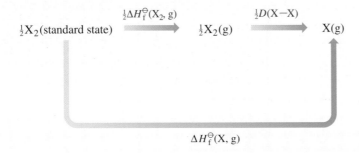

Figure 6.8
A thermodynamic breakdown of the standard enthalpy change for the conversion of elemental halogens into halogen atoms at 25 °C.

In Table 6.4, the values of $\Delta H_f^{\ominus}(X, g)$ in column 4 have been calculated from this equation by adding the values of $\tfrac{1}{2}\Delta H_f^{\ominus}(X_2, g)$ and $\tfrac{1}{2}D(X{-}X)$ given in columns 2 and 3, respectively, and which are derived from data in Table 6.2. Two points are worth making. First, the values are not all that different: they cover a range of less than 45 kJ mol^{-1}, and the values for chlorine, bromine and iodine are especially close. Secondly, the value for fluorine (79 kJ mol^{-1}) lies well below the others: fluorine is the easiest of the elemental halogens to convert into gaseous atoms. Apart from the noble gases, all elements in the Periodic Table are held together by some sort of chemical bond at room temperature. Of these, only caesium and mercury are easier to break up into separate, gaseous atoms than fluorine.

Table 6.4 Thermodynamics of the formation of gaseous halogen atoms and halide ions from the elements in their standard states at 298.15 K

X	$\frac{1}{2}\Delta H_f^{\ominus}$ (X_2, g)	$\frac{1}{2}D(X{-}X)$	$\Delta H_f^{\ominus}$ (X, g)	$-E(X)$	$\Delta H_f^{\ominus}$ (X^-, g)
			kJ mol^{-1}		
F	0	79	79	−328	−249
Cl	0	122	122	−349	−227
Br	15	97	112	−324	−212
I	31	76	107	−295	−188

⬤ What energy term is responsible for this?

⬤ The low value of $D(F{-}F)$; the F—F bond is unusually weak (Figure 6.5).

In Figure 6.9, we go one step further, and add an electron to each halogen atom to get the gaseous X^- ion. The overall reaction is:

$$\tfrac{1}{2}X_2(\text{st}) + e^-(g) = X^-(g) \tag{6.23}$$

where st denotes standard state; because a gaseous halide ion is formed from the standard state of the element, the standard enthalpy change is expressed as $\Delta H_f^{\ominus}$ (X^-, g). From Figure 6.9, $\qquad$ (6.24)

$$\Delta H_f^{\ominus}(X^-, g) = \tfrac{1}{2}\Delta H_f^{\ominus}(X_2, g) + \tfrac{1}{2}D(X-X) - E(X) \tag{6.25}$$

$$= \Delta H_f^{\ominus}(X, g) - E(X)$$

Column 5 of Table 6.4 gives the values of $-E(X)$, and column 6 gives the result of adding them to the values of $\Delta H_f^{\ominus}$ (X, g) in column 4. Again, the figures are not very different.

⬤ Which gaseous halide ion is most readily formed from the elemental state?

⬤ The fluoride ion: $\Delta H_f^{\ominus}$ (X^-, g) is most negative for this element.

Notice that this is true despite the higher electron affinity of chlorine; again, it is a result of the weakness of the F—F bond. If the F—F and Cl—Cl bonds were equally strong, the answer would have been chlorine.

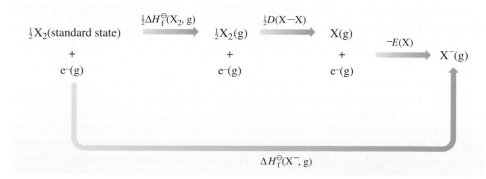

Figure 6.9
A thermodynamic breakdown of the standard enthalpy change for the formation of a gaseous halide ion from the elemental halogen and a gaseous electron.

6.4.1 Halides and oxidation numbers

Halides of an element are often made by the reaction of that element with halogens. For example, sulfur is in Group VI/16 (Figure 6.1), so its highest oxidation number is +6. Which halogen is most likely to lift sulfur to this state? In the lower reaction in Figure 6.10, we imagine the gaseous halide, $SX_6(g)$, being formed from its elements.

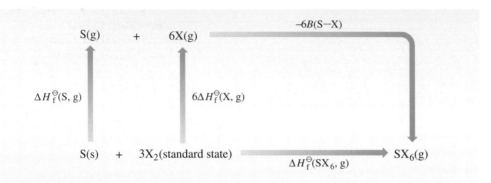

Figure 6.10
A thermodynamic breakdown of the standard enthalpy change for the formation of a gaseous, covalent halide, $SX_6(g)$ from its elements.

The standard enthalpy change is $\Delta H_f^{\ominus}(SX_6, g)$, and it is equal to the sum of the clockwise steps in the cycle:

$$\Delta H_f^{\ominus}(SX_6, g) = \Delta H_f^{\ominus}(S, g) + 6\Delta H_f^{\ominus}(X, g) - 6B(S - X) \tag{6.26}$$

Now look at the terms on the right-hand side of Equation 6.26. $\Delta H_f^{\ominus}(S, g)$ is unaffected by changes in X, so the formation of SX_6 will be most favoured when $\Delta H_f^{\ominus}(X, g)$ is small and the bond energy term, $B(S—X)$, is large, because then $\Delta H_f^{\ominus}(SX_6, g)$ will be at its most negative.

- Which halogen most favours formation of SX_6?

- Fluorine. Because fluorine forms weak bonds with itself, $\Delta H_f^{\ominus}(F, g)$ is small, and because the small fluorine atom usually forms strong bonds with other elements, $B(S—F)$ will be large.

This is correct; if sulfur is heated in fluorine, the gas SF_6 is formed, but if sulfur is heated in excess chlorine, the product is SCl_2; SCl_6 has never been made. Moreover, bromine and iodine should be even less likely to bring out sulfur's highest oxidation number than chlorine. This is because $\Delta H_f^{\ominus}(X, g)$ does not differ greatly for chlorine, bromine and iodine, so the decrease in $B(S—X)$ as the halogen gets bigger will make the formation of the halide less favourable. Not even SBr_2 is formed when sulfur and bromine react together; the product is S_2Br_2 (Structure **6.4**), a dark red liquid. Sulfur and iodine do not react at all. In this and other cases, then, fluorine is best at bringing out high oxidation numbers, followed by chlorine, bromine and iodine in that order.

6.4

Figure 6.10 deals with covalent halides; now we turn to the ionic case. This time silver provides our example. Silver forms many compounds in the +1 oxidation number, but few in the +2. Which halogen is most likely to form a dihalide when heated with silver?

This time we use a cycle of the Born–Haber type (Figure 6.11), from which we obtain

$$\Delta H_f^{\ominus}(AgX_2, s) = \Delta H_f^{\ominus}(Ag^{2+}, g) + 2\Delta H_f^{\ominus}(X^-, g) + L(AgX_2, s) \tag{6.27}$$

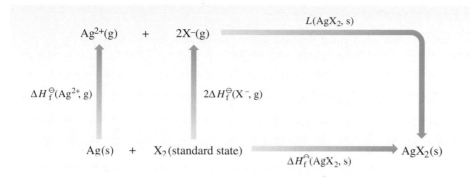

The lattice energy, L,* is approximately given by:

$$L = -\frac{WvZ_+Z_-}{r_+ + r_-} \qquad (6.28)$$

For AgX_2, $v = 3$, $Z_+ = 2$ and $Z_- = 1$, so

$$\Delta H_f^{\ominus}(AgX_2, s) = \Delta H_f^{\ominus}(Ag^{2+}, g) + 2\Delta H_f^{\ominus}(X^-, g) - \frac{6W}{r(Ag^{2+}) + r(X^-)} \qquad (6.29)$$

⬤ Which halogen favours formation of $AgX_2(s)$?

⬤ Fluorine; $\Delta H_f^{\ominus}(AgX_2, s)$ is most negative when $\Delta H_f^{\ominus}(X^-, g)$ is most negative, and when $r(X^-)$ is at its smallest. Both these conditions are fulfilled by fluorine.

This is correct; if silver is warmed with the halogens, the solid products are yellow AgI, cream AgBr, white AgCl and dark brown AgF_2: only fluorine forms a dihalide. Thus, in both covalent and ionic situations, fluorine is the best halogen for bringing out high oxidation numbers.

6.4.2 Oxidizing strength of the halogens in aqueous solution

As noted earlier, the relative oxidizing strengths of the elemental halogens are given by the values of $E^{\ominus}(\frac{1}{2}X_2 \mid X^-)$ in Table 6.1; they run in the order $F_2 > Cl_2 > Br_2 > I_2$. Here we shall only outline the conclusions of a thermodynamic analysis of this trend.

When the elemental halogens act as oxidizing agents, they are converted into $X^-(aq)$ ions. In the process, we imagine first a step in which X_2 is converted into $X^-(g)$. The energy of this step is most favourable for fluorine, but it is quite similar for all the halogens: the values of $\Delta H_f^{\ominus}(X^-, g)$ in Table 6.4 illustrate the point.

There is then a step in which the gaseous halide ions are immersed in water:

$$X^-(g) = X^-(aq) \qquad (6.30)$$

Energy changes for such processes are called **energies of hydration** of ions. As an indication of their size, we give, in Table 6.5, values of $\Delta G_m^{\ominus}$ for Reaction 6.30. Notice that they are large and negative. This is because when halide ions are immersed in water, they become strongly bound by water molecules. X-ray studies of the structure of solutions of metal halides suggest that the interaction is of the type shown in Figure 6.12 (overleaf): as the hydrogen atoms of the water molecules carry a fractional positive charge, they interact with the negatively charged halide ions.

Figure 6.11
A thermodynamic breakdown of the standard enthalpy change for the formation of a solid ionic halide, $AgX_2(s)$, from its elements.

Table 6.5 The standard Gibbs energy changes for the hydration of the halide ions at 298.15 K: $\Delta G_m^{\ominus}$ for Reaction 6.30

	$\Delta G_m^{\ominus}$/kJ mol^{-1}
F$^-$	−475
Cl$^-$	−349
Br$^-$	−324
I$^-$	−289

*Lattice energy calculations are discussed in *Metals and Chemical Change*.[3]

⬤ Why do the hydration energies of the halide ions become markedly less negative between fluorine and iodine?

⬤ As the size of the halide ion increases from fluoride to iodide, the halide–hydrogen distance increases, and so the interaction becomes weaker.

So in passing from $\frac{1}{2}X_2$ to $X^-(aq)$, there is an intermediate step from $\frac{1}{2}X_2$ to $X^-(g)$, whose energy is similar for all the halogens; then there is a subsequent step from $X^-(g)$ to $X^-(aq)$, which becomes much less favourable from fluorine to iodine.

⬤ What effect will this second step have on the change in oxidizing strength of X_2 in descending the Group?

⬤ It makes the ease with which $\frac{1}{2}X_2$ is converted to $X^-(aq)$ diminish; that is, it causes a decrease in the oxidizing strength of X_2.

Thus, the decrease in oxidizing strength of the halogens from fluorine to iodine is primarily due to the decreasing interaction of the halide ions and the solvent as the size of the halide ion increases: the strength of fluorine as an oxidizing agent in aqueous solution is a consequence of the small size of the F^- ion.

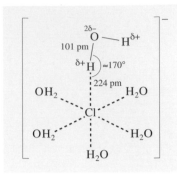

Figure 6.12
The octahedral arrangement of the water molecules around a chloride ion in some aqueous solutions of metal chlorides. The attachment of all the water molecules is like the one shown in detail.

6.5 Summary of Sections 6.1–6.4

1   The halogens are non-metals which, in the elemental state, contain diatomic molecules; at room temperature, fluorine and chlorine are yellow–green gases, bromine is a red liquid, and iodine is a black solid.

2   The elemental halogens are usually obtained by the oxidation of halide ions. Chlorine, bromine and iodine are made industrially by the oxidation of the aqueous halides in brines or seawater. With chlorine, the oxidation is electrolytic; with bromine and iodine, chlorine is the oxidizing agent. Industrial fluorine is made by the electrolytic oxidation of a conducting, liquid blend of KF and HF.

3   On descending the halogen Group, the single-bond covalent radii and the ionic radii of halide ions increase, whereas the ionization energies and electronegativities of the atoms decrease.

4   The bond dissociation energies of the halogen molecules, X_2, run in the order $D(Cl-Cl) > D(Br-Br) > D(F-F) \approx D(I-I)$. The unexpectedly low value for fluorine is attributed to repulsion between non-bonding electrons across the short F—F distance.

5   The halides HF and HCl can be made by the action of concentrated H_2SO_4 on CaF_2 and NaCl, respectively. HBr and HI can be made by direct combination of the elements using a platinum catalyst.

6   $D(H-X)$ and the thermal stability of HX(g) decrease from fluorine to iodine; even at 300 °C, HI(g) contains an appreciable proportion of $H_2(g)$ and $I_2(g)$.

7   The halides HX are all very soluble in water and give acid solutions. In water, HCl, HBr and HI are strong acids; HF is a weak acid because of hydrogen-bonding between H_3O^+ and F^-. Many metal halides can be made by heating metals or metal oxides in hydrogen halide gases.

8   The values of $\Delta H_f^{\ominus}(X, g)$ and $\Delta H_f^{\ominus}(X^-, g)$ are quite similar for all the halogens, but the values for fluorine are the least positive (most negative), primarily because of the low bond dissociation energy of fluorine.

9   Fluorine is the best halogen for bringing out high oxidation numbers. In covalent situations this is because fluorine forms weak bonds with itself and short strong bonds with other elements: only fluorine reacts with sulfur to form a hexahalide. In ionic situations it is because the fluoride ion is small and so forms halides with large negative lattice energies: only fluorine reacts with silver to give a dihalide.

10 The strength of the halogens as oxidizing agents in aqueous solution decreases from fluorine to iodine. This is primarily due to the weakening of the bonding interaction between the halide ion and water molecules as the halide ion gets larger.

QUESTION 6.1

Predict the results, if any, of adding a few drops of bromine to: (i) an aqueous solution of sodium chloride; (ii) an aqueous solution of sodium iodide. Write an equation for any reaction that occurs.

QUESTION 6.2

As Figure 6.13 shows, the strengths of the single bonds that the atoms of Groups IV/14 and VI/16 form with themselves decrease as one descends the Group, except in the case of oxygen, where $B(O-O)$ is unexpectedly low. Explain this anomaly. How does your explanation account for the difference between oxygen and sulfur discussed in Section 4.2?

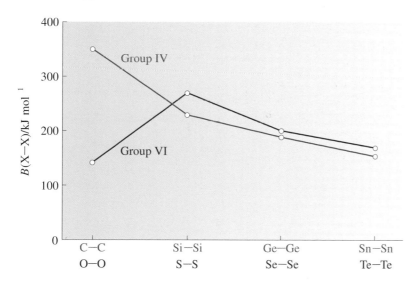

Figure 6.13
Variation in the bond energy terms, $B(A-A)$, in Groups IV/14 and VI/16.

QUESTION 6.3

The average relative molecular mass of the molecules in air is 29. Gases are denser or less dense than air depending on whether their relative molecular masses are greater or less than this figure. Explain how this determines a design feature of the experiment in Figure 6.6 (relative atomic masses can be found in the *Data Book* on the CD-ROMs associated with this Book).

QUESTION 6.4

Write equations for the reactions that occur when (i) beryllium metal and (ii) beryllium oxide are heated in a stream of dry hydrogen fluoride.

QUESTION 6.5 ✻

Halides of an element are often made by heating the element with the halogen at about 250 °C. If this is done with selenium and bismuth (Figure 6.14), one, and only one, of the halogens will convert the elements into halides containing selenium and bismuth in their highest oxidation numbers. Which halogen is it, and what are the formulae of the two halides that are formed?

QUESTION 6.6 ✻

When copper is heated with excess bromine, the black solid $CuBr_2$ is formed; with excess iodine, the product is the white compound, CuI (CuI_2 has never been made). If this difference is analysed by using a cycle of the type shown in Figure 6.11, which term is likely to be primarily responsible for it?

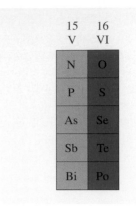

Figure 6.14
The Group V/15 and Group VI/16 elements.

6.6 Higher oxidation states of the halogens

The halogens form compounds and aqueous ions in which their oxidation number is positive; the states +1, +5 and +7 are especially prominent. In this introductory Section, we show how such substances can be obtained via the reactions of the elemental halogens in aqueous solutions. The reaction scheme for chlorine is pictured in Figure 6.15, and you should find it useful to refer to this as we proceed.

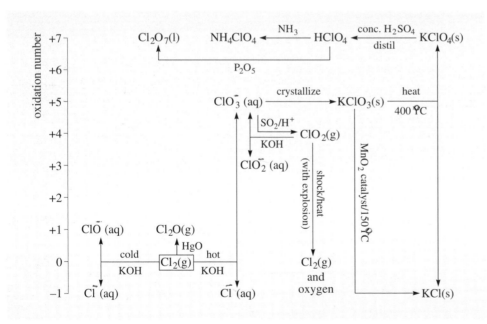

Figure 6.15
A reaction scheme for producing chlorine in its higher oxidation numbers, starting from chlorine gas and using simple experiments.

At 25 °C, a litre of water will dissolve about 0.1 mole of Cl_2 gas in the following equilibrium:

$$Cl_2(aq) + H_2O(l) \rightleftharpoons HOCl(aq) + H^+(aq) + Cl^-(aq) \tag{6.31}$$

Equilibrium tends to the left: most dissolved chlorine is in the form of $Cl_2(aq)$. The substance $HOCl$ is hypochlorous acid and it is written in the undissociated form because the acid is weak. In other words, equilibrium also lies to the left in the equation

$$HOCl(aq) \rightleftharpoons H^+(aq) + ClO^-(aq); \quad K_a \sim 10^{-8} \text{ mol litre}^{-1} \tag{6.32}$$

⬤ What happens to both these equilibria if hydroxide ions are added?

⬤ They both shift to the right: the hydroxide ions displace the equilibrium by combining with the hydrogen ions on the right:

$$H^+(aq) + OH^-(aq) = H_2O(l) \tag{6.33}$$

The overall reaction, the reaction of chlorine with alkali, can then be obtained by adding Equation 6.31, Equation 6.32 and twice Equation 6.33:

$$Cl_2(aq) + 2OH^-(aq) = ClO^-(aq) + Cl^-(aq) + H_2O(l) \tag{6.34}$$

Now the dissolved chlorine is present as both chloride and hypochlorite ions.

⬤ What type of reaction is Reaction 6.34?

⬤ A disproportionation. Chlorine in oxidation number zero is simultaneously oxidized (to oxidation number +1 in ClO^-) and reduced (to oxidation number -1 in Cl^-).

Alkaline hypochlorite solutions are useful for bleaching and sterilizing purposes: they release chlorine on dilution or acidification, when the equilibrium in Reaction 6.34 is displaced to the left. With domestic 'liquid' bleach, the alkali is NaOH.

Reaction 6.34 occurs in the cold, but if the alkali is heated before the chlorine is bubbled into it, the hypochlorite disproportionates in its turn to chlorate and chloride:

$$3ClO^-(aq) = ClO_3^-(aq) + 2Cl^-(aq) \tag{6.35}$$

and the overall reaction becomes:

$$3Cl_2 + 6OH^-(aq) = ClO_3^-(aq) + 5Cl^-(aq) + 3H_2O(l) \tag{6.36}$$

If the alkali is KOH, crystals of $KClO_3$, potassium chlorate, are deposited on cooling because it is sparingly soluble. The chlorate ion is pyramidal (Structure **6.5**)[*] in accordance with VSEPR theory; the chlorine oxidation number is +5. Bromine and iodine behave in a similar way to chlorine, except that BrO^- and IO^- disproportionate much more easily than ClO^- in the alkaline solution. The result is that the reaction of these elements with *hot or cold* alkali produces bromates or iodates.

The standard redox potentials for the process

$$XO_3^-(aq) + 6H^+(aq) + 5e = \tfrac{1}{2}X_2(aq) + 3H_2O(l) \tag{6.37}$$

are a measure of the oxidizing strengths of the halate ions when they are reduced to halogen in acid solution. In Table 6.6, they are compared with the halogen data. The halate values all exceed 1.0 V: in contrast to the halogens, all three halates are strong oxidizing agents. Notice in particular that the bromate and iodate values are substantially larger than their halogen counterparts. This means that bromate in acid solution is thermodynamically capable of oxidizing bromide to bromine. Likewise, iodate can oxidize iodide:

$$IO_3^-(aq) + 5I^-(aq) + 6H^+(aq) = 3I_2(aq) + 3H_2O(l) \tag{6.38}$$

Both reactions occur.

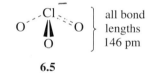

6.5

Table 6.6 Values of $E^{\ominus}(XO_3^-|\tfrac{1}{2}X_2)$ and $E^{\ominus}(\tfrac{1}{2}X_2|X^-)$

	$E^{\ominus}(XO_3^-|\tfrac{1}{2}X_2)$	$E^{\ominus}(\tfrac{1}{2}X_2|X^-)$
	V	
Cl	1.47	1.36
Br	1.50	1.10
I	1.19	0.53

[*]Notice that Structure **6.5** contains two double bonds and one single bond, so it is not consistent with the fact that all three bonds are of equal length. This is because only one of three resonance forms has been shown. Resonance with the two other forms, as for NO_3^- in Question 4.2b, makes all three bonds the same length. The same convention is used in Structures **6.6** and **6.8**.

Solid halates are also strong oxidizing agents and usually evolve oxygen on strong heating. In the presence of a manganese dioxide catalyst, $KClO_3$ decomposes at 150 °C:

$$2KClO_3(s) = 2KCl(s) + 3O_2(g) \qquad (6.39)$$

The role of sodium chlorate as a plant- or weed-killer depends on its oxidizing powers, as does the use of $KClO_3$ in matches (Section 11.3).

$KClO_3$ is also a common oxidizer in fireworks, where it is mixed with salts of strontium, barium or copper to produce red, green or blue colours, respectively. Precocious schoolchildren have occasionally killed or maimed themselves by experimenting with the very dangerous explosive mixture of the weedkiller sodium chlorate (oxidizer) and sugar (reducing agent).

● Is +5 the highest possible oxidation number of the halogens?

● No, the highest possible oxidation number is +7, the same as the Mendeléev Group number.

If potassium chlorate is heated *without any MnO_2 catalyst*, at about 400 °C, it slowly disproportionates:

$$4KClO_3(s) = 3KClO_4(s) + KCl(s) \qquad (6.40)$$

The addition of water removes the highly soluble KCl, leaving the sparingly soluble potassium perchlorate, $KClO_4$, a colourless crystalline solid containing a tetrahedral anion (Structure **6.6**). Its low solubility in water is unusual for an alkali metal salt; $RbClO_4$ and $CsClO_4$ are even less soluble.

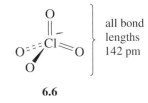

all bond lengths 142 pm

6.6

When potassium perchlorate is distilled under reduced pressure with concentrated sulfuric acid at 90–160 °C, perchloric acid vapour, $HClO_4$, is evolved. It can be condensed to a liquid and used to make other perchlorates such as $NaClO_4$ and NH_4ClO_4; for example

$$NH_3 + HClO_4 = NH_4ClO_4(s) \qquad (6.41)$$

These salts are much more soluble in water than $KClO_4$. Thermodynamically, aqueous perchlorate, $ClO_4^-(aq)$, is a strong oxidizing agent like aqueous halates, but it often seems surprisingly stable because its redox reactions are usually very slow. With solid perchlorates, the oxidizing power is more evident.[*] Ammonium perchlorate is the most common oxidizer in solid-fuel rocket propellants. The two huge booster rockets on the American space shuttle (Figure 6.16) contain a mixture of 70% NH_4ClO_4 and 16% aluminium powder with appropriate binding agents. A typical reaction is

$$6NH_4ClO_4(s) + 8Al(s) = 3N_2(g) + 12H_2O(g) + 3Cl_2(g) + 4Al_2O_3(s) \qquad (6.42)$$

Figure 6.16 At blast-off, the American space shuttle sits on top of a huge rocket powered by a liquid hydrogen/liquid oxygen propellant. To left and right are booster rockets, which are driven by the solid combination Al/NH_4ClO_4, and jettisoned 127 seconds into the flight. The billowing white clouds in their wake consist of fine particles of aluminium oxide. On January 28, 1986, a ring-seal on the right-hand booster failed, and a sideways flame played upon, then ruptured and ignited the main hydrogen fuel tank. Space shuttle *Challenger* blew up 73 seconds into the flight, and the crew of seven died.

[*]The oxidizing power of perchlorates is shown in 'The p-Block elements in action' on one of the CD-ROMs accompanying this Book.

The best method of generating perbromates, which were first made in 1968, involves the passage of fluorine into alkaline bromate:

$$F_2(g) + 2OH^-(aq) + BrO_3^-(aq) = BrO_4^-(aq) + 2F^-(aq) + H_2O(l) \qquad (6.43)$$

Periodates are remarkable for occurring in forms with different iodine coordinations. The most common coordination is octahedral, which occurs in derivatives of periodic acid, H_5IO_6 (Structure **6.7**). There are five replaceable hydrogens although in most instances, only one or two are substituted. For example, if an alkaline solution of sodium iodate is oxidized with chlorine, and cooled, $Na_2H_3IO_6$ is precipitated:

$$IO_3^-(aq) + Cl_2(g) + 2Na^+(aq) + 3OH^-(aq) = Na_2H_3IO_6(s) + 2Cl^-(aq) \qquad (6.44)$$

In the presence of dilute nitric acid, this is converted into $NaIO_4$, in which the iodine coordination is tetrahedral (Structure **6.8**):

$$Na_2H_3IO_6(s) + H^+(aq) = NaIO_4(s) + 2H_2O(l) + Na^+(aq) \qquad (6.45)$$

Whatever their coordination, periodates are very strong oxidizing agents. They will, for example, convert the nearly colourless $Mn^{2+}(aq)$ ion into the deep purple permanganate, $MnO_4^-(aq)$, in acid solution. Only the most powerful oxidizing agents can do this.

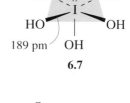

6.6.1 Halogen oxides and oxoacids

Most halogen oxides are liable to decompose, often explosively, on heating or under the influence of shock or impact. This is especially so in the case of chlorine and bromine oxides, all of which have positive values of $\Delta H_f^\ominus$ and $\Delta G_f^\ominus$ at room temperature.

● What does this tell you about their stability?

● Decomposition into their constituent elements is thermodynamically favourable.

Thus, all oxides of bromine decompose at or below room temperature, and have no practical importance. We therefore confine ourselves to the more important oxides of chlorine shown in Figure 6.17 and to I_2O_5 and OF_2.

	Cl_2O	ClO_2	Cl_2O_7
$T_b/°C$	2	11	81
$T_m/°C$	−121	−59	−92

Figure 6.17
The structure, melting temperatures (T_m) and boiling temperatures (T_b) of the three most important oxides of chlorine: Cl_2O, ClO_2 and Cl_2O_7.

6.6.1.1 Dichlorine monoxide, chlorine dioxide and dichlorine heptoxide

You may find it helpful to refer to the network of chlorine reactions shown in Figure 6.15. Dichlorine monoxide, Cl_2O, is a brownish-yellow gas at room temperature but condenses to an orange liquid with a boiling temperature of 2 °C.

It is conveniently made (Figure 6.18) by passing chlorine through dry, freshly prepared yellow HgO and condensing the gas in a tube cooled in an ice/salt freezing mixture:

$$2Cl_2(g) + 2HgO(s) = Cl_2O(l) + HgO.HgCl_2(s) \qquad (6.46)$$

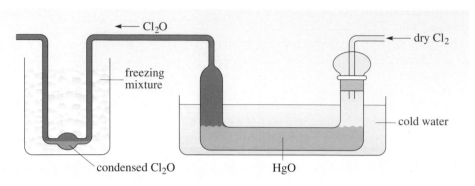

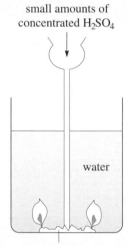

Figure 6.18
The preparation of dichlorine monoxide.

The gas explodes on heating, giving chlorine and oxygen. An especially violent explosion occurs if the gas is mixed with ammonia, or if the liquid makes contact with rubber.

When drops of concentrated sulfuric acid are allowed to fall on powdered potassium chlorate, a crackling sound ensues. In this notoriously dangerous reaction, the noise is the explosive decomposition of evolved chlorine dioxide gas:

$$3KClO_3(s) + 2H_2SO_4(l) = KClO_4(s) + 2KHSO_4(s) + 2ClO_2(g) + H_2O(g) \quad (6.47)$$
$$2ClO_2(g) = Cl_2(g) + 2O_2(g) \qquad (6.48)$$

Figure 6.19 shows a demonstration of the reactivity of chlorine dioxide prepared in this way. A safer preparative method, often used in industry, is the reduction of strongly acidified aqueous chlorate by sulfur dioxide:

$$SO_2(g) + 2ClO_3^-(aq) = SO_4^{2-}(aq) + 2ClO_2(g) \qquad (6.49)$$

ClO_2 is a yellow–green gas, which is very soluble in water. The molecule is unusual in having an odd number of electrons.

⬤ What physical property will this confer on the gas.

⬤ The odd electron must be *unpaired*: the gas will be paramagnetic.

ClO_2 condenses to a dark-red explosive liquid (T_b 11 °C); the gas is also liable to decompose explosively to its elements if its pressure reaches 0.07 atmospheres or more. Despite this, it is extensively used in *dilute* form to bleach pulp in the paper industry where its rival, chlorine is partly consumed in wasteful side reactions such as the chlorination of lignin. Another reason why the use of ClO_2 is growing, in both bleaching and water purification, is that the chlorinated hydrocarbons produced by chlorine (Section 6.1) are not formed in the presence of ClO_2. Both Cl_2O and ClO_2 are V-shaped, but ClO_2 has the shorter Cl—O bond length.

⬤ How is this taken account of in Figure 6.17?

⬤ ClO_2 has been assigned double bonds.

small amounts of concentrated H_2SO_4

water

crystals of $KClO_3$ and small pieces of white phosphorus

Figure 6.19
A demonstration of the violent reactivity of chlorine dioxide; a few crystals of the sparingly soluble $KClO_3$ and two or three fragments of white phosphorus are covered with water. A little concentrated sulfuric acid, which is very dense, is then poured on to the chlorate by using a funnel with a thin-bored stem. The acid/chlorate combination liberates ClO_2, which immediately attacks the phosphorus, causing it to inflame and burn under water.

The oxide Cl_2O_7 can be prepared by the action of phosphorus pentoxide, P_2O_5, on perchloric acid (Figure 6.15). P_2O_5 is a powerful dehydrating agent, which can often abstract the elements of water from substances by forming phosphoric acid:

$$P_2O_5(s) + 3H_2O(l) = 2H_3PO_4(aq) \tag{6.50}$$

If $HClO_4$ is added to P_2O_5 at $-10\,°C$, the acid loses water:

$$2HClO_4(l) - H_2O = Cl_2O_7(l) \tag{6.51}$$

The heptoxide can then be distilled off under reduced pressure at $35\,°C$. It is an oily liquid (T_b $81\,°C$) liable to explode on mechanical shock. The different $Cl-O$ bond lengths in the molecule support the structural formula of Figure 6.17, which contains heptavalent chlorine.

6.6.1.2 Oxoacids of chlorine

Cl_2O_7 dissolves in water, regenerating perchloric acid by a rearranged form of Equation 6.51:

$$Cl_2O_7(l) + H_2O(l) = 2H^+(aq) + 2ClO_4^-(aq) \tag{6.52}$$

It is therefore an **acidic oxide**, and is said to be the **acid anhydride** of perchloric acid, which contains chlorine in the same oxidation number of +7. Cl_2O is the anhydride of hypochlorous acid:

$$Cl_2O(g) + H_2O(l) = 2HOCl(aq) \tag{6.53}$$

The aqueous solution of ClO_2 contains just solvated ClO_2 molecules, but the gas disproportionates with alkalis to generate the anions of chlorous acid, $HClO_2$, and chloric acid, $HClO_3$ (Figure 6.15):

$$2ClO_2(g) + 2OH^-(aq) = ClO_2^-(aq) + ClO_3^-(aq) + H_2O(l) \tag{6.54}$$

The measure of the strength of an acid, HX, is its dissociation constant, K_a (Section 3.2). Table 6.7 shows dissociation constants for the four oxoacids of chlorine. Perchloric acid is very strong, but hypochlorous acid is weak: the higher oxide, Cl_2O_7, is more acidic than the lower, Cl_2O. More precisely, the larger the number of terminal oxygens (those bound just to one chlorine atom) the stronger is the corresponding acid. Linus Pauling suggested that any oxoacid's dissociation constant, K_a, was approximately 10^{5t-8} mol litre^{-1}, where t is the number of terminal oxygens (see Section 11.4.4). Now the structures of the four acids correspond to the formulae HOCl, HOClO, HOClO$_2$ and HOClO$_3$, where the terminal oxygens are written last in each case.

Table 6.7 Approximate acid dissociation constants of the oxoacids of chlorine

	K_a/mol litre^{-1}
HOCl	10^{-8}
HClO$_2$	10^{-2}
HClO$_3$	10^1
HClO$_4$	10^{10}

What would be the values of K_a for HOCl, HClO$_2$, HClO$_3$ and HClO$_4$ if this rule were exact?

10^{-8}, 10^{-3}, 10^2 and 10^7 mol litre^{-1}, respectively, corresponding to $t = 0$, 1, 2 and 3. The agreement is reasonable.

6.6.1.3 Oxygen difluoride and diiodine pentoxide

The oxide F_2O is called oxygen difluoride in deference to the high electronegativity of fluorine. It is the only known long-lived fluorine–oxygen compound at room temperature. Its name implies that, unlike the other halogens, fluorine occurs only in oxidation numbers zero and -1. It is made by passing fluorine quickly through very dilute aqueous alkali:

$$2F_2(g) + 2OH^-(aq) = OF_2(g) + 2F^-(aq) + H_2O(l) \tag{6.55}$$

OF$_2$ (Structure **6.9**) is a colourless gas; it oxidizes water slowly, but explodes in steam:

$$OF_2(g) + H_2O(g) = 2HF(g) + O_2(g) \qquad (6.56)$$

6.9

Although periodic acid is known (Section 6.6), I$_2$O$_7$ is not. The highest known normal oxide of iodine is diiodine pentoxide, I$_2$O$_5$, which can be made by the dehydration of the white crystals of iodic acid, HIO$_3$, which is itself obtained by the oxidation of iodine with nitric acid. Dehydration is achieved by heating in a stream of dry air:

$$2HIO_3(s) = I_2O_5(s) + H_2O(g) \qquad (6.57)$$

I$_2$O$_5$ consists of white crystals containing the molecular unit shown in Structure **6.10**. It will oxidize carbon monoxide quickly and completely at 90 °C:

$$I_2O_5(s) + 5CO(g) = I_2(s) + 5CO_2(g) \qquad (6.58)$$

By finding out how much iodine or CO$_2$ is produced, this reaction can be used to determine how much CO there is in a poisonous atmosphere.

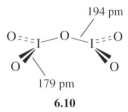

6.10

6.6.2 Interhalogen compounds

The four halogens react with one another to form compounds containing diatomic molecules that comprise different halogen atoms. Six such combinations are possible, but only ClF, ICl and IBr can be obtained in a pure state at room temperature. Only these three are therefore included in Table 6.8. The more electronegative element is assigned the oxidation number of −1, the less electronegative, +1. As Table 6.8 implies, higher oxidation numbers are obtained in fluorides and in I$_2$Cl$_6$. This is usually done by using an excess of the more electronegative halogen, raising the temperature and, if necessary, increasing the pressure. Thus, iodine and fluorine yield IF$_5$ at room temperature and IF$_7$ at 250–300 °C. Likewise, the reaction of chlorine and fluorine at 225 °C gives mainly ClF and, at 300 °C, ClF$_3$; if the fluorine pressure is raised to 250 atm and the temperature to 350 °C, the product is ClF$_5$.

Table 6.8 A selection of interhalogen compounds and their boiling temperatures (T_b) and melting temperatures (T_m)

Oxidation number			
+1	+3	+5	+7
ClF	ClF$_3$	ClF$_5$	
colourless gas; T_b −156 °C	colourless gas; T_b 12 °C	colourless gas; T_b −13 °C	
	BrF$_3$	BrF$_5$	
	yellow liquid; T_b 126 °C	colourless liquid; T_b 41 °C	
ICl	I$_2$Cl$_6$	IF$_5$	IF$_7$
red crystals; T_m 27 °C	yellow solid; T_m 101 °C	colourless liquid; T_b 105 °C	colourless gas; T_b 7 °C
IBr			
black crystals; T_m 41 °C			

The shapes of the molecules agree with the predictions of VSEPR theory, and three of the more important instances are illustrated in Figure 6.20. As Table 6.8 implies, iodine trichloride consists of dimers, I_2Cl_6, which are planar, with the structure shown in Figure 6.21a. The two-coordinate bridging chlorines retain noble gas configurations if we allow them each to form one dative bond and one shared electron pair bond with iodine (Figure 6.21b).

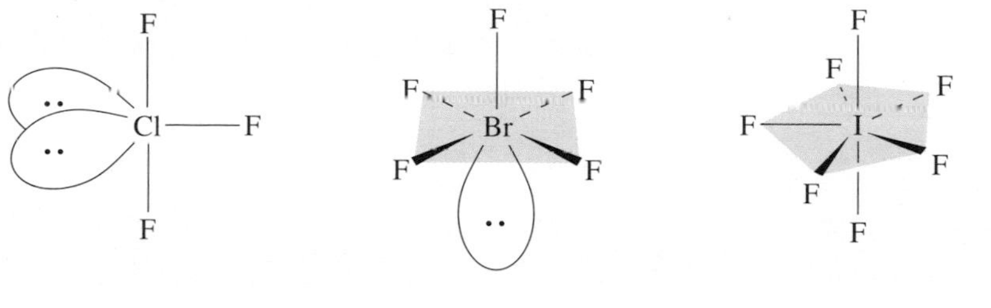

Figure 6.20
The shapes of interhalogen molecules such as ClF_3, BrF_5 and IF_7 are largely in agreement with the predictions of VSEPR theory. The orbital lobes show non-bonding pairs of electrons.

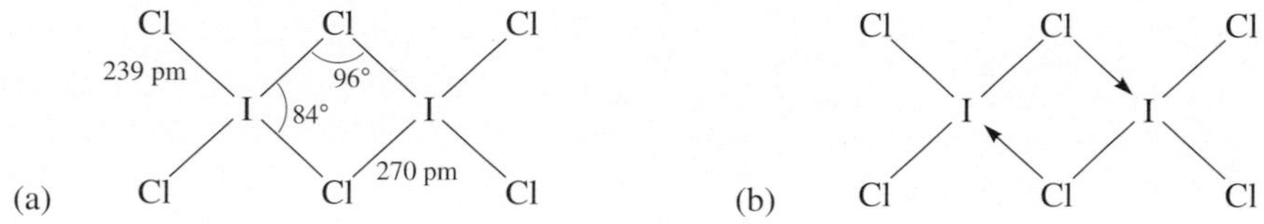

Figure 6.21
(a) The geometry of I_2Cl_6;
(b) a structural formula in which the bridging chlorines retain noble gas configurations although the iodines do not.

The interhalogen fluorides are all powerful fluorinating agents. ClF_3 is especially vigorous, often reacting more quickly than fluorine itself. It can be transported as a liquid in steel cylinders, on which it forms a protective lining of insoluble fluoride. It fluorinates metal oxides and chlorides, often oxidizing them, if that is possible, at the same time:

$$AgCl(s) + ClF_3(g) = AgF_2(s) + \tfrac{1}{2}Cl_2(g) + ClF(g) \qquad (6.59)$$

$$Al_2O_3(s) + 2ClF_3(g) = 2AlF_3(s) + Cl_2(g) + \tfrac{3}{2}O_2(g) \qquad (6.60)$$

ClF_3 is one of the most reactive of all chemical compounds, instantly setting fire to glass wool and asbestos. One drop of ClF_3 sets fire to paper, cloth or wood, and the liquid reacts with water with a noise like the crack of a whip.

6.6.3 Polyhalogen ions

When we assign electrons to bonds in interhalogens prior to applying VSEPR, we find that the number of outer electrons on the central atom in the compound varies. For example, it is 8 in I_2 or ClF, 10 in ClF_3, 12 in BrF_5 and 14 in IF_7. This flexibility is also evident in reactions in which the central atom increases its outer electrons by accepting the non-bonding pair from a halide ligand. A very simple example is the triiodide ion, I_3^-. The solubility of iodine in water is small (about 0.3 g litre^{-1} at 25 °C), so if a spoonful of solid iodine is added to a beaker of water, most of it collects on the bottom of the beaker. But this very quickly dissolves if potassium iodide solution is added, and the familiar red–brown 'iodine' solution is formed. The iodine is made soluble by combining with the aqueous iodide ion:

$$I^-(aq) + I_2(s) = [I-I-I]^-(aq) \qquad (6.61)$$

⬤ What type of reaction is this?

● It is a Lewis acid–Lewis base reaction: the iodine atoms in I_2 each have eight outer electrons. But in one of them this jumps to ten when I^- acts as a Lewis base, donating a non-bonding pair to I_2, which acts as the Lewis acid.

VSEPR theory now predicts that the triiodide ion is linear: an iodine atom has seven outer electrons, so if the negative charge of the I_3^- ion is put on the central iodine, that iodine now has eight outer electrons. Two of these are used to form the terminal bonds to the two other iodines. This leaves six, which become three non-bonding pairs, so there are five repulsion axes disposed in a trigonal bipyramid. In this disposition, the three non-bonding pairs occupy the equatorial positions (Figure 6.22a). X-ray diffraction studies of solids obtained from aqueous solutions of I_3^-(aq) confirm the linear shape (Figures 6.22b and c).

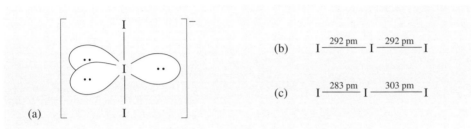

(a)

(b) $I \xrightarrow{\text{292 pm}} I \xrightarrow{\text{292 pm}} I$

(c) $I \xrightarrow{\text{283 pm}} I \xrightarrow{\text{303 pm}} I$

Figure 6.22
(a) The linear shape of I_3^- is correctly predicted by VSEPR theory, the three non-bonding pairs occupying the equatorial positions in a trigonal bipyramid. In solids, the ion is always linear, but sometimes it is symmetrical as in the anion of $[N(CH_3)_4]^+ I_3^-$ (b), and sometimes unsymmetrical as in the anion of $Cs^+ I_3^-$ (c). In either case, the bond lengths are longer than the I–I distance in I_2 (266 pm). This point is taken up later when we discuss bonding in noble gas compounds.

Many other polyhalide anions can be formed by reactions of the type shown in Equation 6.61. If CsF reacts with ClF_3(g) at 100 °C, $CsClF_4$(s) is formed; a similar reaction with BrF_5 gives $CsBrF_6$. In these reactions, ClF_3 and BrF_5 act as Lewis acids, accepting a non-bonding pair from the F^- ion in CsF:

$$F^- + ClF_3 = ClF_4^- \qquad (6.62)$$

● What shape does VSEPR theory predict for ClF_4^-?

◑ Square planar (Figure 6.23): the negative charge is assigned to the central chlorine, giving eight valence electrons in all. Four are used in Cl–F bonds, leaving two non-bonding pairs. The six repulsion axes are octahedrally disposed, the non-bonding pairs being at 180° to each other to minimize non-bonding pair–non-bonding pair repulsions.

In the case of $CsBrF_6$, it appears that the BrF_6^- ion is regular octahedral. This confounds the predictions of VSEPR theory, which indicates seven repulsion axes. However, it is worth remembering that the ion is not free, but in a crystal, where its shape could be influenced by interaction with surrounding ions.

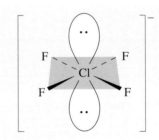

Figure 6.23
The structure of ClF_4^- is in agreement with VSEPR theory.

So far we have considered reactions in which a halogen or interhalogen expands its valence shell by accepting a halide non-bonding pair. But reactions are also possible in which the interhalogen valence shell contracts through loss of a halide ligand. The following reactions occur when ClF_3 and IF_7 react with arsenic pentafluoride:

$$ClF_3(g) + AsF_5(g) = [ClF_2]^+ [AsF_6]^-(s) \qquad (6.63)$$

$$IF_7(g) + AsF_5(g) = [IF_6]^+ [AsF_6]^-(s) \qquad (6.64)$$

Here the interhalogen transfers a fluoride ion to AsF_5, which is a very good fluoride ion acceptor, and ends up as a polyhalogen positive ion.

● When AsF_5 accepts the fluoride ion, which is the Lewis acid, and which is the Lewis base?

The F^- ion carries a non-bonding pair, which it donates to AsF_5: F^- is the Lewis base and AsF_5 is the Lewis acid.

Other positive halogen ions include the blue paramagnetic ion, I_2^+, which is formed when iodine is oxidized by dissolving it in oleum (a solution of sulfur trioxide, SO_3, in concentrated sulfuric acid). In solids containing this ion, the bond length is 256 pm, as compared with 266 pm in the free I_2 molecule. The distance in I_2^+ is shorter because, in moving from a halogen molecule to the ion, an electron is removed from an antibonding orbital, thereby increasing the bond order. Similarly, when Cl_2^+ is formed from Cl_2, the electron is removed from the $2\pi_g$ level.

6.6.4 Summary of Section 6.6

1   In cold aqueous alkalis, Cl_2 disproportionates to $Cl^-(aq)$ and $ClO^-(aq)$; in hot alkali the disproportionation is to $Cl^-(aq)$ and $ClO_3^-(aq)$. Bromine and iodine disproportionate to halide and halate in hot or cold alkali.

2   Halates are strong oxidizing agents; bromates oxidize bromide, and iodates oxidize iodide to the halogens in acid solution. $KClO_3$ evolves oxygen on gentle heating in the presence of MnO_2, leaving KCl.

3   In the absence of MnO_2, solid $KClO_3$ disproportionates on heating to form KCl and sparingly soluble $KClO_4$. Distillation of $KClO_4$ under reduced pressure with concentrated H_2SO_4 yields perchloric acid, $HClO_4$.

4   Perchlorates contain chlorine in its highest oxidation number of +7. They are strong oxidizing agents, but appear stable in aqueous solution because such reactions are usually very slow. They are used in solid rocket propellants as oxidizers, notably in the combination Al/NH_4ClO_4.

5   Periodates can be made by the oxidation of alkaline iodate with chlorine; they occur as salts of both the acid H_5IO_6 and the acid HIO_4.

6   Cl_2O is made by the reaction of chlorine and HgO, ClO_2 by reduction of acidified chlorate with SO_2, and Cl_2O_7 by dehydration of $HClO_4$ with P_2O_5. Cl_2O and ClO_2 are gases; Cl_2O_7 is an oily liquid. All three compounds are liable to explosive decomposition into chlorine and oxygen. Cl_2O and Cl_2O_7 are the anhydrides of hypochlorous and perchloric acid, respectively.

7   Neither Br_2O_7 nor I_2O_7 have been made. The highest known oxide of iodine is I_2O_5, which is obtained by the dehydration of iodic acid, HIO_3.

8   F_2O is called oxygen difluoride; fluorine occurs only in oxidation numbers -1 and zero. F_2O is made by the reaction of fluorine with dilute aqueous alkali.

9   The dissociation constant of an oxoacid is approximately 10^{5t-8} mol litre^{-1}, where t is the number of terminal oxygens.

10 Covalent interhalogen molecules, XY, are obtained by combination of the elements. When Y = F, higher fluorides such as ClF_3, BrF_5 and IF_7 can be made by increasing the temperature and the fluorine pressure. These higher fluorides are powerful fluorinating agents.

11 Polyhalogen ions are formed when halogens or interhalogens accept or donate halide ions. In the first category come $I_3^-(aq)$, ClF_4^- and BrF_6^- in alkali metal salts. Donation is easiest to powerful fluoride ion acceptors such as AsF_5, as in $[ClF_2]^+[AsF_6]^-$.

QUESTION 6.7 ⚹

Write equations for the reactions you would expect to occur between:
(a) bromine and cold aqueous alkali, $OH^-(aq)$; (b) $ClF_5(g)$ and $AsF_5(g)$;
(c) $CsCl(s)$ and $ICl(l)$. Predict the shapes of any halogen species that are formed.

QUESTION 6.8 ⚹

Only two binary *compounds* (compounds formed by just two elements) are
known in which a halogen element is found in an oxidation number equal to the
halogen Mendeléev Group number. What are they? In what set of compounds
are chlorine, bromine and iodine all found in this highest oxidation number?

QUESTION 6.9 ⚹

The anhydride of chloric acid, $HClO_3$, has never been prepared. What formula
would it have? By drawing an analogy with the preparation of $HClO_4$ and its
anhydride, suggest how the anhydride of $HClO_3$ might be obtained from $KClO_3$.
Then identify a reaction in the text which suggests why the preparation might
not be successful.

GROUP VIII/18: THE NOBLE GASES

7

The principal source of the noble gases is the atmosphere Table 7.1 shows the nine most plentiful components of the atmosphere by mass. Over 98% is nitrogen and oxygen, but the noble gas argon comes third, making up over half the residue. It is therefore more abundant than water vapour, and much more abundant than CO_2. Trailing far behind carbon dioxide, come neon, krypton, helium and xenon in that order.

Table 7.1 Composition of the atmosphere

Component	Mass/t	Molar mass/g mol^{-1}
N_2	3.87×10^{15}	28
O_2	1.19×10^{15}	32
Ar	6.59×10^{13}	40
H_2O	1.70×10^{13}	18
CO_2	2.46×10^{12}	44
Ne	6.48×10^{10}	20
Kr	1.69×10^{10}	84
He	3.71×10^9	4
Xe	2.02×10^9	131
Total	5.14×10^{15}	—

7.1 The discovery of the noble gases

The discovery of the noble gases was initiated when, in 1892, Lord Rayleigh noted that atmospheric nitrogen obtained by the removal of oxygen, CO_2 and water vapour from air was slightly denser (1.2572 g litre^{-1}) [*] than chemically prepared nitrogen (1.2505 g litre^{-1}), produced in reactions such as:

$$NH_4^+(aq) + NO_2^- (aq) = N_2(g) + 2H_2O(l) \tag{7.1}$$

where a mixture of solutions of ammonium sulfate, $(NH_4)_2SO_4$, and sodium nitrite, $NaNO_2$, is heated.

● Look at Table 7.1: which undiscovered element do you think was mainly responsible for this discrepancy?

● Of the noble gases, argon is the most abundant. Its molar mass (40) is also higher than that of N_2 (28), so the presence of argon in nitrogen tends to raise the density.

[*] These and subsequent densities are those of the gas at 1 atmosphere pressure and 0 °C.

The idea that Rayleigh's atmospheric nitrogen might contain a hitherto undiscovered gas was taken up by William Ramsay (Figure 7.1) in collaboration with Rayleigh. Ramsay passed a sample repeatedly over heated magnesium (Figure 7.2); its volume gradually decreased as magnesium nitride was formed:

$$3Mg(s) + N_2(g) = Mg_3N_2(s) \qquad (7.2)$$

Figure 7.1
William Ramsay (1852–1916) read classics at Glasgow University, but then frustrated parental expectations by becoming a chemist rather than a clergyman. In 1904, he won the Nobel Prize for Chemistry: 'Rayleigh gave numbers about which there could be no reasonable doubt. I asked him then if he minded my trying to solve the mystery. He thought that the cause of the discrepancy was a light gas in non-atmospheric nitrogen; I thought that the cause was a heavy gas in atmospheric nitrogen. He spent the summer in looking for the light gas; I spent July in hunting for the heavy one. And I have succeeded in isolating it.'

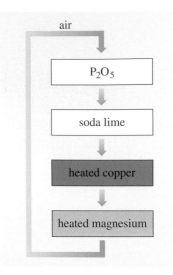

Figure 7.2
Ramsay's scheme for obtaining argon from air: air was continuously cycled over:
(1) phosphorus pentoxide, P_2O_5, which is a powerful drying agent and removes water;
(2) soda lime, a dried blend of sodium and calcium hydroxides, which removes carbon dioxide;
(3) heated copper which removes oxygen by forming CuO;
(4) heated magnesium, which removes nitrogen (Equation 7.2).

As the volume of the residual gas decreased, its density increased. When all the nitrogen had apparently been absorbed, the volume was only about 1/100th of the parent sample of air, and the density had risen to 1.780 g litre^{-1}. The resulting gas was chemically inactive: it was unaffected when heated even with the most reactive substances, including fluorine, so when Rayleigh and Ramsay announced its discovery in 1894, it was named *argon* (Gk 'the lazy one'). Ramsay was also able to show that argon was a *monatomic* gas; that is, its molecules consist of free argon *atoms*.

The appearance of a single new element with unprecedented properties challenged Mendeléev's Periodic Law, because this law implies that elements come in groups or families. Ramsay, however, had a bold answer to this challenge.

🔘 What do you think the answer was?

🔘 He suggested that an *entire family* of new elements awaited discovery.

In 1895, Ramsay found helium in a uranium mineral. Then, in 1898, by liquefying samples of his 'argon' and carefully distilling the liquid, he obtained not just a purer argon, but, in quick succession, krypton, neon and xenon, which were present as very minor contaminants.

Like argon, all these gases were monatomic and seemed quite inactive; together they made up a complete family (Figure 7.3). Thus, the Periodic Law not only survived the argon challenge; it emerged from it with enhanced prestige.

Ramsay's five new elements could be distinguished by their emission spectra. These are observed when the gases are sealed in glass tubes under reduced pressure, and an electric discharge passed through them (Figure 7.4). The differently coloured lines of the emission spectra are seen when the resulting glow is viewed through a prism. The prevailing colours in clear glass are helium (yellow), neon (red–orange), argon (bluish), krypton (green) and xenon (blue).

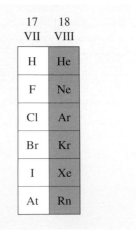

Figure 7.3
The place of the noble gases in the Periodic Table.

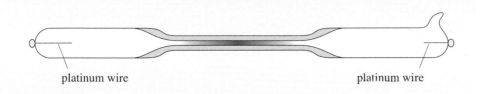

platinum wire   platinum wire

Figure 7.4
Ramsay's apparatus for observing the emission spectra of the noble gases: gas samples at low pressure were sealed in a glass tube pierced at each end by platinum wires. The middle portion is a fine capillary. When a high voltage is applied across the platinum wires, a brilliant glow appears in the capillary section.

7.2 The manufacture and properties of the noble gases

Some properties of the noble gases are shown in Table 7.2. Coming at the end of the Periods, they have high ionization energies because of the build up of nuclear charge. The high values are a measure of the stability of the noble gas configuration with respect to the loss of electrons. Note that they decrease down the Group. At the same time, the melting and boiling temperatures increase.

Table 7.2 Properties of the noble gases

Element	Ionization energy/kJ mol^{-1}	T_m /K	T_b /K
He	2 372	0.95	4.2
Ne	2 081	25	27
Ar	1 520	84	87
Kr	1 351	116	120
Xe	1 170	161	165
Rn	1 037	202	211

⬤ What is the connection between these changes?

⬤ The decrease in ionization energy is an indication of increasing polarizability down the Group as more electrons are added at greater distances from the nucleus, and the electron cloud becomes more 'flabby'. In the solids and liquids, the induced dipole–induced dipole interactions that hold the atoms together, increase with polarizability (Section 4.5). Thus, the melting and boiling temperatures increase down the Group as the ionization energy decreases.

Noble gases are isolated primarily by the **fractional distillation** (Figure 7.5, overleaf) of liquid air. What happens is much influenced by the boiling temperatures of liquid nitrogen (77 K) and liquid oxygen (90 K). A typical scheme is depicted in Figure 7.6 (overleaf) and starts with dry liquefied air from which CO_2 has been removed. The column is more complicated than the one shown in Figure 7.5, but the consequences are the same: temperature decreases from the bottom of the column to the top, and a fraction tapped off nearer the bottom of the column is rich in the less-volatile oxygen; conversely, a fraction accumulating on trays higher up is rich in the more-volatile nitrogen.

⬤ What property in Table 7.2 accounts for the availability of an argon fraction, a liquid rich in argon, between the oxygen and nitrogen fractions in Figure 7.6?

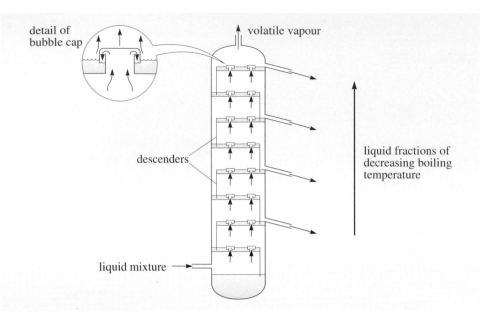

Figure 7.5 In fractional distillation, a liquid mixture enters the bottom of a column at a temperature at which it is volatile. The vapour then ascends the column, which is broken into sections by perforated trays. Above the perforations are bubble caps, where components of the ascending vapour can condense to a liquid, which then becomes part of a shallow layer on the tray. Some liquid flows back through the descenders to lower trays, where its constituents may be re-evaporated and rejoin the ascending vapour. The temperature of the sections decreases in an upward direction. In the warmer lower sections, components of the mixture with the highest boiling temperatures are removed from the vapour by condensation, but as the vapour ascends, components with progressively lower boiling temperatures are liquefied. Thus, on different trays of the column, different mixtures of the components can be run off. These mixtures are called *fractions*.

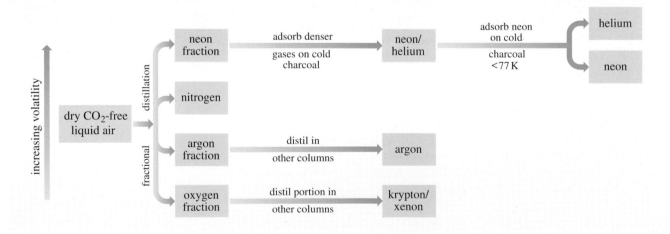

Figure 7.6 A scheme, based primarily on fractional distillation, for obtaining noble gases from liquid air.

● The boiling temperature of argon (87 K) shows that the volatility of argon lies between that of oxygen and nitrogen.

During this initial distillation, the gases that are too volatile to be condensed in the column, notably neon and helium, stream off at the top. Conversely, the tiny amounts of relatively involatile krypton and xenon remain dissolved in the liquid oxygen fractions near the bottom.

Further separation and purification of each noble gas can then be achieved either in further distillation columns, or by differential adsorption on solids such as charcoal which can absorb gases. The less-volatile noble gas is more strongly absorbed. Thus, as Figure 7.6 implies, charcoal below 77 K will take up neon while helium passes on. The neon can later be released by warming.

The kind of process depicted in Figure 7.6 is used to obtain neon, argon, krypton and xenon. Helium is not usually collected, because there is a much more economical source — natural gas, in which helium can accumulate at concentrations up to 7%. As Ramsay's discovery of the gas in a uranium mineral suggests, it consists of accumulated α-particles from the α-decay of radioactive elements in rocks; for example

$$^{238}_{92}U = ^{234}_{90}Th + ^{4}_{2}He \tag{7.3}$$

Radon occurs only in trace amounts, its longest-lived isotope being ^{222}Rn. It is produced by the α-decay of ^{226}Ra, which is itself just one of the radioactive isotopes produced in the chain of decay processes by which ^{238}U is transformed into ^{206}Pb. The ^{222}Rn is itself α-active:

$$^{222}_{86}Rn = ^{218}_{84}Po + ^{4}_{2}He; \text{ half-life} = 3.8 \text{ days} \tag{7.4}$$

α-particles are not an especially penetrating kind of radiation — they are stopped by a thin layer of paper or skin — but they can do very serious damage if they can find a route to relatively vulnerable and unprotected tissue without passing through the skin.

● Why do you think this makes ^{222}Rn an especially dangerous radioactive isotope?

● It is a gas and can therefore be inhaled. Its decay can then release α-particles into the vulnerable tissue that lines the lungs, leading to the development of lung cancer.

Radon concentrations in air are very variable, and are greatest over rocks that are relatively rich in isotopes in the ^{238}U decay chain. As many as 2 500 premature UK cancer deaths per year are said to be attributable to this entirely natural source of radioactivity.

7.3 Uses of the noble gases

The only noble gases produced in large quantities are argon (about 1 million tonnes p.a.) and helium (about 10 000 tonnes p.a.). An important use is as an inert blanket in high-temperature metallurgical processes such as the extraction and refining of titanium and zirconium, where the hot metals would otherwise react with oxygen or nitrogen. The very low boiling temperature of liquid helium (4.2 K) makes it the coldest liquid refrigerant available. It is essential in the study of low-temperature phenomena such as superconductivity and in high-field magnets such as those used in high resolution NMR spectroscopy. Because of the low solubility of helium in blood, a mixture of helium and oxygen is sometimes used instead of compressed air by deep-sea divers. Compressed air is avoided because nitrogen has higher solubility in blood and is the cause of a condition known as the bends (see Section 11.2). Argon, with about 12% nitrogen, is used to fill electric light bulbs (Figure 7.7). All the gases have some use in producing coloured lights in electric discharge tubes, a phenomenon to which the general name 'neon lighting' has been attached. The colours obtained in clear glass for each gas were given earlier, but they can be varied by mixing gases. A neon–argon mixture, for example, gives a deep lavender glow.

Although, as we shall now see, some noble gases do form chemical compounds, these are of no commercial importance. The noble gases are far more useful to us because of their inertness, than through any chemical reactivity they possess.

7.4 Summary of Sections 7.1–7.3

1 The noble gases are monatomic and occur in the atmosphere; argon is by far the most plentiful. Helium occurs in natural gas through the accumulation of α-particles from radioactive decay.

2 In 1894, Rayleigh noted that atmospheric 'nitrogen' had a higher density than chemically prepared nitrogen. Ramsay removed atmospheric nitrogen by combining it with hot magnesium, and obtained argon and other noble gases as a residual gas.

3 The melting and boiling temperatures of the noble gases increase down the Group as the polarizability of the atoms increases. The industrial fractional distillation of liquid air gives initially a fraction containing helium and neon gases, a liquid nitrogen fraction, a liquid argon fraction, and a liquid oxygen fraction containing krypton and xenon. Further separation and purification is achieved by more fractional distillation, or differential adsorption on charcoal.

4 Radon occurs only in trace amounts as ^{222}Rn with a half-life of 3.8 days. Being a gas and an α-particle emitter, it can, when inhaled, damage cells in the unprotected tissues of the lungs, thereby causing cancer.

5 The noble gases are useful because of their inertness. Argon and helium are used as inert atmospheres in high-temperature metallurgical processes; argon and krypton are used as filling gases in light bulbs, where they prevent filament evaporation and allow a higher filament temperature.

QUESTION 7.1

A liquid oxygen fraction rich in krypton and xenon is obtained by a procedure like that in Figure 7.6. The three components are separated by two successive fractional distillations. On the first column, the separation gives one component as a gas, and a liquid fraction, consisting of a mixture of the two others; on the second column, this liquid is separated into one component as a gas, and the other as a liquid. By using boiling temperatures from Table 7.2 and given that the boiling temperature of oxygen is 90 K, draw a diagram like that in Figure 7.6, displaying the process, and the exit points for each component.

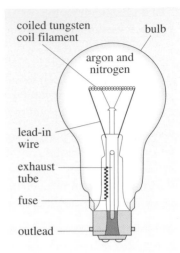

Figure 7.7
A standard electric light bulb contains a helically coiled tungsten filament, which becomes incandescent when heated by the passage of an electric current to about 2 800 K. The space is usually filled by a mixture of roughly nine-tenths argon and one-tenth nitrogen. The gas prevents evaporation from the hot filament, and allows a higher operating temperature (by 300 K), which leads to greater efficiency. Still greater efficiency is achieved by replacing argon with krypton: the more massive and therefore slower-moving krypton molecules conduct heat away from the filament less well, so less electrical energy is needed to maintain the operating temperature.

7.5 Noble gas compounds

In 1895, Ramsay sent a sample of argon to Moissan (Figure 6.2), who reported that fluorine and argon did not react with each other. Other attempts to obtain noble gas compounds were made, notably at the instigation of Linus Pauling in the 1930s. He persuaded D. M. Yost and A. L. Kaye to pass an electric discharge through a mixture of xenon and fluorine, but they did not observe a reaction. The important part played by the stability of the noble gas configuration in elementary theories of chemical bonding also encouraged the idea that the noble gases would not form ionic or covalent bonds.

In 1962, Neil Bartlett (Figure 7.8) was working with the very reactive red gas, platinum hexafluoride, PtF_6. It must be kept out of contact with air or moisture. But in the course of the work, some air accidentally leaked into the apparatus, and at room temperature an orange solid, PtF_6O_2, was formed:

$$PtF_6(g) + O_2(g) = PtF_6O_2(s) \tag{7.5}$$

🔵 What would the usual rules give for the oxidation number of platinum in PtF_6O_2?

🔵 With each oxygen contributing -2, and each fluorine -1, it would be $+10$, an unprecedentedly high value.

Bartlett then investigated the crystal structure. It proved to be very similar to that of $KSbF_6$, an ionic compound written $K^+[SbF_6]^-$. This suggested that PtF_6O_2 also contained a complex ion of the type MF_6^-, and therefore that the positive ion was O_2^+. Reaction 7.5 now becomes:

$$PtF_6(g) + O_2(g) = O_2^+[PtF_6]^-(s) \tag{7.6}$$

Here PtF_6 has acted as an oxidizing agent, and platinum has been reduced to platinum(V); the oxygen molecule has been oxidized to O_2^+. This is remarkable because the ionization energy of the oxygen molecule is so high ($1\,175\,kJ\,mol^{-1}$). Even more remarkable consequences followed when Bartlett noticed that the ionization energy of O_2 was very similar to that of xenon ($1\,170\,kJ\,mol^{-1}$).

🔵 What new experiment do you think this observation encouraged?

🔵 The attempted reaction of xenon with PtF_6, in the hope that a compound $Xe^+[PtF_6]^-$ would be formed.

This argument is analysed in more detail in Question 7.2. Here, it will suffice to say that when Bartlett passed colourless xenon gas into red PtF_6 vapour, the first noble gas compound appeared instantly as an orange–yellow solid on the walls of the containing vessel. He published his discovery under the title 'Xenon hexafluoroplatinate(V), $Xe^+[PtF_6]^-$'. Subsequent investigations have failed to confirm this formula: the product appears to be a mixture that includes the compound $[XeF^+]PtF_6^-$. But after this work, noble gas chemistry was never quite the same again.

7.5.1 The fluorides and oxides of xenon and krypton

After Bartlett's discovery, other chemists quickly re-examined Pauling's suggestion that xenon would combine with fluorine. It was found that by varying the reaction conditions, as shown in Figure 7.9 (overleaf), XeF_2, XeF_4 and XeF_6 could all be prepared. They are colourless solids at room temperature, but melt well below 150 °C.

Figure 7.8
Neil Bartlett was born in Newcastle in 1932, and graduated from the University of Durham. He was briefly a school teacher before moving to a teaching position at the University of British Columbia in Canada, where he synthesized the first true noble gas compound in 1962. Subsequently, he held chemistry chairs at Princeton University and the University of Berkeley in California, where he currently (2002) continues to work on noble gas and fluorine chemistry.

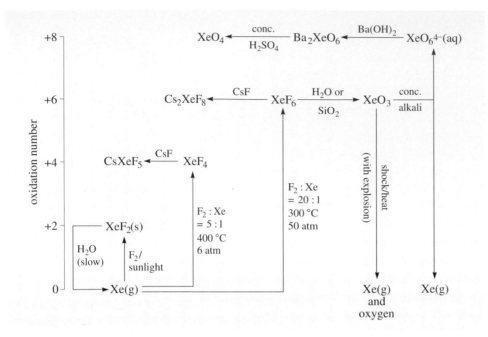

Figure 7.9
A network of the reactions of xenon and its compounds, in which the substances are classified by oxidation number.

All three xenon fluorides are strong oxidizing agents and revert fairly easily to xenon on reaction with the mildest of reducing agents. Thus, XeF_2 slowly oxidizes water with evolution of xenon:

$$2XeF_2(s) + 2H_2O(l) = 2Xe(g) + 4HF(aq) + O_2(g) \qquad (7.7)$$

At 0 °C, this reaction takes many hours for completion. With XeF_6, the oxidation of water, although thermodynamically very favourable, is again slow, and a rapid hydrolysis reaction occurs instead, yielding xenon trioxide:

$$XeF_6(s) + 3H_2O(l) = XeO_3(aq) + 6HF(aq) \qquad (7.8)$$

This oxide is also produced when XeF_6 attacks glass or silica:

$$2XeF_6(s) + 3SiO_2(s) = 2XeO_3(s) + 3SiF_4(g) \qquad (7.9)$$

XeO_3 is a colourless solid. It explodes with a blue flash on heating or under shock, giving xenon and oxygen; in this respect it is much more dangerous than the xenon fluorides. This distinction is confirmed by Table 7.3.

⚫ How is the distinction consistent with the thermodynamic data in Table 7.3?

⚫ For the xenon fluorides, both in the solid and gaseous states, $\Delta H_f^\ominus$ is negative; for XeO_3, it is positive. The decomposition

$$XeO_3(s) = Xe(g) + \tfrac{3}{2}O_2(g) \qquad (7.10)$$

will therefore be exothermic. As gases are produced, $\Delta S_m^\ominus$ will be positive, so $\Delta G_m^\ominus$ must be negative: the oxide, unlike the fluorides, is thermodynamically unstable with respect to its elements.

If XeO_3 is dissolved in strong alkali, it slowly disproportionates to xenon and the perxenate ion, $XeO_6^{4-}(aq)$:

$$2XeO_3(s) + 4OH^-(aq) = Xe(g) + XeO_6^{4-}(aq) + O_2(g) + 2H_2O(l) \qquad (7.11)$$

If the alkali in question is $Ba(OH)_2$, the salt Ba_2XeO_6 can be crystallized from solution.

Table 7.3 Standard enthalpies of formation of some noble gas compounds at 298.15 K

Substance	State	$\dfrac{\Delta H_f^\ominus}{\text{kJ mol}^{-1}}$
XeF_2	s	−163
XeF_2	g	−107
XeF_4	s	−267
XeF_4	g	−206
XeF_6	s	−338
XeF_6	g	−279
XeO_3	s	400
XeO_4	s	643
KrF_2	s	20
KrF_2	g	60

● What is the xenon oxidation number in XeO_6^{4-}?

● +8; this gives the right charge when added to the −12 supplied by the six oxygens.

With cold concentrated sulfuric acid, barium perxenate forms the insoluble barium sulfate, $BaSO_4$, and gaseous xenon tetroxide, XeO_4:

$$Ba_2XeO_6(s) + 2H_2SO_4(l) = 2BaSO_4(s) + XeO_4(g) + 2H_2O(l) \qquad (7.12)$$

The gas must be quickly chilled to prevent decomposition, and it then forms a yellow solid, which melts at about −40 °C. XeO_4 is even more dangerous than XeO_3, and is subject to sudden explosion at −40 °C or above. All these discoveries were made within 10 years of Bartlett's breakthrough: in that short time, xenon moved from being an element that formed no chemical compounds to one found in oxidation numbers, 0, +2, +4, +6 and +8. The chemistry of krypton is much more limited; the only known oxide or fluoride is KrF_2, a volatile white solid.

● Look at Table 7.3; can KrF_2 be made by the combination of krypton and F_2 gases at room temperature? ✳

● The positive value of $\Delta H_f^{\ominus}$ suggests not; formation from the standard states $Kr(g)$ and $F_2(g)$ is thermodynamically unfavourable.

Figure 7.10 shows how this difficulty is circumvented. KrF_2 decomposes slowly at room temperature into krypton and fluorine, confirming the thermodynamic instability implied by Table 7.3.

Radiochemical studies show that radon does form chemical compounds, but because of the short half-life of ^{222}Rn, they have not been made in visible amounts.

7.5.1.1 Noble gas fluorides as fluoride donors and acceptors

Like the higher fluorides of chlorine, bromine and iodine, noble gas fluorides can act as fluoride ion donors in the presence of good fluoride acceptors such as AsF_5; for example

$$XeF_2(g) + AsF_5(g) = XeF^+[AsF_6]^- \qquad (7.13)$$

Fluoride ion acceptor behaviour also occurs. XeF_6 will combine with alkali metal fluorides; at 80 °C the reaction is

$$2CsF(s) + XeF_6(g) = Cs_2XeF_8(s) \qquad (7.14)$$

Cs_2XeF_8 is the most thermally stable noble gas compound yet made: it decomposes only above 400 °C. The ion $[XeF_8]^{2-}$ has the shape of a square antiprism (Structure 7.1). A fluoride ion has also been forced on to XeF_4. At 190 °C, the following reaction occurs:

$$CsF(s) + XeF_4(g) = CsXeF_5(s) \qquad (7.15)$$

The ion XeF_5^- is pentagonal (Structure 7.2).

7.5.1.2 XeF₂ and KrF₂ as fluorinating agents

Both substances are very convenient fluorinating agents because the reaction by-product is xenon or krypton, which is easily swept out of the reaction vessel. Because XeF_2 is thermodynamically stable with respect to xenon and fluorine,

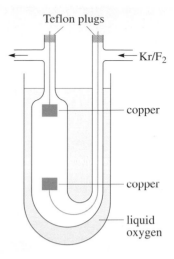

Figure 7.10
Krypton difluoride can be made by subjecting mixtures of krypton and fluorine to an electrical discharge. The reaction vessel is immersed in liquid oxygen at about 90 K, so solid KrF_2 condenses on to the walls before it has time to decompose. As KrF_2 is thermodynamically unstable with respect to its elements, $Kr(g)$ and $F_2(g)$ in their ground states do not react together. The electric discharge converts the reactants to excited, higher-energy states, from which the reaction is favourable.

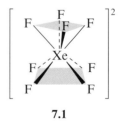

7.1

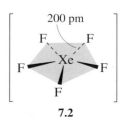

7.2

it is a milder fluorinating agent than fluorine. It reacts more quickly if dissolved in liquid HF; a typical reaction is

$$S(s) + 3XeF_2(hf) = SF_6(g) + 3Xe(g) \qquad (7.16)$$

By contrast, KrF_2, being thermodynamically unstable with respect to its elements, is a more powerful oxidizing agent than fluorine. It oxidizes gold to its highest-known oxidation number of +5 by forming gold pentafluoride in a two-stage process, the first of which involves the $[AuF_6]^-$ ion:

$$7KrF_2(g) + 2Au(s) = 2KrF^+[AuF_6]^-(s) + 5Kr(g) \qquad (7.17)$$

On heating the solid at 60 °C, the red AuF_5 is formed:

$$[KrF]^+[AuF_6]^-(s) = AuF_5(s) + Kr(g) + F_2(g) \qquad (7.18)$$

Gold(V) is known only as AuF_5 or the $[AuF_6]^-$ ion, and fluorine alone will not convert gold to AuF_5.

7.5.2 Other xenon compounds

In nearly all known xenon compounds that exist at room temperature, xenon atoms are bound chiefly to the highly electronegative fluorine and oxygen atoms. If xenon is bound to other atoms such as carbon, they are usually part of a group made electronegative by the presence of fluorine atoms. Thus, if the gas hexafluoroethane is irradiated so that the $F_3C{-}CF_3$ bond is broken and $CF_3\cdot$ radicals are generated, a white waxy solid, $Xe(CF_3)_2$, is formed when the gas is passed over XeF_2. Here, the three fluorine atoms bound to the carbon atom make it electronegative. Even so, $Xe(CF_3)_2$ decomposes completely in about 4 hours at room temperature.

Consider, however, a solution of XeF_2 in an excess of the strong fluoride ion acceptor, liquid antimony pentafluoride, SbF_5. The XeF_2 donates a fluoride to the solvent, and forms XeF^+(apf), where apf denotes SbF_5. If xenon gas is now brought into contact with the solution at 2–3 atmospheres pressure, a green solution of the paramagnetic ion Xe_2^+ is formed:

$$3Xe(g) + XeF^+(apf) + SbF_5(l) = 2Xe_2^+(apf) + SbF_6^-(apf) \qquad (7.19)$$

Thus, the xenon forces the XeF^+ to pass the second of the XeF_2 fluorines to the SbF_5, and Xe_2^+ results. Here, xenon is bound to itself, and not oxygen or fluorine, although Xe_2^+ is in a fluorinated environment. If liquid HF is present in the solution, then at −30 °C, dark green crystals of the compound $Xe_2^+Sb_4F_{21}^-$ appear:

$$Xe_2^+(apf) + 4SbF_6^-(apf) = Xe_2^+Sb_4F_{21}^-(s) + 3F^-(apf) \qquad (7.20)$$

X-ray crystallography shows that the Xe—Xe bond length in Xe_2^+ is 309 pm. This is very long, and it suggests that the bond is weak.

7.5.3 Helium, neon and argon chemistry

Currently, no compounds of helium, neon or argon have been characterized at room temperature and pressure. However, molecules containing argon have been identified at very low temperatures. Quite recently (2000), changes were observed in a frozen mixture of argon and HF when it was irradiated by UV light at temperatures below 20 K. The infrared spectrum of the sample contained three bands with major peaks at $1\,970\,cm^{-1}$, $687\,cm^{-1}$ and $436\,cm^{-1}$. These suggest the presence of HArF molecules. The first and third bands were assigned to vibrations that are predominantly Ar—H and Ar—F stretching modes, respectively. The band at 687 cm⁻¹ was assumed to be a bending vibration. The HArF molecules decomposed above 27 K, but their characterization holds out the possibility that, for argon at least, compounds like those of krypton and xenon may yet be obtained at room temperature.

7.5.4 The structure of noble gas compounds

When we apply VSEPR theory to XeF_2, we use two of the eight outer electrons on the xenon atom to form two Xe—F electron-pair bonds. The remaining six xenon electrons are grouped into three non-bonding pairs. This gives five repulsion axes distributed towards the corners of a trigonal bipyramid, and the non-bonding pair–non-bonding pair repulsions are minimized when these pairs occupy the equatorial positions; this correctly predicts the linear shape of XeF_2 (Figure 7.11a).

○ What shape does the same reasoning predict for XeF_4?

○ Square planar: four xenon electrons form four Xe—F bonds, leaving two non-bonding pairs, which occupy opposed positions within the six octahedrally disposed repulsion axes (Figure 7.11b).

An important test of this theory emerges when we turn to gaseous XeF_6. This time there are seven repulsion axes: six Xe—F bonds and one non-bonding pair, so the molecule should *not* be a regular octahedron. We noted that in the analogous case of the ion BrF_6^- in $CsBrF_6$, the shape *was* regular octahedral: the theory failed. However, we noted that the ion might be subject to overriding external forces from within the crystal. But with $XeF_6(g)$, we are in the gas phase, and there is no such excuse. So is the theory successful in this crucial case?

The answer is a qualified yes. You should note first that no molecule is rigid; the atoms are in constant motion with respect to each other: the molecule vibrates. The shapes of molecules, therefore, represent the *time-averaged* positions of the moving atoms. Now the movements in $XeF_6(g)$ seem highly unusual. They are consistent with a model in which there is a *distorted* octahedron with a non-bonding pair protruding from the centre of one face (Figure 7.12). The distortion arises because the three fluorines at the corners of that face are pushed apart. The vibration of the molecule then continuously moves the non-bonding pair and accompanying distortion from face to face of the octahedron. This deviation from the normal behaviour of an octahedral molecule is consistent with VSEPR theory. When we consider the oxides, XeO_3 and XeO_4, the shapes are again those predicted by VSEPR theory (Figure 7.13); you will encounter further successes in Question 7.4.

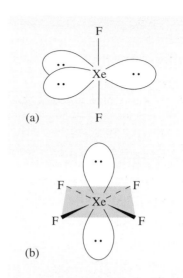

(a)

(b)

Figure 7.11
The shapes of (a) XeF_2 and (b) XeF_4 molecules are correctly predicted by VSEPR theory.

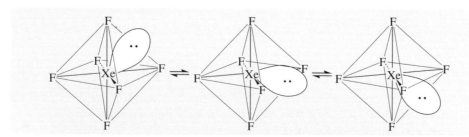

Figure 7.12 Studies of the structure and motion of the XeF_6 molecule are consistent with a *distorted* octahedral structure, of the type predicted by VSEPR theory, in which the non-bonding pair shifts from face to face of the octahedron via an edge.

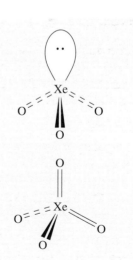

Figure 7.13
The shapes of the XeO_3 and XeO_4 molecules are correctly predicted by VSEPR theory when two of the eight xenon electrons are used to form each xenon–oxygen bond.

As in the case of the polyhalides, however, the situation with regard to *ions in crystals* is not quite so satisfactory. Consider first the case of the pentagonal ion, [XeF₅]⁻ (Structure **7.2**), encountered in Section 7.5.1.1 (p. 93).

⬤ Does VSEPR theory predict this shape?

⬤ Yes; if the overall negative charge of the ion is added to the central xenon, there are nine xenon valence electrons. Five are used to form Xe—F bonds, leaving two non-bonding pairs. Hence there are seven repulsion axes distributed towards the corners of a pentagonal bipyramid. The non-bonding pairs minimize their repulsions by occupying opposed, axial positions, and the ion is pentagonal (Figure 7.14).

This success is balanced, however, by the case of [XeF₈]²⁻ (Section 7.5.1.1). The square antiprismatic shape (Structure **7.1**) is one expected from eight repulsion axes: by twisting the top face of an [XeF₈]²⁻ cube through 45°, the distance between the eight Xe—F bonds is increased, and the repulsions are reduced. However, the procedure just applied to [XeF₅]⁻ predicts nine repulsion axes — that is, eight Xe—F bonds and one non-bonding pair. As Structure **7.1** shows, there is no stereochemical evidence to support the presence of the non-bonding pair, so this must be regarded as a failure of VSEPR. The case resembles that of BrF₆⁻ where we suggested that in a crystal, shapes might be affected by forces acting *between* as well as *within* ions.

7.5.5 The bonding in noble gas compounds

When using VSEPR theory, we assume xenon forms two single, two-electron bonds in XeF₂. Now tellurium comes two places before xenon in the Periodic Table, and has the outer electronic configuration $5s^2 5p^4$ (**7.3**). When a tellurium atom forms two single bonds with hydrogen or a halogen in H₂Te or TeX₂, the two unpaired 5p electrons in the box diagram each pair up with an unpaired electron on a hydrogen or halogen atom.

⬤ Why cannot this happen in XeF₂?

⬤ The outer electronic configuration of xenon is $5s^2 5p^6$ (**7.4**); it has no unpaired electrons to pair with an unpaired electron on fluorine.

One solution is to assume that, when bonding to fluorine, the xenon atom *promotes* an outer electron to some empty, higher energy orbital. The one usually chosen is one of the five 5d orbitals, which normally become occupied only in the next Period in the Periodic Table. Promotion of, say, a 5s electron gives the configuration shown in **7.5**.

Now two two-electron bonds can be formed from unpaired electrons in the 5s and 5d orbitals.

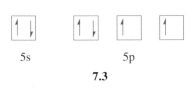

Figure 7.14
The shape of the ion [XeF₅]⁻, according to VSEPR theory.

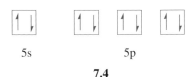

Thus, when we apply VSEPR theory to xenon fluorides, we commit ourselves to both Xe—F single bonds, and to using higher energy orbitals, such as 5d, to form those bonds. But this use of higher-energy orbitals is controversial, and alternative treatments have been proposed.

One alternative uses molecular orbital theory. Suppose we take the F—Xe—F axis in XeF_2 to be the z axis. Then we can avoid the use of higher-energy orbitals by forming molecular orbitals from a $2p_z$ orbital on each of the fluorines, and the $5p_z$ orbital of xenon.

How many molecular orbitals will result, and of what type will each one be?

Three atomic orbitals generate three molecular orbitals. There will be a bonding orbital, which must have an antibonding partner. The remaining, unpartnered molecular orbital must be non-bonding.

The energy-level diagram is shown in Figure 7.15, along with the atomic orbital composition of the three molecular orbitals. Note that the orbitals extend over three atoms; in other words we are dealing with *three-centre bonding* (cf. Section 8.14).

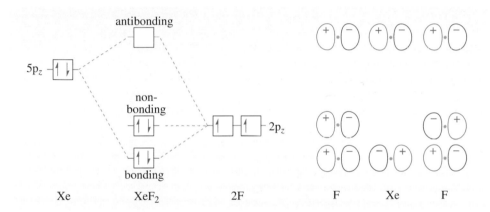

Figure 7.15 Partial orbital energy-level diagram showing the formation of molecular orbitals for XeF_2 from the xenon $5p_z$ orbital and the fluorine $2p_z$ orbitals. The combination of atomic orbitals for each molecular orbital is shown on the right: three atomic orbitals give three molecular orbitals. The xenon orbital is used to give the lower bonding orbital, with matching signs of the wavefunction in the overlap regions, and the upper antibonding orbital with mismatched signs. This leaves the third non-bonding orbital to be formed from just the two fluorine orbitals; as these two orbitals have the same orientation along the z axis in the bonding and antibonding orbitals, they have opposite orientations in the non-bonding one. Note that the non-bonding orbital that is formed from fluorine orbitals alone is fully occupied; this implies that, consistent with its high electronegativity, there is considerable electron density on the fluorine atoms in XeF_2. By adding an exercise of this kind along, say, the y axis, a molecular orbital treatment of XeF_4 can be obtained.

Because the ionization energy of fluorine is larger than that of xenon, the 2p orbitals of fluorine lie below the 5p orbital of xenon. There are two electrons in the $5p_z$ orbital of the xenon atom, and one electron in each of the $2p_z$ orbitals of the fluorine atoms. Thus, there are four electrons to feed into the diagram, two occupying the bonding orbital, and two occupying the non-bonding one.

● Remembering that the bonding is spread over two Xe—F bonds, what is the Xe—F bond order?

● One-half; the filled bonding orbital provides a total bond order of one, and this is divided by two.

More sophisticated methods of assessment give a higher bond order, but the value is still much less than one. It is therefore considerably less than that assumed in VSEPR theory. Which fits the facts better? Iodine comes just before xenon in the Periodic Table, and the molecule IF is written with a single bond. The bond length is 191 pm. This is close to typical values (185–195 pm) for the Xe—F distance in the FXe^+ cation, which, like IF, contains a single bond (Structure **7.6**). These values suggest that the Xe—F single bond length should be about 190 pm. Now look at Figures 7.16a and 7.16b.

● What bond order would you judge the Xe—F bonds in XeF_2 and XeF_4 to be closer to: one or one-half?

● They seem closer to one: the bond lengths are only slightly longer than our suggested single bond length. A bigger difference would be expected if the bond order were considerably less than one.

Thus, the description of the bonding derived from VSEPR theory fits the bond lengths rather better than a molecular orbital treatment that uses just p orbitals in the bonding. In other situations, however, this may not be the case. With I_3^-, for example (Figure 6.22), the lengths of the bonds relative to that in I_2 suggest that a three-centre molecular orbital treatment giving a bond order much less than one is the better model.

7.6

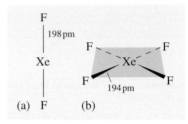

(a)   (b)

Figure 7.16
The shapes and bond lengths in (a) the XeF_2, and (b) the XeF_4 molecules.

7.5.6 Summary of Section 7.5

1   Oxygen and PtF_6 react to form $O_2^+[PtF_6]^-$. The first true noble gas compound was made in 1962 by the reaction of xenon and PtF_6 after Bartlett noticed that the ionization energies of Xe(g) and O_2(g) were very similar.

2   Currently, helium, neon and argon still appear to form no chemical compounds at room temperature and pressure, but molecules of HArF have been observed spectroscopically at very low temperatures.

3   Xenon and fluorine react in sunlight to give XeF_2. By increasing the F_2 : Xe ratio, the gas pressure and the temperature, XeF_4 and XeF_6 can also be made. The three fluorides are volatile white solids at room temperature.

4   The xenon fluorides are strong oxidizing and fluorinating agents, often acting like a mixture of xenon and fluorine. Thus, XeF_2 oxidizes water to oxygen, with the formation of gaseous xenon.

5   KrF_2 is the only known binary krypton compound. Unlike the xenon fluorides, it is unstable with respect to its elements. It is therefore a more powerful fluorinating agent than fluorine, and is made by passing an electric discharge through a Kr/F_2 mixture.

6   Noble gas fluorides act as fluoride donors in the presence of good fluoride acceptors such as AsF_5. On the other hand XeF_6 and XeF_4 act as fluoride acceptors, forming $Cs_2[XeF_8]$ and $CsXeF_5$ with CsF.

7   XeO_3 is formed by the reaction of XeF_6 with water or silica. In strong alkali it disproportionates, giving a solution containing XeO_6^{4-}(aq), from which

Ba_2XeO_6 can be crystallized. With concentrated H_2SO_4, the latter yields $XeO_4(g)$, which must be quickly chilled to prevent decomposition. Both xenon oxides are dangerously explosive, endothermic compounds.

8   Molecules of noble gas compounds have shapes in excellent agreement with VSEPR theory, although to form two-electron bonds, xenon must promote valence electrons to higher-energy orbitals such as 5d. This model can be avoided by a molecular orbital treatment which, in XeF_2, forms three-centre bonds from xenon 5p and fluorine 2p orbitals. The Xe—F bonds then have an order much less than one. The experimentally determined bond lengths are in better agreement with the two-electron, two-centre bonds of VSEPR theory.

QUESTION 7.2

Bartlett undertook the reaction of xenon and PtF_6 because Reaction 7.6 occurred:

$$O_2(g) + PtF_6(g) = O_2^+[PtF_6]^-(s) \tag{7.6}$$

and because the ionization energies of xenon and O_2 were similar. It therefore seemed possible that the reaction of Xe and PtF_6 to give $Xe^+[PtF_6]^-$ might be equally favourable. Construct a thermodynamic cycle around Reaction 7.6, in which there is an ionization step for $O_2(g)$. What other energy term in the cycle changes when O_2 is replaced by Xe? Are there reasons for believing that this too would be little altered by the substitution?

QUESTION 7.3

Write likely balanced equations for reactions between the following substances:

(i) $XeF_4(g)$ and $H_2(g)$;

(ii) $KrF_2(g)$ and water;

(iii) $XcF_6(g)$ and $AsF_5(g)$;

(iv) magnesium powder and $XeO_3(s)$;

(v) $Xe(g)$ and excess $KrF_2(g)$.

QUESTION 7.4

XeO_4 reacts with XeF_6 as follows:

$$XeF_6 + XeO_4 = XeOF_4 + XeO_3F_2 \tag{7.21}$$

Use VSEPR theory to predict the shapes of the oxohalide molecules $XeOF_4$ and XeO_3F_2.

QUESTION 7.5

In Section 6, we discussed the preparation of the compound KHF_2, which contains the linear HF_2^- ion.

(a) Draw a partial orbital energy-level diagram for the [FHF] *molecule* along the lines of Figure 7.15, in which three-centre molecular orbitals are formed from the 1s orbital of hydrogen and the $2p_z$ orbitals of fluorine. Draw the combination of atomic orbitals for each molecular orbital. How does the occupancy of the molecular orbitals change in moving from the HF_2 molecule to the HF_2^- ion?

(b) What is the bond order of each H—F bond in (a) the HF_2 molecule and (b) the HF_2^- ion? The bond length is 112 pm compared with 92 pm in HF(g). How consistent is this difference with your estimated bond order?

GENERAL OBSERVATIONS ON SECOND- AND THIRD-ROW ELEMENTS AND PERIODIC TRENDS

8

Our review of the properties of the p-Block elements began with a description of Groups VII, the halogens, and VIII, the noble gases, elements that display extreme non-metallic properties. Their chemistry is dominated by covalent compounds, and our discussion was set in a framework of covalent bonding models. These elements contrast with those at the left side of the Periodic Table (Figure 8.1), which are the typical metals of Groups I and II. Useful insights into their chemistry can be gained by application of the ionic model.[*] We now turn our attention to the elements of Groups III to VI, focusing especially on the second and third Periods: boron and aluminium, carbon and silicon, nitrogen and phosphorus and oxygen and sulfur. These are elements of great social and industrial importance. The chemistry of sulfur is taken up again in the Case Study *Acid Rain: sulfur and power generation*, where the topic of acid rain is discussed, and the industrial importance of these elements is evident in the overview of industrial inorganic chemistry presented in the *Industrial Inorganic Chemistry* Case Study.

In Groups III to VI, we observe the transition from metallic to non-metallic behaviour, and so we can anticipate diverse chemistry. We naturally expect similarities between elements in the same Group, such as nitrogen and phosphorus, but there are important differences as well. In particular, we shall examine similarities and differences between second- and third-Period elements across the entire width of the elements shown in Figure 8.1. This subject is taken up later, but to put the intervening descriptive chemistry into perspective we make a few general observations on it now.

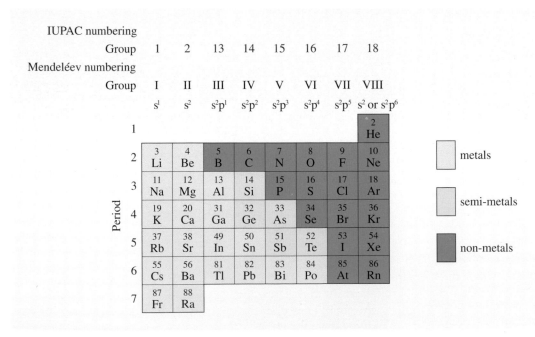

Figure 8.1 The typical or Main Group elements of the Periodic Table.

[*] The ionic model is described in Section 20 of *Metals and Chemical Change*.[3]

8.1 Single and multiple bonds

One difference between second- and third-row elements is that substances containing multiple bonds are more easily made or preserved in the second row than they are in the third. Some important examples can be found among the structures of the elements themselves (Figure 8.2).

IUPAC numbering Group	1	2	13	14	15	16	17	18
Mendeléev numbering Group	I	II	III	IV	V	VI	VII	VIII
	Metallic		Mostly extended structures		Covalent diatomic			Monatomic
1							H—H single bonds	He
2	Li	Be	B	C	$N{\equiv}N$	$O{=}O$	F—F	Ne
					multiple bonds			
3	Na	Mg	Al	Si	P (also P_4)	S_8 rings	Cl—Cl	Ar
4	K	Ca	Ga	Ge	As	Se_∞ chains	Br—Br	Kr
5	Rb	Sr	In	Sn	Sb	Te_∞ chains	I—I	Xe
6	Cs	Ba	Tl	Pb	Bi	Po	At?	Rn
7	Fr	Ra						

(Period — along the left side)

Figure 8.2 The structure of the elemental forms of the typical elements.

● Why do the structures of elemental oxygen and sulfur provide such an example?

● The second-row element oxygen consists of O_2 molecules containing a *double* bond (O=O), whereas the commonest allotrope of sulfur, rhombic sulfur, contains S_8 rings held together by S—S *single* bonds (Figure 8.3).

Group IV provides a further example. Graphite is the commonest form of carbon, and consists of hexagonal rings of carbon atoms, which share sides to form sheets (Figure 8.4).

The broken circles within the hexagons mark the delocalization of π electrons, which gives rise to a bond order greater than one (one and one-third). By contrast, elemental silicon occurs only in a diamond structure, with single Si—Si bonds. Nor is this difference between the second and third Periods confined to the elemental states. In Group IV, for instance, carbon dioxide consists of small discrete molecules, O=C=O, containing two carbon–oxygen double bonds. But SiO_2, the dioxide of the third-row element silicon, contains only Si—O single bonds, and this is achieved by an extended covalent structure (Figure 8.5), in which the coordination to the oxygens around each silicon is tetrahedral, and the coordination to the silicons around each oxygen is V-shaped. Thus, SiO_2 is a giant molecular solid: its high boiling temperature (about 2 800 °C) reflects the strength of the Si—O single bonds that must be broken before it can volatilize as smaller units.

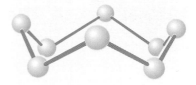

Figure 8.3
The structure of rhombic sulfur.

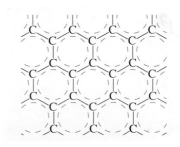

Figure 8.4
The structure of graphite.

The great differences in the volatility and physical states of CO_2 and SiO_2 are therefore a direct consequence of the preference for single bonding in the third Period, and for multiple bonding in the second. The theoretical reasons for this preference will be discussed later.

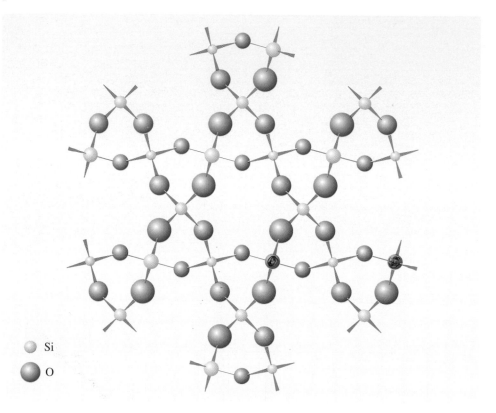

Si

O

Figure 8.5
The structure of one form of SiO_2, β-quartz.

8.2 Third-row elements: a case for expansion of the octet?

The most elementary theories of covalent bonding are founded on two concepts, namely G. N. Lewis's two-centre electron-pair bond and the stability of the octet. Some very common compounds of the third and later rows of typical elements prove these concepts wanting. The fluorides of nitrogen and phosphorus provide one example. The highest fluoride of nitrogen is NF_3. This is a gas at room temperature, and it contains NF_3 molecules (Structure **8.1**) in which N—F electron-pair bonds provide both nitrogen and fluorine with octet configurations. However, phosphorus, a third-row element, forms *two* gaseous fluorides at room temperature, PF_3 and PF_5. The explanation of the bonding in PF_3 is similar to that of NF_3: three electron-pair bonds yield octets on the fluorine and the Group V atoms. The case of PF_5 is different.

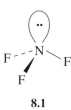

8.1

How many outer electrons does the phosphorus atom in PF_5 acquire, if it forms shared electron-pair bonds?

Ten; each one of its five outer electrons is paired with a fluorine electron in a P—F bond.

This allocation is the one used in VSEPR theory: the repulsions between the five bond pairs of the five P—F bonds give rise to a PF_5 molecule with a trigonal-bipyramidal shape (Structure **8.2**). Forced to choose between the octet and the two-centre electron-pair bond, we retain the latter and abandon the former. An important consequence of this choice emerges when we consider which phosphorus orbitals must be used to form the five electron-pair bonds. The outer electronic configuration of phosphorus is $3s^2 3p^3$ (Figure 8.6): five unpaired electrons are needed to form the five bonds, and only three are available.

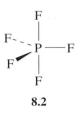

8.2

● What assumption might be made to put this right?

● The same one that we made in one of the treatments of the bonding in XeF_2 (Section 7.4): we assume that, when bonding to fluorine, the phosphorus atom promotes one electron from its $3s^2$ pair to an empty, higher-energy orbital.

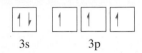

Figure 8.6
The outer electronic configuration of phosphorus.

The one usually chosen is one of the five 3d orbitals (Figure 8.7). These normally become occupied only when we reach the next row of the Periodic Table. The energy needed to promote the electron is then recouped by forming five bonds instead of three.

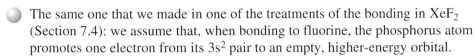

Figure 8.7 Promotion of an electron to a 3d orbital in phosphorus.

One reason why the 3d orbitals might be sufficiently low in energy, and therefore accessible for participation in bonding, is that they have the same principal quantum number ($n = 3$) as the normal phosphorus valence electrons. This then supplies a possible explanation for why nitrogen seems unable to form a pentafluoride: the outer electronic configuration of nitrogen is $2s^2 2p^3$, and there are no such things as 2d orbitals. The most accessible unoccupied orbital is 3s, whose principal quantum number is one greater than that of the valence electrons, and is too high in energy. But despite this apparent plausibility, the involvement of phosphorus 3d orbitals in the bonding in PF_5 is controversial. Although these orbitals have the same n value as the $3s^2 3p^3$ valence electrons, they are still rather high in energy. This has prompted alternative molecular orbital treatments in which σ-bonding molecular orbitals are formed from a 2p orbital on each fluorine, and from just the 3s and three 3p orbitals on phosphorus.

● How many molecular orbitals will be formed from such a combination?

● Nine; there are nine atomic orbitals, which must generate nine molecular orbitals. Four of these molecular orbitals are bonding, and there are four matching antibonding ones; this leaves one non-bonding orbital (Figure 8.8, overleaf).

● Pencil the appropriate number of electrons into Figure 8.8. What is the *average* bond order of a P—F bond according to this scheme?

● Four-fifths: you have ten electrons (five from phosphorus and one from each fluorine 2p orbital) to assign; these fill the four bonding orbitals, and the one non-bonding orbital, giving a total bond order of four; which is distributed over five bonds.

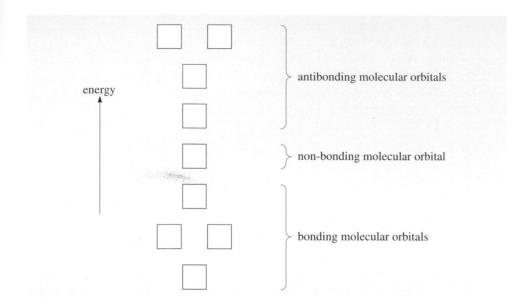

Figure 8.8
The sequence of σ-bonding molecular orbitals in a treatment of the bonding in PF$_5$, which uses a 2p orbital on each fluorine, and the 3s and 3p orbitals on phosphorus.

When we choose this option then, the P—F bond order drops below one. We put forward an analogous molecular orbital treatment when we dealt with XeF$_2$ (Section 7.4): the bond order per Xe—F bond was less than the value of one implied by VSEPR theory. But although the choice of bonding treatment may be a matter of opinion, there is no dispute about the experimental facts: third- and later-row elements of Groups V–VII, such as phosphorus, sulfur and chlorine, form higher fluorides and oxides than their counterparts among the second-row elements, nitrogen, oxygen and fluorine. You will see more examples of this trend in the following Sections of this book.

8.3 Trends in the Periodic Table

In the ensuing Sections on Groups III to VII, the transition across the Periodic Table, from metals to non-metals, illustrates the diverse chemistry of these Groups. Thus aluminium in the third Period of Group III is a metal, and sulfur in Group VI is a typical non-metal. Similar gradations of behaviour are observed in vertical comparisons in the Periodic Table. In Group IV, we see the transition from the non-metal carbon, to the metals tin and lead. In the remainder of this Section we draw attention to some of these trends that will be apparent in Groups III to VII and which will be summarized in Section 13.

8.3.1 Trends across the Periods

Ionization energies, electronegativities and covalent radii

The first ionization energies and electronegativities of the typical elements show overall increases across the Periodic Table (Figures 8.9 and 4.4). By contrast, the single-bond covalent radii show a decrease. All three trends reveal the importance of the increase in the nuclear charge as successive protons are added to the nucleus of the atoms. Such an increase causes the outer electrons to be bound more and more tightly across the row. They are trends in the properties of atoms, and as such provide possible explanations of other trends in physical and chemical properties, which we examine in the following Sections.

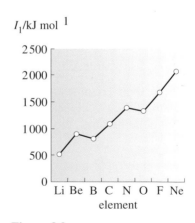

I_1/kJ mol^{-1}

Figure 8.9
The first ionization energies of the elements of the second Period.

Metals, semi-metals and non-metals

At the beginning of a row of Figure 8.1, the nuclear charge and ionization energies are relatively low. Electrons are then more easily lost to, or become part of, an 'electron pool' or 'electron gas', which, according to the simplest theory of metallic bonding, holds the metal together through its interaction with the residual positive ion cores (Figure 8.10). The increase in nuclear charge therefore accounts for the tendency for metals to give way first to semi-metals, and then to non-metals, across a row of Figure 8.1. Examine this tendency, as it borne out by the structures of the elements described in the following Sections.

The structures of halides, hydrides and oxides

At the left-hand side of a row in Figure 8.1, the electronegativity of an element is low, much lower than that of a halogen, oxygen or hydrogen. Halides, oxides and hydrides form ionic structures; thus, the halides and hydrides of lithium, sodium, potassium and rubidium all crystallize with the structure of NaCl. Similarly, the halogens, with high electronegativity, engage in these ionic structures as anions, but show strong tendencies to form covalent compounds, for example, in the halogens and interhalogens (Section 6). In the following Sections note the type of structure taken by the halides, hydrides and oxides of the elements of Groups III to VI and identify the general trends across the Periodic Table. It is also worth noting the chemical properties such as the acid–base character of their hydrides and oxides.

8.3.2 Trends down the Groups

Ionization energies, electronegativities and size

The first ionization energies and electronegativities of the typical elements tend to decrease down a Group. By contrast, measures of size, such as single-bond covalent radii or the ionic radii of similarly charged ions, tend to increase. Figures 8.11 and 8.12 (overleaf) show the contrasted examples of ionization energy and ionic radii.

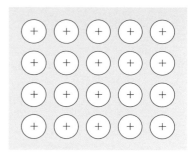

Figure 8.10
In the simplest theory of metallic bonding, the pool of electrons (blue) holds the metal together through interaction with the positive cores.

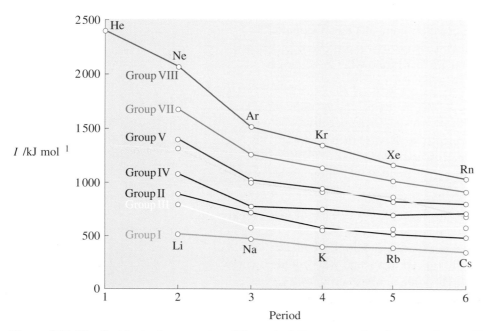

Figure 8.11 The first ionization energies of the typical elements tend to decrease down a Group.

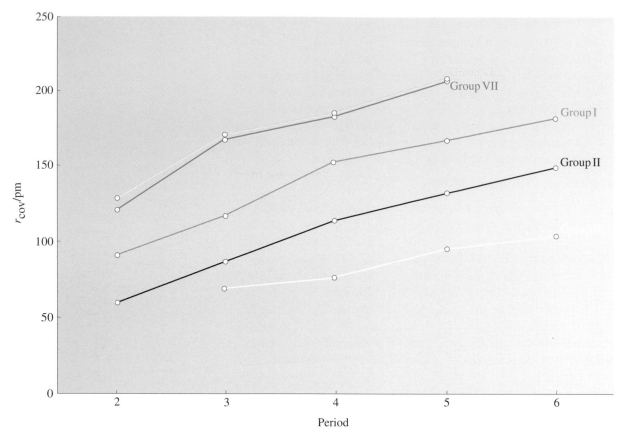

Figure 8.12 The ionic radii of similarly charged ions of the typical elements tend to increase down a Group.

All these trends reveal the importance of the increase in distance of the outer electrons from the nucleus as the successive electron shells build up. This increasing separation weakens the hold that the nucleus has on the outer electrons.

Ionization energy is also the principal factor determining the metallic and non-metallic properties of the elements. It is instructive to explore how these properties of the elemental forms of Groups III to VI follow the trend, down the Group, as illustrated in Sections 9 to 12.

Similarly, within Groups of Figure 8.1, electronegativity is highest at the top. On descending a Group it decreases, so that differences in electronegativity between any element and oxygen or halogen (high electronegativity) then increase. The Oxides and halides of the elements are therefore expected to become more ionic on descending a Group. Again, note examples of this trend in the following Sections on Groups III to VI.

Finally, bond formation between like elements depends on a number of factors that are dependent on the size of atoms. In particular, the ability of elements to engage in $\pi-\pi$ bonding with each other results in the formation of multiple bonds. Examples of the tendency to form π bonds occur in Groups III to VI, and you should note how this is affected by the covalent radii, which increase as we descend each Group.

8.4 Summary of Section 8

1   Substances containing multiple bonds are more easily made or preserved when they contain second-row rather than third-row typical elements.

2   Third-row typical elements of Groups V–VII form higher fluorides and oxides than their counterparts in the second row.

3   The bonding in these compounds can be described as the formation of a number of conventional two-electron bonds equal to the classical valency (e.g. five in PF_5). This requires involvement of d orbitals in bonding. Alternatively, there are molecular orbital treatments that require only the s and p orbitals of the third-row element, and imply lower bond orders.

4   Many trends in chemical behaviour are related to the variation of physical properties down the Groups and across the Periods of the Periodic Table:

   • down the Groups, ionization energy and electronegativity show an overall decrease and covalent radius increases;

   • across the Periods, ionization energy and electronegativity both increase, and covalent radius decreases.

QUESTION 8.1 ⚡

The commonest form of phosphorus is white phosphorus, a waxy solid consisting of discrete P_4 molecules, in which the phosphorus atoms occupy the corners of a tetrahedron (Figure 8.13). What is the covalency of the phosphorus atoms in this structure? How does this compare with the covalency of nitrogen in elemental nitrogen?

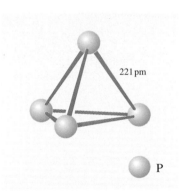

221 pm

P

Figure 8.13 The structure of white phosphorus, P_4.

QUESTION 8.2 ⚡

The highest fluoride of oxygen is the gas OF_2; the highest fluoride of sulfur is the gas SF_6, which contains octahedral SF_6 molecules.

(i)   If we assume that each sulfur–fluorine bond in SF_6 is a single bond, how many of the five 3d orbitals on sulfur must be involved in the bonding?

(ii) A molecular orbital treatment of the bonding in SF_6 that uses a p orbital on each fluorine atom, and just the 3s and 3p orbitals on sulfur generates four bonding orbitals, four antibonding orbitals and two non-bonding orbitals. Explain why there are ten molecular orbitals in all. What is the bond order of each sulfur–fluorine bond according to this scheme?

THE GROUP III/13 ELEMENTS

9

The typical elements shown in the mini-Periodic Table (Figure 9.1) divide into Groups I and II (the metallic s-Block elements) and Groups III to VIII (the p-Block elements). Having considered the chemistry of Groups VII and VIII, the essentially non-metallic Groups, we now proceed to a Group in which the properties are more diverse: Group III/13, shown in Figure 9.2.

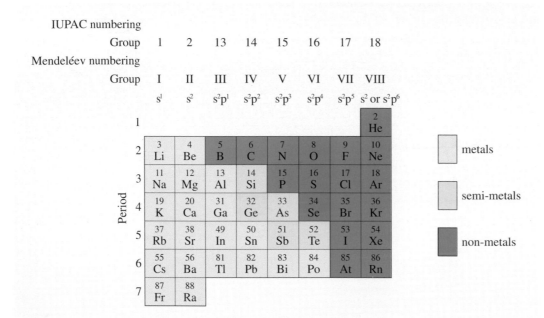

Figure 9.1
Mini-Periodic Table showing the typical elements.

The electrical conductivity of boron shows it to be a non-metal; aluminium, gallium, indium and thallium are metals.

We acknowledge this greater variation within the Group by dealing separately with boron and aluminium. Then we shall take gallium, indium and thallium together, a move that can be justified not only chemically, but also by reference to electronic configurations. Table 9.1 shows that, as expected, the atoms of the Group III elements all have outer electron configurations of the type s^2p^1.

Figure 9.2
The Group III/13 elements: boron is a non-metal, the others are metals.

Table 9.1 Electronic configurations of Group III/13 atoms

Atom	Electronic configuration
B	[He] $2s^2\,2p^1$
Al	[Ne] $3s^2\,3p^1$
Ga	[Ar] $3d^{10}\,4s^2\,4p^1$
In	[Kr] $4d^{10}\,5s^2\,5p^1$
Tl	[Xe] $4f^{14}\,5d^{10}\,6s^2\,6p^1$

● What distinguishes the configurations of gallium, indium and thallium from those of boron and aluminium?

◐ There are inner d shells (and, in the case of thallium, f shells) of lower principal quantum number between the valence electrons and the filled shells of the preceding noble gas.

As we shall see, this presence of filled inner shells has important chemical consequences.

9.1 Boron: occurrence and extraction

Boron occurs naturally as borates — substances containing oxoanions in which the oxidation number of boron is +3. There are large deposits in California and Turkey. Borax, often written $Na_2B_4O_7.10H_2O$, is a good example (but see Section 9.1.5). Mined borax is purified by recrystallization from water. Treatment of its hot solution with a strong mineral acid, such as sulfuric acid, causes white, crystalline boric acid, H_3BO_3, to crystallize on cooling.

$$\begin{array}{c} OH \\ | \\ B \\ HO \quad\quad OH \end{array}$$

9.1

However, the structure of the acid molecule (Structure **9.1**) shows that $B(OH)_3$ is a better formulation. When boric acid is heated, boric oxide, B_2O_3, is formed:

$$2B(OH)_3 = B_2O_3 + 3H_2O(g) \tag{9.1}$$

9.1.1 The boron atom

Table 9.1 shows that the boron atom has three outer, valence electrons. Suppose that it uses them to form electron-pair bonds.

● How many bonds can be formed in this way? In the process, does the atom gain a noble gas configuration?

◐ Three single bonds are formed, thereby increasing the number of outer electrons to six — that is, two short of the electronic configuration of neon.

This is a unique feature of boron chemistry: it is the only non-metal or semi-metal that cannot gain an octet solely by using its outer electrons to form shared electron-pair bonds. Much boron chemistry can be regarded as processes in which the atom gains a share in a greater number of electrons. Many of these reactions of boron can be classified as interactions between a Lewis acid and a Lewis base.

9.1.2 Lewis acidity in the boron halides

The boron halides have the expected formula, BX_3. Boron trifluoride can be made by heating boric oxide, calcium fluoride and concentrated sulfuric acid:

$$B_2O_3(s) + 3CaF_2(s) + 3H_2SO_4(l) = 2BF_3(g) + 3CaSO_4(s) + 3H_2O(l) \tag{9.2}$$

BCl_3 and BBr_3 can be obtained by direct combination of the elements, and BI_3 by heating BCl_3 with HI. At 25 °C, BF_3 and BCl_3 are gases, BBr_3 is a liquid and BI_3 is a solid (Table 9.2). All four halides are easily hydrolysed by water, boric acid being one of the products.

Table 9.2 Properties of the boron trihalides

	BF_3	BCl_3	BBr_3	BI_3
$T_m\,/°C$	−127	−107	−46	50
$T_b\,/°C$	−100	13	91	210
B—X distance/pm	130	175	187	210

Now both experimental measurements and VSEPR theory agree that the BX_3 molecules in these substances are trigonal planar (Structure **9.2**). As noted in Section 9.1.1, boron then has six outer electrons.

9.2

● How can boron increase this number to eight?

● The trihalide can act as a Lewis acid, acquiring a share in the non-bonded pair of a Lewis base, such as ammonia or trimethylamine; for example:

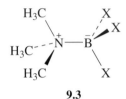

9.3

$$(CH_3)_3N(g) + BX_3(g) = (CH_3)_3\overset{+}{N}-\overset{-}{B}X_3(s) \qquad (9.3)$$

In the white solid product, the geometry around both the donor nitrogen atom and the acceptor boron atom is tetrahedral (Structure **9.3**); the structure of the complex is analogous to that of a substituted ethane, C_2H_6. It is isoelectronic with $(CH_3)_3C-CX_3$. The Lewis base may also be a halide ion; thus when BF_3 is heated with solid KF, the tetrahedral tetrafluoroborate ion (Structure **9.4**) is formed:

$$KF(s) + BF_3(g) = KBF_4(s) \qquad (9.4)$$

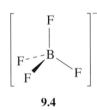

9.4

● Using the electronegativity of the halogen as a criterion, which trihalide should form the more stable complex in Reaction 9.3, BF_3 or BBr_3?

● BF_3; the more electronegative fluorine atoms will exercise a stronger attraction on the incoming electron pair of the Lewis base than the less electronegative bromine atoms.

In fact, experiment shows the opposite: the bromide complex is the more stable; for example, $\Delta H_m^{\ominus}$ for Reaction 9.3 becomes more negative from BF_3 to BBr_3, suggesting that the stability of the $(CH_3)_3NBX_3$ complexes is in the order F < Cl < Br. We can begin to explain the stability order by considering the bonding in the boron trihalides, which is illustrated in Figure 9.3.

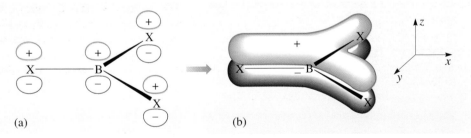

(a)   (b)

Figure 9.3 The bonding in the boron trihalides. The three two-electron bonds in the *xy* plane use a combination of the boron 2s, $2p_x$ and $2p_y$ orbitals. This leaves an *empty* boron $2p_z$ orbital above and below the plane, as in (a). Because the molecule is planar, this overlaps with *filled* p_z non-bonding orbitals on the halogen atoms, resulting in partial delocalization of halogen non-bonding electrons into the boron $2p_z$ orbital. In other words, the B—X bonds are strengthened by π bonding via a π orbital extending over all four atoms, as in (b).

Evidence for the π bonding is provided by bond lengths: in free BF₃, for example, the B—F distance is 130 pm (Table 9.2); in complexes like those shown in Structures **9.3** and **9.4**, where the planar geometry has been destroyed, and the possibility of π bonding is much diminished, it is longer (about 140 pm) and so the bonds are presumed weaker. The π bonding is believed to be most marked in BF₃, where the small fluorine atoms are closer to the central boron than are the halogen atoms in the other boron halides, which allows better overlap with the vacant boron $2p_z$ orbital. Because this π bonding largely disappears in reactions such as Reaction 9.3, BF₃ is more reluctant to undergo the reaction than the other trihalides, which accounts for the order of stability of the (CH₃)₃NBX₃ complexes mentioned above: F < Cl < Br.

This Section has shown you two ways in which a trivalent boron atom can increase the number of electrons that it shares with other atoms: boron compounds can act as Lewis acids, or, if a boron atom is bound to halogen or oxygen atoms with non-bonding electrons, it can participate in π bonding.

9.1.3 Boron and the borides

Boron of purity 95–98% can be made by reducing B₂O₃ in a furnace with magnesium. High-purity boron can be obtained by the decomposition of the triiodide at 1 000 °C:

$$2BI_3(s) = 2B(s) + 3I_2(g) \tag{9.5}$$

Boron is a hard inert black solid of very low electrical conductivity (5×10^{-5} S m⁻¹), which melts at 2 180 °C. Elemental boron occurs in several forms, but they all consist of interconnected B₁₂ units, in which the twelve boron atoms are positioned at the corners of a regular icosahedron, a solid figure with twenty triangular faces (Figure 9.4).

🔵 Could the lines between the boron atoms in Figure 9.4 represent conventional single bonds?

🔵 No; each boron is joined to *five* others, but its three outer electrons can form a maximum of three electron-pair bonds.

In the simplest form of elemental boron, each B₁₂ icosahedron is linked to six others by B—B bonds formed by the six darker coloured atoms in Figure 9.4. These have a length of 171 pm, nearly identical with the B—B distance in gaseous B₂Cl₄ (**9.5**). They can therefore be regarded as orthodox single electron-pair bonds. Now each boron atom has three valence electrons, so there are 36 valence electrons in each icosahedron. As six of these must be used in the six B—B inter-icosahedral single bonds, that leaves 30. These 30 must largely be dispersed within the icosahedron so as to bind the 12 boron atoms together. They may be thought of as occupying bonding molecular orbitals, which extend over the whole B₁₂ unit. Alternatively, they can be regarded as an electron gas, which creates a metallic type of bonding that is localized within and around the icosahedron. Because of this 'localized delocalization', electrons cannot easily be transferred from icosahedron to icosahedron in an electric field, so boron is a non-conductor. Localized delocalization, or the formation of multicentre molecular orbitals, is a further means by which boron atoms acquire a share in a larger number of electrons than that made available by the formation of localized, shared electron-pair bonds.

177 pm

Figure 9.4 The different forms of elemental boron consist of interconnected B₁₂ icosahedra. Each icosahedron is linked to others. The core of the icosahedron is a pair of parallel pentagons of boron atoms with opposite orientations; these are shaded. The icosahedron is completed by boron atoms above and below the pair of pentagons. If we ignore slight irregularities in the real crystal, the unit has 20 faces, each consisting of an equilateral triangle of boron atoms of side 177 pm.

170 pm · Cl · Cl · Cl · B—B · 175° · Cl · Cl

9.5

The localization can be broken down when boron forms metallic borides. These are often metallic conductors, and can be thought of as alloys. Titanium diboride, TiB_2, for example, is best made by heating boric and titanium oxides with carbon:

$$TiO_2(s) + B_2O_3(s) + 5C(s) = TiB_2(s) + 5CO(g) \tag{9.6}$$

In this substance (Figure 9.5), there is very strong metallic bonding within and between layers; the material is very hard, melts close to $3\,000\,°C$, and has a conductivity five times that of titanium metal. It is used as a protective coating for metals, such as tungsten, in high-temperature applications.

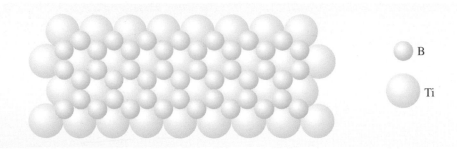

Figure 9.5 The structures of the diborides of titanium and magnesium. Both metals have a hexagonal close-packed structure, ABABAB, etc., in which the spheres in a B-type layer sit on alternate sockets (hollows) in close-packed A-type layers. In the diborides, the B-type layers are replaced by layers of boron atoms, which sit on *every* socket in the A-type layers of the metals. As the number of sockets in a layer is twice the number of spheres, the empirical formula is MB_2.

Magnesium diboride, which has the same structure as TiB_2, can be made by heating magnesium and boron together in the correct proportions. In 2001, it was found to be a superconductor below 39 K. This is easily the highest superconducting temperature observed in any element or binary compound. It may lead to important applications. The Meissner effect, the standard test for a superconductor, is shown in Figure 9.6.

Figure 9.6 A small cylindrical magnet floats above a disc made of a material that has been chilled below the temperature at which it becomes superconducting. The magnet creates a magnetic copy of itself within the superconductor – a copy by which it is repelled.

9.1.4 The boron hydrides

The extraordinary nature of boron hydrides (Box 9.1) was first revealed by the German chemist, Alfred Stock (Figure 9.7).

BOX 9.1 Alfred Stock and boron hydrides

Virtually nothing was known about the boron hydrides in 1909, when Alfred Stock began his work at what was then Breslau in Germany (now Wroclaw, Poland). As the hydrides inflame spontaneously in air, he devised the now-standard technique of vacuum line chemistry, including special grease-free valves that were necessary when handling such reactive compounds. He isolated B_2H_6, B_4H_{10}, $B_{10}H_{14}$ and, subsequently, B_5H_9, B_6H_{10} and B_5H_{11}. Liberal use was made of mercury in pumps and manometers, and during this time, he suffered headaches, vertigo and numbness, culminating, in 1923, with almost total loss of memory and hearing. Having recognized those symptoms as indications of mercury poisoning, he then spent much of the remainder of his working life on this subject, often using himself as a guinea-pig, and publishing numerous valuable warnings and precautionary advice. His health deteriorated further in the 1930s, and he had political difficulties with the Nazis. By 1943, hardening of the muscles drove him to retire to Silesia, but as the Russians closed in, he and his wife became refugees; in 1946, he died in obscurity at Aken on the Elbe, after a life of much accomplishment and suffering.

Figure 9.7
Alfred Stock (1876–1946).

🔵 From what you know already, what should be the molecular formula of the simplest boron hydride?

🔵 As there are halides, BX_3, one is entitled to expect a hydride, BH_3.

In fact, the simplest hydride to have been isolated is diborane, B_2H_6. It is made industrially by heating BF_3 with sodium hydride:

$$6NaH(s) + 2BF_3(g) = 6NaF(s) + B_2H_6(g) \qquad (9.7)$$

The formula is analogous to that of ethane, but the structure (Figure 9.8) is different. To interpret the bonding we assume that there are four B—H single bonds in the blue plane. This leaves one electron on each boron atom, and one electron from each of the two bridging hydrogens, a total of four in all. These four electrons are assigned to two three-centre B—H—B bonds, one being above the blue plane and one being below it. The two electrons that we assign to each of these bonds occupy a bonding molecular orbital formed from a hydrogen 1s orbital and two boron 2p orbitals (Figure 9.9).

As each bonding pair is spread over two B—H distances, the B—H bond order is much less than one. This agrees with Figure 9.8, where the boron–hydrogen distance in the bridge is 14 pm longer than the single B—H bond in the plane.

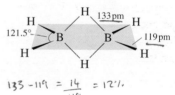

$$133 - 119 = \frac{14}{119} = 12\%$$

Figure 9.8
The structure of diborane, B_2H_6; the coordination around each boron is roughly tetrahedral.

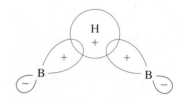

Figure 9.9
The atomic orbital composition of the bonding molecular orbital in one of the two B—H—B bent three-centre bonds that lie above and below the horizontal plane in Figure 9.8.

113

Diborane decomposes on heating into other, more complicated boron hydrides, and over 20 have been isolated by carefully controlling the heating conditions. For example, at 80–90 °C and 200 atmospheres, B_4H_{10} is produced; in the presence of further hydrogen, B_5H_9 is formed in a quick pass through a reaction vessel at 200–240 °C. In B_5H_9 (Figure 9.10), the boron atoms lie at the corners of a square pyramid, the four at the base being linked by B—H—B three-centre bonds. In addition, each boron atom is bound to a hydrogen atom by a conventional B—H single bond. The five borons and nine hydrogens supply 24 valence electrons. Ten are used in the five B—H terminal single bonds, and eight in the four B—H—B three-centre bonds. This leaves six to be delocalized over, and to hold together, the cluster of five boron atoms. Although the boron hydrides and their derivatives are many and various, they invariably feature at least two of the following: (i) B—H or B—B conventional single bonds; (ii) three-centre B—H—B bonds; (iii) multicentre bonds in a cluster of boron atoms.

Figure 9.10
The structure of B_5H_9.

9.1.5 Boron–oxygen compounds

Boron–oxygen compounds account for some of the major industrial uses of boron. In boric acid, $B(OH)_3$, and boric oxide, B_2O_3, boron forms three coplanar, triangularly disposed bonds (Structure **9.6**). Crystals of boric acid are white, flaky and transparent. The $B(OH)_3$ units are linked in layers by hydrogen bonds (Figure 9.11); the interlayer forces are of the weak van der Waals type.

9.6

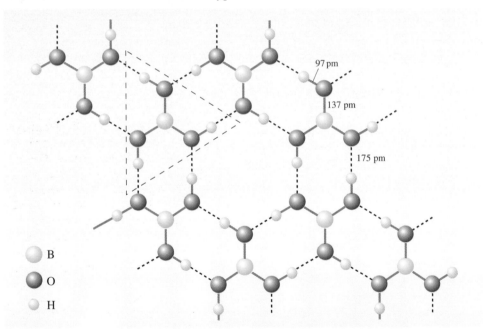

B
O
H

Figure 9.11
The structure of solid boric acid. The hydrogen bonds holding the $B(OH)_3$ units together are 175 pm long. One $B(OH)_3$ unit is indicated by a broken-line triangle.

● Will $B(OH)_3$ be a strong or weak acid in aqueous solution?

◗ Very weak; the strength of oxoacids increases with the number of terminal oxygens in the structural formula, and $B(OH)_3$ has no terminal oxygens.

This is quite correct. However, the ionization of $B(OH)_3$ in water is unusual: $H^+(aq)$ is generated by abstracting OH^- from a water molecule:

$$B(OH)_3(aq) + H_2O(l) = H^+(aq) + [B(OH)_4]^-(aq) \qquad (9.8)$$

Here, boric acid acts as a Lewis acid, accepting a non-bonded pair from the Lewis base OH⁻, and forming four tetrahedrally disposed B—O bonds (Structure **9.7**). Thus, boron can occur in three- or four-fold coordination with oxygen atoms.

At normal pressures, B_2O_3 (T_m 450 °C) consists of a three-dimensional network of BO_3 groups (Structure **9.6**) in which the blue triangles share vertices. Borates can be made by opening up the B_2O_3 network using metal oxides, by reactions such as the one shown in Figure 9.12. This reaction explains the use of B_2O_3 in the formation of borosilicate glasses, such as Pyrex.

9.7

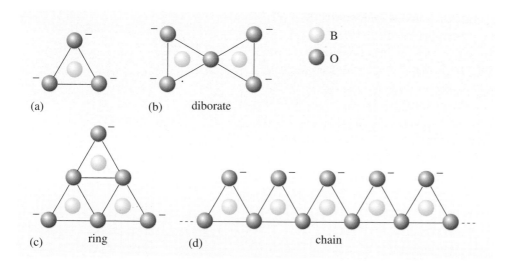

$Ca^{2+} O^{2-}$ +

Ca^{2+}

Figure 9.12 Metal oxides react with boric oxide to form borate glasses containing giant disordered molecular anions, or borate compounds. An O^{2-} ion is transferred to the B_2O_3 network, opening up a B—O—B link, and replacing it with two B—O⁻ bonds.

When the proportions of the oxides are right, crystalline metal borates are formed. If the network is fully opened up, discrete BO_3^{3-} ions are formed (Figure 9.13a).

(a)

(b) diborate

B
O

(c) ring

(d) chain

Figure 9.13 Borate anions that can be made by opening up the B_2O_3 network with metal oxides. These examples contain only three-coordinate boron. BO_3 triangles share zero, one or two vertices with —O⁻ sites at any vertex that is unshared: (a) discrete BO_3^{3-} ions (no shared vertices); (b) diborate, $B_2O_5^{4-}$, ions (each BO_3 triangle shares one vertex); (c) $B_3O_6^{3-}$ rings (each BO_3 triangle shares two vertices); (d) as (c) but the triangles are arranged in a chain.

Progressively less opening is marked by pairs, rings and chains of BO_3 triangles, joined at vertices in all cases (Figure 9.13b, c and d). All these compounds contain just three-coordinate boron, but in some cases, four-coordinate atoms may be present as well. Thus, in the mineral borax, written $Na_2B_4O_7.10H_2O$ in Section 9.1, the anion is $[B_4O_5(OH)_4]^{2-}$ (Figure 9.14, overleaf), in which there are two three-coordinate and two four-coordinate boron atoms.

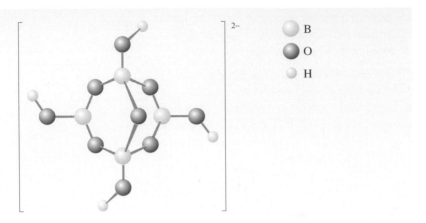

Figure 9.14
The ion $[B_4O_5(OH)_4]^{2-}$, which is found in the mineral borax.

○ What would be a more correct formula for the compound?   *Na₂B₄O₇.10H₂O   ANION is [B₄O₅(OH)₄]²⁻*
2 THREE COORDINATE
2 FOUR COORDINATE

○ $Na_2[B_4O_5(OH)_4].8H_2O$; two of the ten water molecules in the usual formula must be incorporated into the borate anion.

9.1.6 Summary of Section 9.1

1   Boron occurs naturally as borax, $Na_2[B_4O_5(OH)_4].8H_2O$, in which boron is both trigonally and tetrahedrally coordinated by oxygen. Sulfuric acid precipitates boric acid from an aqueous solution of borax, and boric acid can be thermally decomposed to boric oxide, B_2O_3.

2   Boron falls short of an octet when it uses its valence electrons to form electron-pair bonds. In the planar halides, BX_3, the number of shared electrons is enlarged by π-bonding between the empty boron $2p_z$ orbital, and filled non-bonding orbitals on the halogens. Alternatively, boron can achieve an octet of electrons when the halides act as Lewis acids; this reaction is least favourable for BF_3, where the weakening of the π-bonding in the halide is usually the largest.

3   Elemental boron is a black solid with a high melting temperature made by magnesium reduction of B_2O_3, or by heating BI_3. It consists of interconnected B_{12} icosahedra. There is localized metallic-type bonding (multicentre molecular orbitals) within the icosahedra, but not between them, so boron itself is a non-conductor. Borides (e.g. TiB_2), however, are often metallic.

4   In B_2H_6, two BH_2 groups are linked by bridging hydrogens in two B—H—B, three-centre, two-electron bonds. Other boron hydrides, such as B_5H_9, contain B—H—B bonds, but often also involve multicentre bonds in clusters of boron atoms.

5   Boric acid is a weak acid, which generates protons by abstracting OH⁻ from water. B_2O_3 consists of triangular BO_3 groups linked through their vertices. Metal oxides open up the network, replacing B—O—B linkages by two B—O⁻ bonds. This can give rise to glasses, or to borates, in which BO_3 groups are joined at only zero, one or two vertices. Some borates also contain four-coordinate boron.

QUESTION 9.1 ✸

The halides BX_3 can be isolated as chemical compounds; the hydride BH_3 cannot. What stabilizing factor, usually assumed to be present in the boron halides, would be missing in BH_3?

QUESTION 9.2 ✳

The structure of the hydride B_4H_{10} is shown in Figure 9.15. The B(1)—B(3) distance is 171 pm; the B(2)—B(4) distance is 280 pm. Identify the B—B single bonds, B—H single bonds and B—H—B three-centre bonds, and then find out if your selection uses up all the valence electrons in the compound.

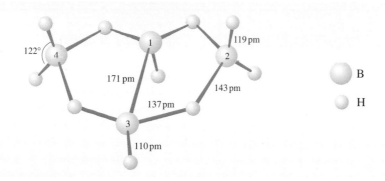

Figure 9.15
The structure of the hydride B_4H_{10}.

QUESTION 9.3 ✳

(a) Represent the structure of B_2O_3 by using triangles of the type used in Figure 9.13. Explain why, even though the triangles consist of one boron atom surrounded by three oxygens, the formula is B_2O_3.

(b) The borate whose anion has the chain structure shown in Figure 9.13d is made by heating mixtures of CaO and B_2O_3 in the right proportions. What is its empirical formula?

9.2 Aluminium

Aluminium is the most abundant metallic element in the Earth's crust. It is extracted by purifying bauxite, an impure hydrated oxide, which may be represented as $Al_2O_3.3H_2O$, and the subsequent electrolytic extraction of the metal. Many of the uses of aluminium — in canning, aircraft construction, and overhead transmission lines — depend on its low density and resistance to corrosion in air and water. This resistance is a kinetic effect, caused by a coherent oxide film, which protects the underlying metal. Some of the important aluminium chemistry that we discuss is summarized in Figure 9.16.

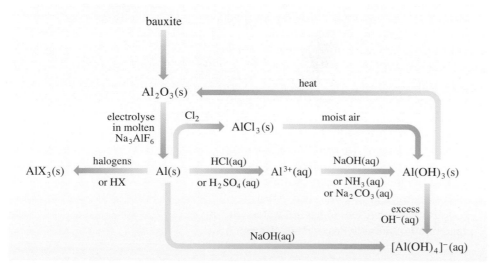

Figure 9.16
Some significant aluminium chemistry.

9.2.1 Aqueous chemistry

Like most metals, aluminium, in contrast to the non-metal boron, forms an aqueous cation. Hydrates of aluminium sulfate, $Al_2(SO_4)_3$, for example, can be made by dissolving bauxite in sulfuric acid and evaporating the solution. When such sulfates are dissolved in water, $Al^{3+}(aq)$ is formed:

$$Al_2(SO_4)_3(s) = 2Al^{3+}(aq) + 3SO_4^{2-}(aq) \qquad (9.9)$$

Careful addition of aqueous sodium hydroxide to this solution will first precipitate insoluble aluminium hydroxide:

$$Al^{3+}(aq) + 3OH^-(aq) = Al(OH)_3(s) \qquad (9.10)$$

The oxide is produced by filtering and heating the hydroxide:

$$2Al(OH)_3(s) = Al_2O_3(s) + 3H_2O(g) \qquad (9.11)$$

Both oxide and hydroxide are unusual in being amphoteric. Thus, they will dissolve in and neutralize acids:

$$Al(OH)_3(s) + 3H^+(aq) = Al^{3+}(aq) + 3H_2O(l) \qquad (9.12)$$

⬤ What else will an amphoteric oxide/hydroxide do?

⬤ It will dissolve in and neutralize alkalis.

In this case, if *excess* sodium hydroxide is added to the precipitate that is initially formed in Reaction 9.10, the precipitate dissolves to form the tetrahydroxyaluminate ion:

$$Al(OH)_3(s) + OH^-(aq) = [Al(OH)_4]^-(aq) \qquad (9.13)$$

As Figure 9.16 implies, the bases ammonia and sodium carbonate are not strong enough to bring about this dissolution. The resemblance of $Al(OH)_3$ to $Be(OH)_2$, which is also amphoteric, is very close.

Because the protective oxide film on the metal is soluble in both acids and alkalis, aluminium, again like beryllium, dissolves in both hydrochloric acid and sodium hydroxide, liberating hydrogen and giving clear, colourless solutions:

$$2Al(s) + 6H^+(aq) = 2Al^{3+}(aq) + 3H_2(g) \qquad (9.14)$$

$$2Al(s) + 6H_2O(l) + 2OH^-(aq) = 2[Al(OH)_4]^-(aq) + 3H_2(g) \qquad (9.15)$$

9.2.2 Aluminium sulfate and water treatment

The Group I and Group II metals form carbonates, those of Group II being insoluble in water. Aluminium carbonate, $Al_2(CO_3)_3$, however, cannot be prepared; if aluminium sulfate is added to a solution containing carbonate or hydrogen carbonate ions, $Al(OH)_3$ is precipitated and carbon dioxide is evolved:

$$2Al^{3+}(aq) + 3CO_3^{2-}(aq) + 3H_2O(l) = 2Al(OH)_3(s) + 3CO_2(g) \qquad (9.16)$$

$$2Al^{3+}(aq) + 3HCO_3^-(aq) = Al(OH)_3(s) + 3CO_2(g) \qquad (9.17)$$

Large amounts of aluminium sulfate are used in the water industry to clear water of fine suspensions, such as clay particles, which are otherwise difficult to filter off. The tiny particles usually carry surface negative charges, which repel each other and prevent coagulation. The positively charged aluminium ions get between the negative particles, counteracting the repulsion and encouraging aggregation. Then, when Reaction 9.17 occurs because of the HCO_3^- ions usually present in natural waters, the particles are carried down with the precipitate of aluminium hydroxide that is formed.

Reactions 9.16 and 9.17 are of interest because acids liberate CO_2 from carbonate and hydrogen carbonate solutions, so here, $Al^{3+}(aq)$ plays the part of an acid. This becomes more apparent when you consider that $Al^{3+}(aq)$, like $Mg^{2+}(aq)$, is shorthand for an octahedral aquo complex (Structure **9.8**). Then, Equation 9.17 becomes

$$[Al(H_2O)_6]^{3+}(aq) + 3HCO_3^-(aq) = Al(OH)_3(s) + 3CO_2(g) + 6H_2O(l) \qquad (9.18)$$

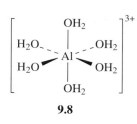

9.8

🔵 Why can $[Al(H_2O)_6]^{3+}$ be said to act as a Brønsted–Lowry acid in this equation?

🔵 Three of the six water molecules that were attached to the aluminium have been lost, but the other three have acted as *proton donors*, leaving aluminium associated with hydroxide ions rather than with water molecules.

A more obvious sign of the acidic character of $[Al(H_2O)_6]^{3+}$ is the fact that aqueous solutions of aluminium sulfate, unlike those of, say, sodium sulfate, are acidic:

$$[Al(H_2O)_6]^{3+}(aq) + H_2O(l) = [Al(H_2O)_5(OH)]^{2+}(aq) + H_3O^+(aq) \qquad (9.19)$$

Again, a water molecule coordinated to Al^{3+} is transformed into an OH^- ligand; at the same time, hydrated protons, H_3O^+, are generated.

The reactions in this section are relevant to the disaster at Camelford in Cornwall in July 1988. A relief driver arrived at the Lowermoor water treatment works with 20 tonnes of aluminium sulfate solution, and the key to the storage tank where it was to be deposited. Unfortunately, the key also fitted another tank where the water was held prior to discharge into the public supply mains, and it was into this tank that the driver discharged his load. Around 9 p.m. on the same evening, South West Water began receiving complaints: milk curdled when it was added to tea or coffee, because the aluminium sulfate coagulated milk in the way that it coagulates suspensions in natural waters. By the time the problem had been diagnosed, 30 000 fish had been killed in the rivers Camel and Allen, and the aluminium concentrations in Camelford drinking water briefly reached 200 times the EC limit (Figure 9.17).

Whether the townsfolk have suffered long-term damage to their health remains a matter of dispute. This brings us to the question of the toxicity of aluminium.

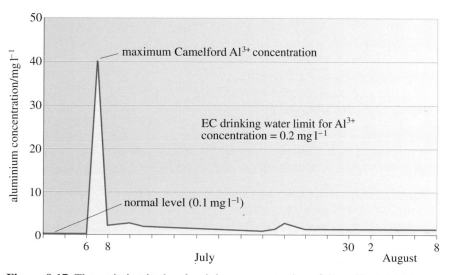

Figure 9.17 The variation in the aluminium concentration of Camelford drinking water in July and August 1988.

9.2.3 Aluminium toxicity

In the 1980s, research into the plaques and fibres in the brains of people with Alzheimer's disease suggested that aluminium might be partly responsible for the condition. This now seems less likely. The World Health Organization sets a tolerable intake of aluminium for a 60 kg adult at 60 mg per day. For most people, the mass actually ingested daily is about 10 mg. Aluminium is mostly excreted in the faeces. That which passes across the gastrointestinal barrier into the blood stream is dealt with by the kidneys (Figure 9.18).

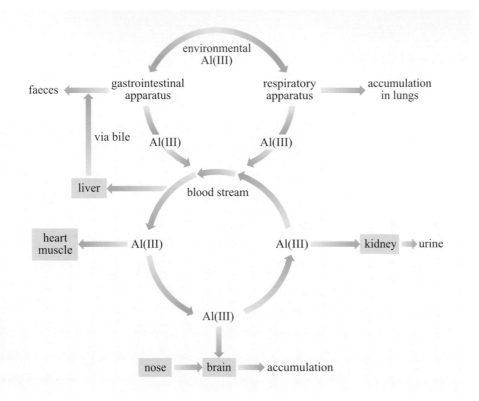

Figure 9.18
The metabolism of aluminium in humans. Most is eliminated in the faeces and kidneys, but there is a small accumulation in the whole body, including the brain and lungs.

However, there is no doubt that aluminium can damage people whose kidney function is impaired. The condition called dialysis dementia was first noticed in patients who had received long-term haemodialysis for renal failure. Its symptoms included speech disorders, memory loss, convulsions and seizures, followed, in some cases, by death within a year. The incidence of the disease was highest when the municipal water used in the dialysis contained high concentrations of aluminium. Aluminium is, therefore, a potential neurotoxin.

The mechanisms of aluminium toxicity remain uncertain. Here we describe one of several proposals.

Glycolysis is a nearly universal pathway in biological systems. It includes a key step in which adenosine triphosphate (ATP) loses its terminal phosphate group to a glucose molecule and becomes adenosine diphosphate (ADP) (Figure 9.19).

The process is catalysed by an enzyme called hexokinase. Before the ATP can be accepted by the enzyme, it must become complexed to a magnesium ion, Mg^{2+}, through oxygen atoms attached to phosphorus atoms 2 and 3 (Structure **9.9**). The exact way in which the magnesium ion bridges the two terminal phosphates is

9.9
(A = adenosine)

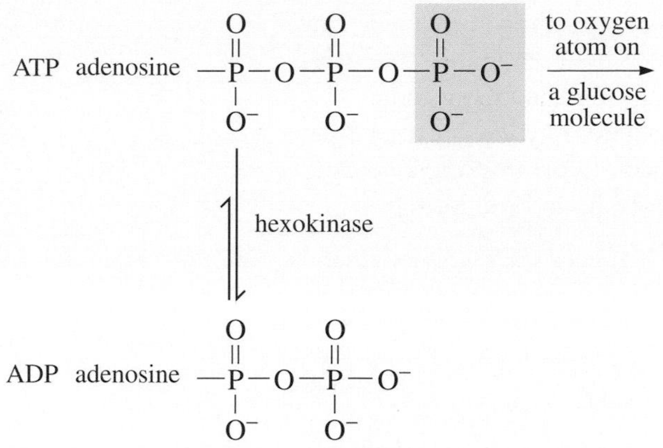

ATP adenosine

$$\text{adenosine} - \underset{\underset{O^-}{|}}{\overset{\overset{O}{\|}}{P}} - O - \underset{\underset{O^-}{|}}{\overset{\overset{O}{\|}}{P}} - O - \underset{\underset{O^-}{|}}{\overset{\overset{O}{\|}}{P}} - O^- \longrightarrow$$

to oxygen atom on a glucose molecule

hexokinase

ADP adenosine

$$\text{adenosine} - \underset{\underset{O^-}{|}}{\overset{\overset{O}{\|}}{P}} - O - \underset{\underset{O^-}{|}}{\overset{\overset{O}{\|}}{P}} - O^-$$

Figure 9.19
The conversion of ATP into ADP in the first stage of glycolysis.

uncertain, but in this complexed state, the ATP is accepted by the enzyme. The magnesium ion then moves from the bridging position between phosphate groups 2 and 3 to one between groups 1 and 2. This exposes the terminal phosphate group to removal, and transfer to glucose.

The change of bridging position is possible because Mg^{2+} does not bind particularly strongly to the oxygen donor sites. But suppose substantial amounts of dissolved aluminium are present.

⬤ What ion is then available for binding to the oxygen donors?

⬤ Al^{3+}; this binds more strongly than Mg^{2+} to oxygen donor ligands, as simple electrostatic arguments would predict.

The result is that Al^{3+} will bridge phosphate groups 2 and 3 like Mg^{2+}, but because of the stronger binding, it is reluctant to shift on the enzyme to bridge groups 1 and 2. Loss of the terminal phosphate is therefore impaired.

Whether this explanation of aluminium neurotoxicity is correct or not, it points to a useful general idea about poisons. Their key feature is a resemblance to some chemical species (in this case Mg^{2+}) that is essential to a biological process, and which is close enough to gain access to a metabolic process. But once the poison has been incorporated, then differences from the essential species intrude and the metabolic process is blocked.

A further example of this is provided by the effect of acid rain on the concentration of aluminium in natural waters. As discussed in the *Acid Rain* Case Study, when the hydrogen ions of acid rain fall upon soils that are naturally quite acidic, they replace other positive ions bound into the soil structure. One of these ions is Al^{3+}, so acid rain leads to an increase in the concentration of dissolved aluminium in natural waters. This is a major contributor to the decline of fish stocks in lakes. The Al^{3+} ion appears to bind to oxygen-donor ligands at the surface of fish gills, in a manner similar to that shown in Structure **9.9**. This makes the membrane of gill cells more permeable to further Al^{3+}, which can then enter the cells and replace Ca^{2+} ions on key proteins. The regulation of the cell concentrations of other ions, such as Na^+, is disrupted. At the same time, the increased concentration of aluminium in gill cells leads to precipitation of aluminium hydroxide, mucus formation and breathing difficulties.

9.2.4 Aluminium halides

Aluminium forms solid trihalides when it is heated with the halogens or hydrogen halides. Unlike the boron trihalides, these substances do not contain discrete MX_3 molecules at room temperature. Thus, AlF_3 melts at 1 290 °C and has a three-dimensional structure, characteristic of an ionic compound, in which each aluminium atom is surrounded by six fluorines, and each fluorine atom is nearly linearly coordinated to two aluminiums. $AlCl_3$, however, has a layer structure (Figure 9.20), which resembles that of $MgCl_2$.

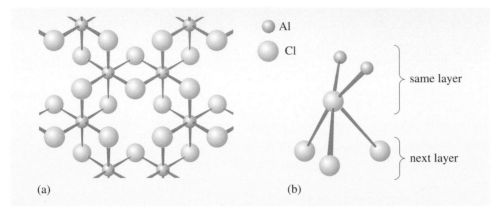

(a)   (b)

Figure 9.20
(a) One of the layers of the $AlCl_3$ structure, viewed from above; the layer has three decks: the top deck consists of chlorines, the middle deck of octahedrally coordinated aluminiums, and the bottom deck of chlorines. (b) The environment of each chlorine atom: there are two aluminiums on one side, but on the other side there are three chlorines, much farther away in an adjacent layer.

Here, the halogen environment is very different from what would be expected in a structure composed of ions. The trichloride melts under slight pressure at only 194 °C, and both the liquid and vapour contain discrete Al_2Cl_6 molecules with a structure similar to that of B_2H_6 (Figure 9.21).

● How does the percentage difference in the lengths of the terminal and bridge bonds compare with that in B_2H_6?
    ~ P₀ 113

● Comparison with Figure 9.8 shows that it is much smaller; in Al_2Cl_6 the bridge bond is longer by about 7%; in B_2H_6 the difference is about 12%.

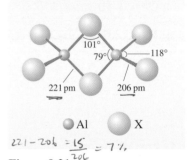

● Al   ● X

221 − 206 = 15 ≈ 7%
             206

Figure 9.21
The structure of the Al_2X_6 molecules found in liquid and gaseous $AlCl_3$, and in solid $AlBr_3$ and AlI_3. The bond length and bond angle data refer to the chloride.

This difference means that a three-centre bond treatment in which the bridge bonds have an order much less than one is less appropriate for Al_2Cl_6; the molecule is better represented by allowing each bridging chlorine to form one shared electron-pair bond, and one dative bond as in Structure **9.10**; all bonds then have order one. A bridging chlorine can do this because it can use more than one of its orbitals to form the two bonds in the bridge; hydrogen, by contrast, has only one orbital available – the 1s.

9.10

Solid $AlBr_3$ and AlI_3 both consist of Al_2X_6 molecules of the type shown in Figure 9.21. Thus, from AlF_3 to AlI_3, we see signs of increasingly covalent character in the transition from a three-dimensional ionic structure in AlF_3, to a layer structure in $AlCl_3$, and then to molecular structures in $AlBr_3$ and AlI_3.

9.2.5 Two observations about aluminium chemistry

All the compounds and complexes that we have considered so far show that aluminium chemistry is dominated by just one oxidation number, +3, the value equal to the Mendeléev Group number. No compounds in other oxidation numbers have been characterized at room temperature. However, if the metal and trichloride are heated together in a sealed container to 1 000 °C, a gaseous monochloride, AlCl, is formed:

$$2Al(l) + AlCl_3(g) = 3AlCl(g) \qquad (9.20)$$

On cooling, the compound disproportionates, regenerating the metal and the trichloride.

A second general feature of aluminium chemistry is the tendency of the element to form the oxide or hydroxide compounds. This is especially clear when comparisons are made with, say, sodium. The tendency is made evident in the insolubility of $Al(OH)_3$ in water and its solubility in alkali, in the way in which the non-existent carbonate decomposes to Al_2O_3, and in the acidity of the aqueous ion (Equation 9.19). Again, unlike NaCl, $AlCl_3$ fumes in moist air because hydrogen chloride gas is formed along with aluminium hydroxide:

$$AlCl_3(s) + 3H_2O(g) = Al(OH)_3(s) + 3HCl(g) \qquad (9.21)$$

Finally, the strong tendency of the metal to form the oxide provides the protective film and corrosion resistance without which the metal would be much less useful.

9.3 Gallium, indium and thallium

In Groups I and II, sodium and magnesium are less readily oxidized than the metals beneath them. One sign of this is the less negative values of $E^{\ominus}(Na^+|Na)$ and $E^{\ominus}(Mg^{2+}|Mg)$ compared with those of the elements below them. Table 9.3 compares values for Group II metals with those for Group III metals, and you can see that the trends are in opposite directions: thermodynamically speaking, aluminium is more readily oxidized to oxidation number +3 than gallium, indium or thallium.

[handwritten margin note: or more positive values / lower value]

Table 9.3 Values of $E^{\ominus}(M^{2+}|M)$ for the Group II metals, and values of $E^{\ominus}(M^{3+}|M)$ for the corresponding Group III metals

Gp II	$E^{\ominus}(M^{2+}|M)/V$	Gp III	$E^{\ominus}(M^{3+}|M)/V$
Mg	−2.36	Al	−1.68
Ca	−2.87	Ga	−0.53
Sr	−2.90	In	−0.34
Ba	−2.91	Tl	+0.72

[handwritten margin notes: More (−) the value the more readily the element is oxidized. Goes with the lower the 1st ionization energy value. Higher the value the more difficult to make the ion. This contributes to the resistance to oxidation]

A possible reason for the reversal emerges from a comparison of the first ionization energies of the Group II and Group III elements (Figure 9.22, overleaf).

From boron to aluminium, there is the usual drop from the second row to the third row of the Group, but thereafter, the values remain unexpectedly high, most notably at thallium whose first ionization energy exceeds that of aluminium. In Section 8, you saw that there is a steep drop in ionization energy when a new Period begins, followed by an *overall* increase across a Period as the nuclear charge builds up.

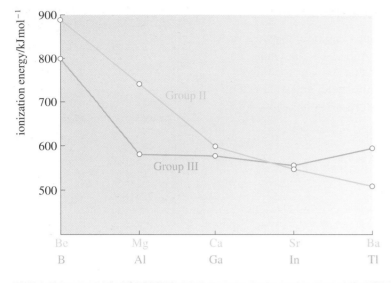

Figure 9.22
The first ionization energies of the Group II and Group III elements. There is a marked decrease between magnesium and barium, which is not matched by that between aluminium and thallium.

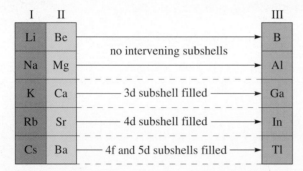

Figure 9.23
In the lithium and sodium Periods, no intervening subshells are filled between Groups II and III; in the potassium and rubidium Periods, however, the 3d and 4d subshells must be filled between these Groups, and in the caesium Period, both the 4f and 5d subshells are filled.

Now look at Figure 9.23. Why might the ionization energies of gallium, indium and, especially, thallium be unexpectedly raised relative to those of aluminium?

Aluminium follows immediately after a Group II element, but prior to gallium and indium, d shells must be filled first. This leads to a more prolonged build-up of nuclear charge, so the ionization energies of gallium and indium are raised. The effect is magnified at thallium, where prior filling of 5d *and* 4f subshells occurs.

Unexpectedly high ionization energies for gallium, indium and thallium make conversion of the metals into ions more difficult. They are a major contribution to the greater resistance to oxidation revealed in Table 9.3.

Such effects, however, do not obliterate the strong resemblance of gallium, indium and thallium to aluminium that their presence in the same Group implies. All three elements are metals, which react with fluorine or chlorine to form trihalides, all of which are solids at room temperature. The metals also dissolve in dilute acids, evolving hydrogen and forming aqueous ions. With gallium and indium, these ions are $Ga^{3+}(aq)$ and $In^{3+}(aq)$; we return to thallium in a moment. Addition of alkali precipitates colourless $Ga(OH)_3$, which is amphoteric, and $In(OH)_3$, which is not.

● What will be the effect of adding excess NaOH to the two suspensions?

● $In(OH)_3$ will be unaffected; $Ga(OH)_3$ dissolves according to the following reaction:

$$Ga(OH)_3(s) + OH^-(aq) = [Ga(OH)_4]^-(aq) \qquad (9.22)$$

In descending Group III from B_2O_3 to In_2O_3, there is therefore a decrease in acidity, and an increase in the basic character of the oxides/hydroxides: B_2O_3 and $B(OH)_3$ are acidic, Al_2O_3 and Ga_2O_3 are amphoteric, and In_2O_3 is basic. Tl_2O_3 confirms this trend, since it too is basic.

Finally, a few comments on thallium. Look at Table 9.3, remembering that $E^{\ominus}(H^+|\frac{1}{2}H_2) = 0$ V.

● When thallium metal dissolves in dilute acids, with evolution of hydrogen, can the product be $Tl^{3+}(aq)$?

● No; $E^{\ominus}(Tl^{3+}|Tl) > E^{\ominus}(H^+|\frac{1}{2}H_2)$; thallium going to $Tl^{3+}(aq)$ cannot reduce $H^+(aq)$ to hydrogen gas.

In fact, the product is $Tl^+(aq)$; thallium, more than any other Group III metal, has a prominent +1 oxidation number. The build-up of nuclear charge in the preceding 4f and 5d block elements leaves thallium's ionization energies higher than they would otherwise be. The higher oxidation number is therefore harder to attain, and the state most stable to oxidation or reduction is +1.

● Write down the electronic configuration of Tl^+.

● $[Xe]\,4f^{14}\,5d^{10}\,6s^2$; the outer 6p electron has been lost, leaving two outer electrons in a full 6s shell.

The emergence at the bottom of Groups III–V of a stable lower oxidation number, two fewer than the Group number is sometimes called the **inert pair effect**, because the outer electronic configuration of the ion is a filled s^2 subshell, which is presumed to be hard to remove during oxidation. The effect increases down the Group: AlCl, AlBr and AlI do not exist at room temperature, but the corresponding compounds of gallium and indium can be made by cooling a heated mixture of the metals and their trihalides:

$$2M(s) + MX_3(s) = 3MX(s) \qquad (9.23)$$

They all, however, decompose in water, either evolving hydrogen or disproportionating to the metal and $M^{3+}(aq)$. Only in the case of thallium does a long-lived $M^+(aq)$ ion exist. The ionic radius of Tl^+ (160 pm) resembles that of K^+, and, like potassium, thallium forms, unusually, a soluble, alkaline carbonate and hydroxide, Tl_2CO_3 and TlOH. When ingested, thallium seems to follow potassium in its metabolism, and it probably interferes with vital roles played by potassium in the nervous system. Thallium, and especially thallium(I), is extremely poisonous, as some well-publicized murder cases testify (Figure 9.24). Again, this testifies to the general explanation of toxicity given in Section 9.2.3: the nervous system is deceived into accepting thallium instead of potassium. Essential biological processes are then blocked because thallium is an inadequate substitute.

(a)

(b)

Figure 9.24
Two notorious thallium poisoners: (a) Caroline Grills despatched four relatives and family friends in Sydney, Australia during the late 1940s. During the subsequent life-sentence, her fellow inmates christened her 'Aunt Thally'. (b) Graham Young had already done time in Broadmoor for poisoning when, in 1971, he began doctoring his workmates' tea. Two men who each received a total of about 1 g of thallium(I) acetate, subsequently died, and the police found that Young had kept a careful record of the doses and symptoms of these and other victims. He was returned to Broadmoor, where he died in 1990.

9.3.1 The inert pair effect

As noted in Section 9.3, the Group III elements supply an excellent example of the inert pair effect. Their atoms have outer electron configurations of the type [FIS] $ns^2 np^1$, where n is the period number, and FIS is short for filled inner subshells. In compounds, the elements occur in two oxidation states: a lower state, +1, and a higher state, +3, which is equal to the group number. If we use an ionic model, compounds in the +1 oxidation state contain ions with the configuration [FIS] ns^2, and those in the +3 state contain ions of configuration [FIS]. Thus oxidation of the +1 to the +3 state involves the loss of an outer pair of s electrons. *As we descend the Group, the difficulty experienced in removing the s pair increasingly exceeds what we would expect.* Consequently, the +1 state becomes increasingly more stable, and at thallium, its stability to oxidation is very marked. Our explanation for this in Section 9.3 was rooted in the nature of the filled inner subshells.

Note well the italicized sentence. It implies that the inert pair effect is a relative phenomenon. It emerges only through a comparison with other Groups. For example, the atoms of the Group II elements have configurations of the type [FIS] ns^2, and in the +2 oxidation state, their ionic configuration is [FIS]. For a Group II element, the process

$$M(g) = M^{2+}(g) + 2e^-(g) \qquad (9.24)$$

involves the loss of the outer s pair, and its enthalpy change is $(I_1 + I_2)$, the sum of the first and second ionization energies.

● Rewrite this last sentence so that it applies to the loss of the outer s pair in Group III.

● For a Group III element, the process

$$M^+(g) = M^{3+}(g) + 2e^-(g) \qquad (9.25)$$

involves the loss of the outer s pair, and its enthalpy change is $(I_2 + I_3)$, the sum of the second and third ionization energies.

Figure 9.25 shows the change in these ionization energy sums as one descends Groups II and III. To make the comparison easier, the scales for each plot have been adjusted so that the slopes of the lines between Be and Mg, and between B and Al, are the same.

As one descends Group II, the expected fall in ionization energy is observed as the outer shell becomes more remote from the nucleus: it gets progressively easier to remove the s pair. But in Group III, the slope of the decrease changes abruptly at aluminium. Indeed, between Al and Ga, and In and Tl, there are *increases*. Relative to the situation in Group II, it gets progressively harder to remove the s pair as one descends Group III. It is for this reason that in Group III, the +1 oxidation state becomes much more stable as one moves from Al to Tl whereas, as Table 9.3 (p. 123) shows, in Group II, the zero oxidation state becomes *less stable* as one moves from Mg to Ba.

We have spent some time on the inert pair effect because isotopes of some of the elements 113–116 have recently been made. Because of the 'island of stability', their half-lives are relatively long and it may be possible to study their chemistry. We expect them to be typical elements lying beneath the elements Tl, Pb, Bi, Po, At and Rn in the Periodic Table. As the inert pair effect increases down Groups III, IV and V, it should influence the chemistry of these new elements in an especially marked way.

STUDY NOTE

The inert pair effect is considered on one of the CD-ROMs accompanying this Book.

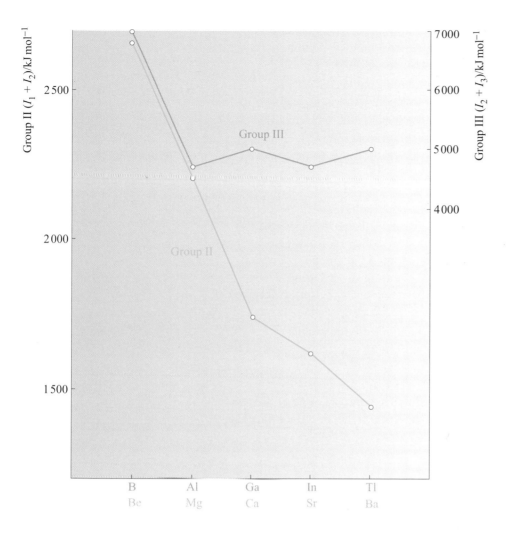

Figure 9.25 The values of $(I_1 + I_2)$ for the Group II elements (scale on left axis) and of $(I_2 + I_3)$ for the Group III elements (scale on right axis). The two scales are different.

9.3.2 Summary of Sections 9.2 and 9.3

1 Al_2O_3 and $Al(OH)_3$ are amphoteric. Thus, the protective film on aluminium metal can be dissolved by both acids and strong alkalis, so the metal dissolves in either aqueous NaOH or HCl, with evolution of hydrogen.

2 The strong tendency of aluminium to become bound to oxide or hydroxide is evident in the precipitation of $Al(OH)_3$ by hydroxide from aqueous solutions of aluminium(III), in the solubility of $Al(OH)_3$ in excess $OH^-(aq)$, in the instability of the non-existent $Al_2(CO_3)_3$, in the acidity of $[Al(H_2O)_6]^{3+}(aq)$, in the hydrolysis of $AlCl_3$ in moist air, and in the formation of a protective oxide film on the metal.

3 Aluminium sulfate solution can be made by dissolving bauxite in sulfuric acid; it finds application in the water industry. The highly charged positive ions coagulate negatively charged suspended matter, and carry it down with the precipitate of $Al(OH)_3$, formed by the reaction with HCO_3^- ions present in natural waters.

4 Most ingested aluminium is eliminated in the faeces, or through the kidneys, but it is potentially toxic to people with an impaired kidney function. This may be because Al^{3+} forms a rather stable complex with ATP, and prevents Mg^{2+} from fulfilling its role in glycolysis.

5 The structures and melting temperatures of the aluminium trihalides show signs of increasingly covalent character from three-dimensional AlF_3, through the layer structure of $AlCl_3$, to molecular Al_2X_6 structures in $AlBr_3$ and AlI_3.

6 With the exception of the metal and its alloys, aluminium occurs only in oxidation number +3 at normal temperatures: the monohalides, formed by heating the trihalides with the metal, disproportionate on cooling.

7 The basic character of oxides/hydroxides increases down Group III, changing from acidic at boron, to amphoteric at aluminium and gallium, to basic at indium and thallium.

8 The ionization energies of gallium, indium and thallium are raised relative to those of the lower elements of Groups I and II by the build-up of nuclear charge as d and f shells are filled in the preceding elements. This makes the elements more electronegative, and, thermodynamically, less readily oxidized to oxidation number +3. It also leads to the inert pair effect: a gradual increase in the stability of oxidation number +1 towards the bottom of the Group. Thus, $GaCl$, $GaBr$, GaI, $InCl$, $InBr$ and InI have all been made, and $Tl^+(aq)$ is formed when the metal dissolves in acids.

9 Tl^+ and K^+ have similar sizes; they both form soluble carbonates and hydroxides. Thallium(I) compounds are very poisonous.

QUESTION 9.4 ✳

Both gallium and indium metals dissolve in dilute HCl, but only one of them in dilute NaOH. Which one? Explain your reasoning, and write an equation for the reaction with the alkali.

QUESTION 9.5 ✳

Which of the two sulfates, Na_2SO_4 and $Al_2(SO_4)_3$, will decompose more readily on heating to give a metal oxide and oxides of sulfur?

QUESTION 9.6 ✳

How can ionic radii and/or electronegativities be used to explain the structural changes in the solid aluminium halides from AlF_3 to AlI_3?

QUESTION 9.7

Which element, and which oxidation numbers, should be most stabilized by the inert pair effect in each of Groups IV and V of the Periodic Table?

THE GROUP IV/14 ELEMENTS

10

10.1 Structures and properties of the elements

As in Group III, the elements of Group IV show a transition from non-metallic to metallic properties with consequent diverse chemical behaviour (Figure 10.1). Electronic configurations of the Group IV elements are shown in Table 10.1.

The non-metal carbon is the basis of organic chemistry, yet most of the carbon on Earth is present as chalk, limestone and fossil fuels. However, all the deposits derive originally, via some form of life, from atmospheric carbon dioxide. Carbon exists in several allotropic forms, including diamond and several newly discovered structures. The abundant element silicon, which occurs in many rocks, is the basis of the modern electronics industry, owing to its semiconducting properties. Another semi-metal, germanium, is relatively rare, and has been widely used in transistors. The facile isolation of metallic tin accounts for its use for thousands of years in alloys such as bronze. Its resistance to corrosion by air and water led to its use in the plating of steel food containers known as 'tins'. New alloys of tin are now used in superconducting magnets. The soft toxic metal lead has a long history of use as a workable waterproof material for roofs and plumbing. Its toxicity has resulted in the decline of its use in tetraethyl-lead, the antiknock additive to petrol.

Carbon atoms form strong single bonds with each other and are also able to form multiple bonds via $p\pi$–$p\pi$ bonding. Both types of bonding appear in the allotropes of carbon.

⬤ What are the common allotropes of carbon?

⬤ Under ambient conditions, graphite is the stable form of carbon (Figure 10.2) whereas diamond is metastable (Figure 10.3, overleaf).

Other, recently discovered allotropes include the fullerenes, the best known being C_{60}, buckminsterfullerene (Figure 10.4, overleaf), and the buckytubes.

Silicon and germanium also have the diamond structure; they are **isostructural** (same crystal structures). Tin (symbol Sn, from the Latin *stannum*) is polymorphic. Grey tin, a semi-metal with the diamond structure, is the stable form below the transition temperature of 13 °C. Above this temperature, white tin, a metal, is the stable form. Tin is also the principal constituent of solder. During Scott's expedition to the South Pole, the petrol cans were found to leak: it is thought that the very low temperatures in the Antarctic caused the tin to change phase and the solder to disintegrate. In white tin, each Sn atom is approximately six coordinate, and four other tin atoms are only a little farther away. The last member of Group IV is lead (symbol, Pb, from the Latin *plumbum*), and it has a typical close-packed metal structure.

Figure 10.1
The Group IV/14 elements: carbon is a non-metal, silicon and germanium are semi-metals, and tin and lead are metals.

Table 10.1 Electronic configurations of Group IV/14 atoms

Atom	Electronic configuration
C	[He] $2s^2\,2p^2$
Si	[Ne] $3s^2\,3p^2$
Ge	[Ar] $3d^{10}\,4s^2\,4p^2$
Sn	[Kr] $4d^{10}\,5s^2\,5p^2$
Pb	[Xe] $4f^{14}\,5d^{10}\,6s^2\,6p^2$

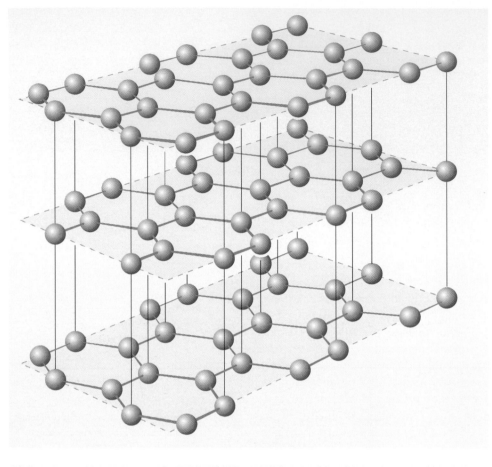

Figure 10.2
The crystal structure of graphite.

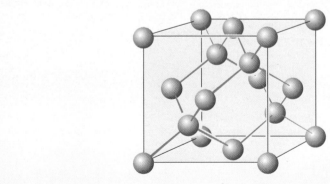

Figure 10.3
A unit cell of the diamond structure.

How does the coordination number for the stable form of the elements at room temperature vary down Group IV?

graphite	silicon	germanium	white tin	lead
3	4	4	6	12

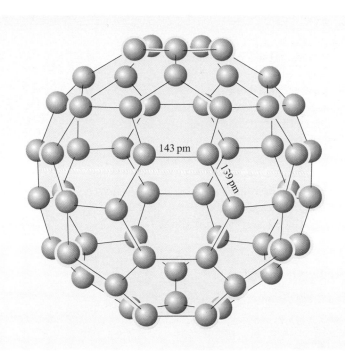

143 pm

159 pm

Figure 10.4
The structure of
buckminsterfullerene, C_{60}.

Notice that the structures of the Group IV elements gradually change down the Group as the elements become more metallic in character: from the giant covalent structures of graphite and diamond, silicon, germanium and grey tin, to the metallic structures of white tin and lead. As we might expect, the electrical properties of the isostructural forms also show a gradation down the Group. Diamond is an insulator, crystalline silicon and germanium, which have metallic lustres, are intrinsic semiconductors, and grey tin conducts almost as well as some metals.

- What can we say about the band gaps between valence band and conduction band for these elements?

- The band gaps of the isostructural forms of the elements *decrease* down the Group.

White tin (metallic) is a much better conductor than grey tin.

Graphite displays anisotropic conductivity. In the plane of the layers, graphite conducts electricity better than either silicon or germanium, but is a poor conductor in the perpendicular direction.

The homonuclear bond enthalpy terms decrease down the Group: C—C, 347 kJ mol^{-1}; Si—Si, 226 kJ mol^{-1}; Ge—Ge, 188 kJ mol^{-1}; Sn—Sn, 152 kJ mol^{-1}. There is also a fall in melting temperature down the Group: 4 100 °C for diamond; 1 420 °C for silicon; 945 °C for germanium; grey tin changes to white tin before melting (white tin melts at 232 °C); lead melts at 327 °C. The isostructural forms show diminishing hardness down the Group: diamond is the hardest known substance, silicon and germanium are somewhat softer and grey tin is a powder, although it can be crystallized. On the Mohs scale, by which the hardness of a substance is determined by its ability to scratch other substances, diamond registers 10, silicon 7 and grey tin 1.5.

10.2 Carbon

With four valence electrons and moderate electronegativity, carbon forms mainly covalent compounds. When carbon is in its standard thermodynamic state, it has the graphite structure (Figure 10.2). This **anisotropic structure** consists of sheets of regular hexagons of carbon, like benzene rings fused together. The bond distance in the hexagons (142 pm) is intermediate between the values observed for normal single and double bonds (154 pm in ethane and 134 pm in ethene). The distance between the sheets (335 pm), however, is about double the van der Waals radius of carbon. Graphite, therefore, has less than two-thirds the density of diamond, so that, although graphite is the stable form at normal pressures, it can be turned into diamond at high pressures and temperatures (more than 105 atmospheres at 2 500 °C). We can account for the bond distance in the graphite sheet in terms of the number of σ and π electrons.

⬤ Of the four valence electrons, how many are used in σ bonding?

⬤ Three, to form an electron-pair bond with each of the three neighbours.

This leaves one π electron per carbon atom (as in ethene or benzene). Imagine an infinite π orbital covering the whole sheet, formed from the $2p_z$ atomic orbitals, one on each carbon (perpendicular to the sheets). But since there are three bonds to carbon in the graphite sheet, the π bond order in each is about one-third. The total bond order is thus about $1\frac{1}{3}$, intermediate between a single and a double bond.

Both graphite and diamond have extended covalent structures, and in their chemical reactions they behave similarly; they are unreactive, infusible and insoluble because their crystals are held together by strong covalent bonds. Their very different structures, however, can explain their rather different physical properties.

Graphite is black, conducts electricity in the hexagonal planes, and is used as a lubricant and as the 'lead' in pencils. Its high melting temperature, 3 570 °C, makes it useful as a crucible material for metal casting. Carbon fibres (graphite) are used to strengthen plastics, finding application in the frames of tennis racquets, for example. In graphite, only van der Waals forces hold the layers together, and they are relatively far apart. This explains graphite's relatively low density and its anisotropy. The crystals shear parallel to the planes with ease. However, contrary to a common misconception, this is not the reason why graphite makes a useful lubricant. The lubricant properties of graphite are dependent on the presence of adsorbed layers of water vapour or oxygen: if these are lost due to low pressure or high temperature, then the lubricant properties disappear. For high-vacuum conditions, such as outer space, surface additives are incorporated in order to preserve the low-friction, low-wear properties. It is the mobility of electrons in the giant π orbital that is responsible for the electrical conductivity in the layers, like a two-dimensional metal.

The diamond structure is **isotropic** (the same in all principal directions), with tetrahedrally coordinated carbon atoms. The carbon–carbon distance is what we would expect for single bonds (154 pm). Diamond is transparent, an electrical insulator, refractory, the hardest substance known, and is used in cutting tools. Diamond's isotropic three-dimensional structure explains its hardness; it is an electrical insulator, but interestingly it has the highest thermal conductivity of any known substance (approximately five times that of copper), which is why diamond cutting tools used for drilling do not overheat. A deposit of a thin layer of diamond can be used as a protective layer on microchips, in order to conduct away the heat generated in the circuits, but not interfere with the electrical performance.

Other forms of carbon that you may have come across are *charcoal* and *soot*. These are forms of microcrystalline carbon. **Activated charcoal**, made by heating charcoal in steam at high temperatures to remove impurities, is very porous and highly adsorbent[*].

In 1985 Robert Curl, Harry Kroto and Richard Smalley discovered new and unexpected forms of pure carbon now known as the **fullerenes**. The most famous of these molecules is C_{60}, **buckminsterfullerene**, with the shape of a soccer ball (Figure 10.4). By 1990 fullerenes could be made in large enough amounts to confirm the structural predictions of the earlier discoveries: they are made by passing an electric arc between two graphite electrodes in a partial atmosphere of helium. The fullerenes produced in this way consist mostly of the two most stable forms, C_{60} (75%) and C_{70} (23%), together with a few of higher molecular mass, including two forms of C_{120} (Figure 10.5a, b) as well as C_{60} units linked by a single carbon (C_{121}) and by two carbon atoms (C_{122}). Smaller fullerenes can also be made, down to C_{28}. Research in the inorganic chemistry of carbon has been dominated by fullerenes since their discovery.

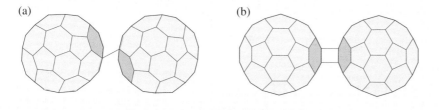

(a) (b)

Figure 10.5
The structure of two fullerene dimers in which the C_{60} units are connected by (a) one single bond and (b) two bonds.

Buckminsterfullerene (a truncated icosahedron) is the most symmetrical of the molecules; it consists of 12 pentagons joined to 20 hexagons.

⬤ Examine the structure of C_{60} (Figure 10.4) and describe the environment of neighbouring carbon atoms. What would you expect to observe in the ^{13}C NMR spectrum of a solution of C_{60}?

◐ Each carbon atom sits at the junction of three rings: two hexagons and one pentagon; so each carbon atom is in an identical environment. The ^{13}C NMR spectrum is therefore expected to be a single resonance. In fact this is what is observed, and was one of the first experimental observations that confirmed the proposed icosahedral structure.

The carbon–carbon distances common to adjacent hexagons are 139 pm. The carbon–carbon linkages shared by hexagons and pentagons are 143 pm, similar to the carbon–carbon distance in the graphite hexagons of 142 pm. We can think of these structures as a carbon sheet with graphite-type delocalized bonding, which bends back on itself to form a polyhedron. The fullerene molecules all have 12 pentagons of carbon atoms linking together different numbers of hexagons: C_{70} has 25 hexagons, and its shape has been likened to that of a rugby ball (Figure 10.6, overleaf).

[*] A substance is said to be *adsorbed* on a surface when there is bonding between the molecule and the surface; this may be weak van der Waals bonding or stronger covalent bonding.

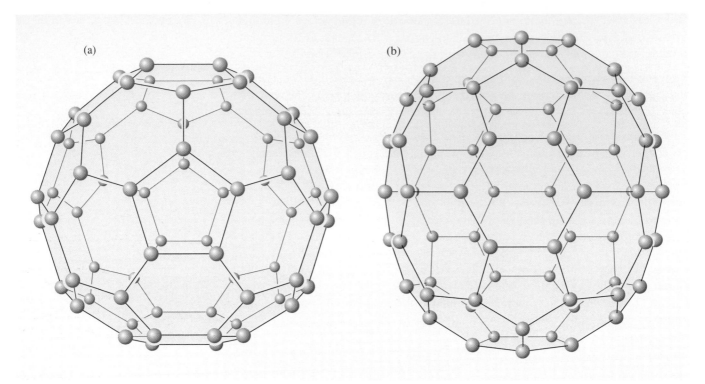

(a) (b)

Figure 10.6 The structure of C_{70} seen from two different perspectives.

The C_{60} buckyballs (as they are often known) close-pack together in the crystals, forming a face-centred cubic array of balls about 100 pm apart. It has been found that fullerenes can be formed around atoms of metals such as potassium, caesium, lanthanum and uranium. Salts of fullerenes have also been formed, containing such ions as $C_{60}{}^{n-}$, where n can have values from 1 to 5. For example, potassium buckide, K_3C_{60} (Figure 10.7) is a metallic crystal with a cubic close-packed array of buckyballs, in which potassium ions occupy both the octahedral and the tetrahedral holes. Fluorination of the fullerenes yields a variety of compounds including the fully fluorinated $C_{60}F_{60}$. Partial fluorination results in distortion of the spherical shape of C_{60}. Thus in $C_{60}F_{18}$ the fluorine atoms are bound to one hemisphere causing flattening of that half of the molecule (Figure 10.8). By contrast, in $C_{60}F_{20}$ the fluorine atoms occupy an equatorial belt that results in flattening of the whole molecule (Figure 10.9).

Many compounds of fullerenes have been synthesized, but they have yet to find commercial exploitation.

In 1980 Sumio Iijima at the NEC laboratory in Japan observed, in a similar preparation to that used for fullerenes, onion-like structures in which the innermost sphere has the size of a C_{60} molecule. In 1991, using an electron microscope, he imaged concentric tubes of carbon atoms, resembling rolled up graphite sheets (Figure 10.10, p. 136). These are called carbon **nanotubes** (CNTs), and those shown in Figure 10.10 are multi-walled (MWCNTs). Typically, the innermost tubes have diameters as small as 0.5 nm, about the same as a C_{36} fullerene, and can be grown to lengths of about 100 μm.

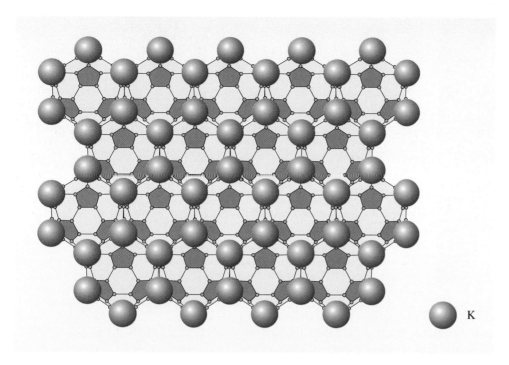

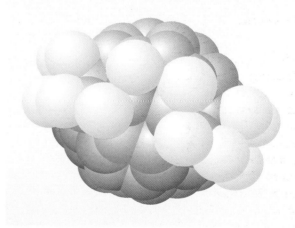

Figure 10.7
The structure of potassium buckide, K_3C_{60}. The layer of K^+ ions forms two rows, one occupying octahedral holes, the other tetrahedral holes. A third row, also occupying tetrahedral holes, and lying above this layer, makes up the complement of $3K^+$ ions.

K

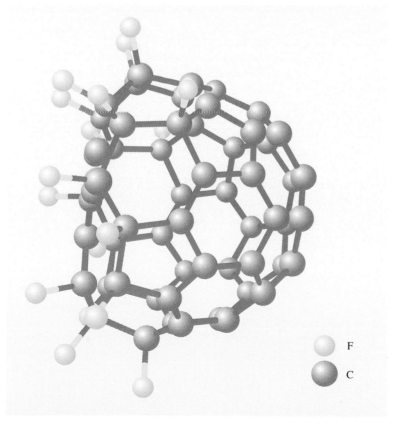

Figure 10.9
The flattened structure of $C_{60}F_{20}$, a molecule that has been called saturnene, owing to its belt of F atoms.

F

C

Figure 10.8
The structure of $C_{60}F_{18}$ in which one-half of the molecule is flattened.

Figure 10.10
Images of carbon nanotubes.

Figure 10.11
A scanning tunnelling image of a single walled carbon nanotube. In this case the tube is open, but nanotubes are also observed in which the ends are closed with hemispherical carbon structures resembling half of a fullerene.

More recently, Iijima also synthesized single cylinders called single-walled nanotubes (SWNTs), an example of which is shown imaged in Figure 10.11. The properties of these nanotubes indicate many promising applications, and so it is interesting to consider their structure.

A SWNT can be considered as a single sheet of graphite (a so-called graphene sheet) rolled up to form a tube by joining the carbon atoms at the edges of the sheet. Examples of this are shown in Figure 10.12, where the sheet is rolled up (a) in the x direction and (b) in the y direction. But note that this rolling up of a graphene sheet is not believed to be the mechanism by which nanotubes are formed.

The x and y directions are only two of the many axes about which a graphene sheet may be rolled up. Each direction would generate a different type of nanotube. We can therefore envisage a large number of such nanotubes defined by a variety of parameters, such as their diameter and the direction of roll-up of the graphene sheet, each representing a different allotrope of carbon. Evidently, carbon has a diversity of allotropic structures that was unsuspected until very recently.

Current research interest in carbon nanotubes is intense owing to their potential application in a variety of roles. The most promising of these is in molecular electronics. For four decades the miniaturization of computer chips has followed an empirical law (Moore's law), by which the number of transistors that can be fabricated on an integrated circuit approximately doubles every two years. It is predicted that, for theoretical and financial reasons, the limit of this trend will be reached by about the year 2015. Further development will then depend on the replacement of lithographic technology, and efforts are being made to synthesize molecular components for computers, which have the potential to reduce the size of computer components by factors of up to 10^5. SWNTs have been found to possess many of the properties essential to this task; the structure shown in Figure 10.12a has the conductivity of a metal, and that in Figure 10.12b is a semiconductor and can be made to work as a transistor. The combination of two crossed SWNTs can act as a rectifier. Molecular electronics is undoubtedly a subject of intense interest, but major obstacles concerning the reproducible synthesis of nanotubes and their chemical linking to form circuits remain.

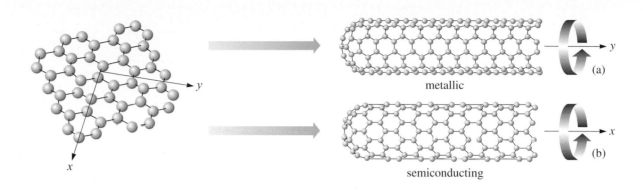

Figure 10.12 Some ways in which a graphene sheet can roll up to form a SWNT.

Amongst their other useful properties is field emission, the emission of electrons when subjected to an electric field, a feature that has already been exploited to produce a flat-panel display. Carbon nanotubes show the ability to absorb small molecules reversibly, within the cavity of the tube. This property may be exploited for the storage of hydrogen as a fuel, especially for transport (Section 5.3). With strength similar to steel, ropes of SWNTs are also promising strong materials, which may rival graphite fibres. Thus carbon nanotubes are fulfilling much of the promised application that it was hoped that fullerenes would deliver following their discovery.

10.2.1 Carbides — molecular, salt-like and interstitial

Carbides are binary compounds in which carbon is the more electronegative partner. We shall come across silicon carbide, SiC, again later when we look at the chemistry of silicon; this is a very hard material, manufactured in large amounts as an abrasive known as carborundum. Its hardness can be accounted for by the fact that its crystal structure is the same as that of diamond, with alternate carbon atoms replaced by silicon (Figure 10.13). A major industrial use of SiC is in steel refining where its addition to the molten metal removes metal oxide impurities with production of CO and a silicate slag, which is skimmed off.

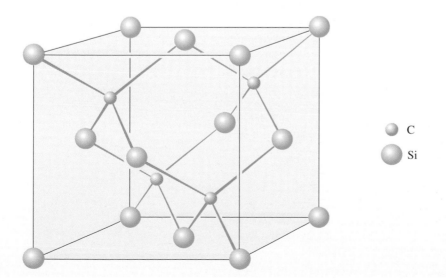

C

Si

Figure 10.13
The structure of SiC.

Salt-like carbides, such as Be_2C, are sometimes called methanides because they produce methane on hydrolysis. Acetylides (methynides) are also salt-like carbides, and contain the $(C{\equiv}C)^{2-}$ ions. The use of calcium carbide (calcium ethynide), CaC_2, to produce acetylene (ethyne), C_2H_2, on treatment with water, has been superseded by production from petrochemicals.

Interstitial carbides are so called because some of the holes in a close-packed metal structure may be occupied by 'guest' atoms; that is, the carbon atoms are in interstices rather than lattice sites. Some of these materials have great industrial importance; steel is the best-known example. The interstitial atoms are primarily carbon, but there are also some boron, nitrogen, silicon, phosphorus and other atoms. Steels, which contain 0.5–1.5% of carbon, look and behave like metals, but are harder, higher melting and more corrosion-resistant than pure iron. Many transition elements form such non-stoichiometric carbides. Other transition metals, such as tungsten and titanium, also form interstitial carbides with a much higher proportion of carbon, which renders these materials very hard. This enables them to be used in cutting tools.

10.2.2 Oxides of carbon

The common oxides of carbon are carbon monoxide, CO, and carbon dioxide, CO_2. Carbon monoxide is a flammable gas; it is formed when carbon is burnt in a limited supply of oxygen; in an excess of oxygen, carbon dioxide forms. Carbon monoxide is colourless, odourless and highly poisonous. The gas is almost non-polar and is one of very few neutral oxides. Its molecule has a very short interatomic distance (Structure **10.1**) and high molar bond enthalpy ($1\,076\,kJ\,mol^{-1}$); the strong bond is due to the π overlap of the 2p orbitals.

Carbon monoxide very readily forms coordination compounds, particularly with transition metals in low oxidation numbers. One important industrial use of these compounds is found in the **Mond process** for purifying nickel (developed by L. Mond in 1899). The impure metal is treated with CO to give the volatile **carbonyl compound**, nickel tetracarbonyl, $Ni(CO)_4$ (Structure **10.2**). The $Ni(CO)_4$ vapour is subsequently decomposed at higher temperature to give pure nickel.

The toxicity of carbon monoxide is a consequence of its strong coordinating properties. Part of our respiration cycle involves the coordination, and subsequent release, of oxygen, O_2, to the haemoglobin in the blood. The oxygen molecule coordinates an iron atom in the centre of the haemoglobin molecule (Figure 10.14a), and the CO molecule can bind at the same place (Figure 14b), reducing the supply of oxygen in the bloodstream.

112.8 pm

$C{\equiv}O$

10.1

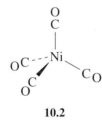

10.2

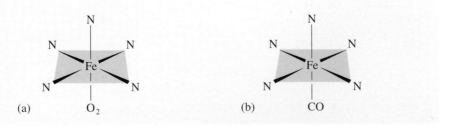

    (a) O_2 (b)   CO

Figure 10.14 (a) The location of O_2 bound to iron in oxyhaemoglobin; (b) the location of CO in carboxyhaemoglobin. The four equational N atoms are in a porphyrin ring, and the axial N atom is in an amino acid that forms part of the protein backbone.

QUESTION 10.1

Describe what you understand by a metal complex or coordination compound. Does $Ni(CO)_4$ fit your description?

QUESTION 10.2 ✳

Would you expect to observe an infrared spectrum for CO?

QUESTION 10.3 ✳

What shape does VSEPR theory predict for the CO_2 molecule? What is the bond order?

QUESTION 10.4 ✳

How many normal modes of vibration do you predict for CO_2? Will they be active in the infrared or Raman or both?

QUESTION 10.5 ✳

According to the harmonic oscillator model, would you expect the stretching frequency of the carbon–oxygen bond in CO to be higher or lower than for a carbon–oxygen bond in CO_2? How would isotopic substitution to form ^{13}CO affect the stretching frequency?

Although it is also a colourless, odourless gas, carbon dioxide has very different chemistry from that of carbon monoxide. It can be made in the laboratory by dropping strong acid on to a carbonate, and on an industrial scale as a byproduct of the production of hydrogen (Equation 10.1) needed for the synthesis of ammonia.

$$CH_4 + 2H_2O = 4H_2 + CO_2 \qquad (10.1)$$

CO_2 is a linear molecule containing two C=O bonds (Structure **10.3**). When CO_2 is cooled to $-78\,°C$ at normal pressure, it turns straight into a solid without going through a liquid phase. This solid is known as **dry ice**; it is used as a refrigerant especially in keeping food frozen. It is also commonly used to produce 'smoke' effects in theatrical productions: it cools the air leading to the formation of mist.

116.3 pm

O=C=O

10.3

At sufficiently high pressure, CO_2 forms a supercritical fluid, a state in which there is no transition from gas to liquid and in which it shows some properties of both the gaseous and the liquid states. The lowest temperature at which this occurs is $31\,°C$ at a pressure of 73 bar (atmospheres). The combination of gas-like diffusion rates and liquid-like density makes supercritical CO_2 a versatile solvent for non-polar substances. It is used for decaffeinating coffee and in many organic syntheses, including radical polymerization, where its use circumvents the problem of entrapping the radical site in a cage of solvent. It has the further advantage that, in contrast to more conventional solvents, it is non-polluting.

Carbon dioxide is only slightly soluble in water, forming a weakly acidic solution containing the hydrogen carbonate ion, HCO_3^-:

$$CO_2(aq) + H_2O(l) = H_2CO_3(aq) = HCO_3^-(aq) + H^+(aq) \qquad (10.2)$$

For many years it was claimed that carbonic acid, H_2CO_3, did not exist. However, carbonic acid has recently been synthesized at low temperature, as a gas and a solid, and studied by infrared spectroscopy and mass spectrometry. In water at room temperature and saturated with CO_2, the ratio of H_2CO_3 to CO_2 is 1 : 100, because the equilibrium in Equation 10.2 lies to the left.

The exchange of CO_2 between the blood and the lungs depends on an enzyme, carbonic anhydrase, which catalyses the combination of CO_2 and OH^- to form HCO_3^- directly The proven existence of H_2CO_3 as a solid or gas suggests that it could be present in the Earth's upper atmosphere as well as on Mars, in comets, in the colder outer bodies of the Solar System and in interstellar grains.

Carbon dioxide also finds everyday uses: it is bubbled into soft drinks under pressure to make them fizzy; soda water tastes only faintly acidic because only a small proportion of the dissolved gas is present as hydrogen carbonate ions: most of it is CO_2 hydrogen-bonded to water (Equation 10.2).

The Earth's atmosphere contains about 0.03% by volume of CO_2, and the equilibrium in Equation 10.2 means that rain falling through even unpolluted skies will be slightly acidic. CO_2 is produced by respiration of plants and animals, but is used by plants in the photosynthesis of carbohydrates. The fossilization of plants in the Carboniferous Period (354–290 million years ago) was responsible for the production of coal. Marine organisms use forms of calcium carbonate (calcite and aragonite) in their exoskeletons. An example of the role of this biomineral in protection is shown in Figure 10.15. Some of these organisms live in coral reefs, but others swim about in the sea, and when they die their shells fall to the sea bed. These shells eventually become sedimentary rocks such as limestone, which may subsequently be metamorphosed into marble.

The CO_2 dissolved in groundwater plays an important part in the weathering of both carbonate and silicate rocks. Insoluble carbonates such as chalk or limestone are slowly taken into solution as soluble hydrogen carbonates. This occurs because an 'insoluble' carbonate, such as $CaCO_3$, is in equilibrium in water with a very small concentration of its ions:

$$CaCO_3(s) = Ca^{2+}(aq) + CO_3^{2-}(aq) \qquad (10.3)$$

What will happen if one of these ions is removed, for instance by reaction with hydrogen ions derived from Equation 10.2:

$$CO_3^{2-}(aq) + H^+(aq) = HCO_3^-(aq) \qquad (10.4)$$

The equilibrium is disturbed, and more will dissolve to take its place (Le Chatelier's principle).

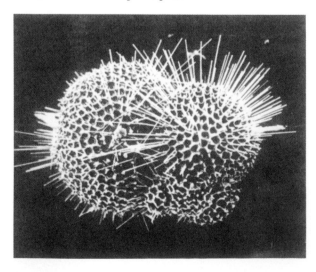

Figure 10.15
An example of biomineralization: the calcite exoskeleton of a foraminiferan, a marine organism.

Silicate rocks are very slowly decomposed or weathered. The alkali metals are removed as soluble carbonates and the alkaline earth metals as soluble hydrogen carbonates. Some silicic acid is also removed in solution, leaving a solid residue of $SiO_2(s)$ as quartz, and clays (which are aluminosilicates).

Carbon dioxide is produced in the respiration (and fermentation and decay) of animals and plants. The pH of blood is influenced by the concentration of dissolved CO_2 (produced by the oxidation of carbohydrates, etc. in the tissues), and this controls the activity of the lungs. CO_2 is also produced by the combustion of fuels. The CO_2 content of the atmosphere is thought to have increased by some 10% in the last 150 years, owing to an increasing human population with increased use of fossil fuels, and also owing to the clearing of forests, which has decreased the involvement of CO_2 in photosynthesis. There is international concern that this increased concentration of CO_2 will contribute to global warming because of the **greenhouse effect,** carbon dioxide being one of the main greenhouse gases. The visible and ultraviolet radiation reaching us from the Sun is absorbed by the Earth, but re-radiated in the *infrared* part of the spectrum. As we saw in Question 10.4, CO_2 absorbs infrared radiation, and so CO_2 effectively traps this energy in the atmosphere. The debate on global warming concerns the effects of a temperature rise, such as a partial melting of the polar ice-caps and consequent rise in the levels of the oceans.

10.2.3 Summary of Section 10.2

1   Carbon forms three main allotropes. Diamond is the hardest known substance; it contains tetrahedral carbon atoms covalently bound to four other carbon atoms in an infinite array. Graphite contains flat layers of carbon hexagons; the 2p orbitals perpendicular to the layers interact to form an extended π orbital. Graphite conducts electricity in the plane of the layers. There are newly discovered forms of carbon. Buckminsterfullerene, C_{60}, has a truncated icosahedral structure, featuring linked pentagons and hexagons of carbon. Carbon nanotubes are rolled single graphite-like sheets, which display promising electronic and tensile properties.

2   Carbon forms binary carbides in which it is the more electronegative element. SiC is a hard abrasive substance with the same atomic arrangement as diamond. The salt-like metal carbides, such as Be_2C and CaC_2, form methane and acetylene, respectively, on hydrolysis. Interstitial carbon atoms in metals form alloys such as steel; the carbon has a profound effect on the physical properties of the metal.

3   Carbon, as a second-row element, has the ability to form multiple bonds by the π overlap of its 2p orbitals.

4   Carbon monoxide, CO, is a poisonous gas, important because of its ability to form coordination compounds: the formation of the volatile carbonyl $Ni(CO)_4$ is used in the purification of nickel.

5   Carbon dioxide, CO_2, is produced naturally by respiration. It forms a weakly acidic solution with water.

10.3 Silicon

Silicon is the second most abundant element in the Earth's crust (26% by mass).

● What is the most abundant element?

● Oxygen; this has an abundance of 46% by mass.

Silicon always occurs in combination with oxygen and other elements, mostly as silica, SiO_2, or silicates. Silica is converted into many useful silicon-containing products, as shown in Figure 10.16.

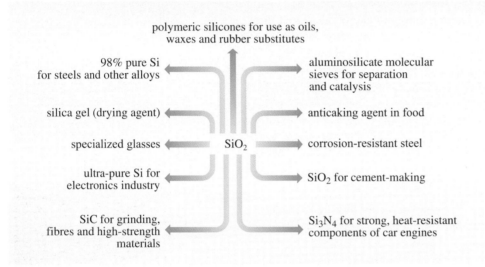

Figure 10.16
Uses of silicon and its compounds.

Reduction of silica with carbon yields elemental silicon. This is carried out on a large scale in huge electric arc furnaces, and the product is 98–99% pure silicon.

QUESTION 10.6 ✳

Figure 10.17 shows the Ellingham diagram [*] for SiO_2 and CO. At what temperature does carbon reduce SiO_2? Calculate the temperature T at which $\Delta G_m^\ominus$ for Reaction 10.5 becomes zero, ignoring the phase change $Si(s) = Si(l)$, which occurs at 1 683 K. Use any values you need from the *Data Book*.

$$\tfrac{1}{2}SiO_2(s) + C(s) = CO(g) + \tfrac{1}{2}Si(s) \tag{10.5}$$

Unpurified silicon (98% pure) is suitable for the formation of silicon-based chemicals such as silicones and silicon alloys. Silicon can form compounds and alloys with many metals. These materials, known as **silicides**, have a wide range of stoichiometries and properties, ranging from metallic to covalent. In particular, silicon is an important minor component of many types of steel and of aluminium alloys, such as are employed in light-alloy car engines. Silicon is an intrinsic semiconductor. It is required in an extremely pure form by the electronics industry, with levels of key impurities, such as phosphorus and boron, of one atom in 10^{10} atoms of Si. High-purity silicon is also used to produce high-purity silica for optical fibres. Methods for the production of pure silicon and for crystallizing the large

[*] Ellingham diagrams are discussed in *Metals and Chemical Change*.[3]

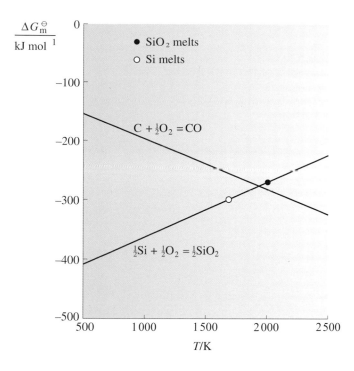

Figure 10.17
Ellingham diagram for SiO_2 and CO.

single crystals for the electronics industry are described in the *Industrial Inorganic Chemistry* Case Study later in this Book. The scale of microchips that are etched onto a **silicon wafer** of a single crystal can be seen in Figure 10.18. To make silicon of sufficiently high purity, unpurified silicon is first converted into $SiHCl_3$ (Equation 10.6), which is then decomposed to pure silicon.

$$Si(s) + 3HCl(g) = SiHCl_3(g) + H_2(g) \tag{10.6}$$

10.3.1 Bonding in silicon compounds

 What is the electronic configuration of the Si atom?

$[Ne]\,3s^2\,3p^2$

Silicon in its compounds is usually four coordinate. Unlike carbon, however, it can increase its coordination number to five or six, if electrons are donated by atoms of another molecule or ion, thus forming a Lewis acid–Lewis base complex: for instance, salts of the complex anions SiF_5^- and SiF_6^{2-} are known.

QUESTION 10.7 ⁒

Using VSEPR theory, predict the shapes of the SiF_5^- and SiF_6^{2-} anions.

In SiF_5^- and SiF_6^{2-} we can see again the idea that not all compounds obey the 'stable octet' rule. It is common for elements of the third row to increase their coordination number to five or six (and higher in later rows), and we shall meet more examples in the chemistry of phosphorus and sulfur.

Compounds of silicon in oxidation number +2, of general type SiX_2, are called **silylenes**; they are mostly unstable at room temperature, and are important as reaction intermediates.

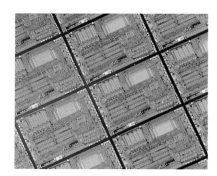

Figure 10.18
A silicon wafer with microchip circuits etched on it.

What would you predict to be the shape of the silylenes?

VSEPR theory predicts that the two bonding pairs of electrons and the non-bonded pair will give bent molecules, and indeed this is observed for the structures that have been determined.

Table 10.2 compares some bond enthalpy terms for single bonds in carbon compounds and in silicon compounds.

Table 10.2 Bond enthalpy terms for C and Si

Bond	$B/\text{kJ mol}^{-1}$	Bond	$B/\text{kJ mol}^{-1}$
C—C	347	Si—Si	226
C—H	413	Si—H	318
C—O	358	Si—O	466
C—F	467	Si—F	597
C—Cl	346	Si—Cl	400
C—Br	290	Si—Br	330
C=O	770	Si=O	638
C=C	612	Si—C	307
C≡C	838		

Note that the Si—Si and Si—H bond enthalpies are significantly lower than those of C—C and C—H. However, with the more electronegative elements from Groups VI and VII, the silicon bond enthalpies are *higher* than the carbon bond energies: the silicon bonds are more polarized than the carbon bonds because silicon is less electronegative than carbon, and the attraction between the differently polarized ends of the bonds leads to extra strength of the silicon bonds. We can use Pauling's equation (Section 4.3) to obtain an estimate of the extra bond strength that we expect from the difference in electronegativity:

$$C(\chi_A - \chi_B)^2 = B(A—B) - \tfrac{1}{2}[B(A—A) + B(B—B)] \tag{10.7}$$

The electronegativities of silicon and fluorine are 1.8 and 4.0, respectively, and so the ionic resonance energy of the Si—F bond is given by

$$C(\chi_F - \chi_{Si})^2 = 96.5(4.0 - 1.8)^2 = 467 \text{ kJ mol}^{-1}$$

If we compare the value for the bond enthalpy term of SiF in Table 10.2 with the average of the Si—Si and F—F values (*Data Book*), the measured additional bond strength is 405 kJ mol^{-1}. (The discrepancy arises from the fact that the values derive from an empirical equation and the electronegativity values are averaged from many data.)

The contrasting chemistry of carbon and silicon is illustrated by their tetrachlorides, CCl_4 is a rather pungent liquid that is immiscible with water and does not react with it. By contrast, $SiCl_4$ is a reactive liquid, which immediately fumes in contact with air, producing white clouds of hydrochloric acid.

QUESTION 10.8

Using values from the *Data Book,* calculate values of $\Delta G_m^{\ominus}$ for the following reactions:

$$CCl_4(l) + 2H_2O(l) = CO_2(g) + 4HCl(g) \tag{10.8}$$

$$SiCl_4(l) + 2H_2O(l) = SiO_2(s) + 4HCl(g) \tag{10.9}$$

−236.2

−143.8

As your calculations should show, both Reactions 10.8 and 10.9 are favourable on thermodynamic grounds, so, clearly, the stability of CCl_4 with respect to hydrolysis must be due to the fact that its hydrolysis is immeasurably slow; that is, it is a kinetic effect due to the mechanism of the reaction. So, a comparison of bond enthalpies between carbon and silicon by itself is a poor guide to relative chemical reactivity. This is because many carbon compounds of the type CX_4 are extremely inert to reactions that would require an expansion of the octet around carbon, whereas many SiX_4 compounds can readily increase their coordination number by adding an electron-pair donor, such as H_2O. This gives the opportunity for subsequent reactions, such as loss of HX and formation of SiO_2. The reactivity of silicon compounds accounts for the fact that, in nature, silicon is found combined only with oxygen, whereas carbon is found in kinetically stable compounds with hydrogen, nitrogen, sulfur and chlorine, as well as with oxygen.

10.3.2 Silicon–oxygen compounds

We have already noted the great difference in physical properties between CO_2 and SiO_2, oxides of elements in the same Group of the Periodic Table: one is a colourless gas, the other a hard crystalline solid. In CO_2 the $2p_x$ and $2p_y$ orbitals on C and O overlap to give pπ–pπ bonding; the result is a discrete molecule containing strong C=O double bonds, O=C=O (see Table 10.2). In SiO_2 the greater strength of the Si—O single bond makes it thermodynamically preferable for Si to form Si—O—Si bridges rather than Si=O double bonds. This leads to the crystalline structure of silica, in which each silicon atom is tetrahedrally coordinated by oxygen atoms, and each oxygen atom is shared with a neighbouring tetrahedron, as in cristobalite (Figure 10.19) one of the most common silica minerals. Silica occurs in several crystalline forms, quartz (or rock crystal; Figure 10.20) being another common mineral.

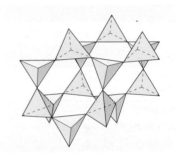

Figure 10.19
Schematic representation of cristobalite, one of the crystalline forms of silica, which is based on tetrahedral SiO_4 groups joined at the apices. Quartz, the more common form of SiO_2, has a similar, but more condensed, structure.

Figure 10.20 Natural crystals of quartz, SiO_2.

Amorphous silica occurs in sponges and many plants, especially in grasses, where it is found in the cell walls and contributes to the strength of the plant. The tips of the needles of stinging nettles are made of hollow tubes of silica, which is why they are so brittle, breaking off after puncturing the skin (Figure 10.21). In this way silica acts as a defence mechanism for the plant.

Quartz is one of the commonest minerals on Earth, occurring as sand on the seashore, as a constituent of granite and flint, and, in less pure form, as agate and opal. Silica behaves like an acid in that it combines on heating with the basic oxides and carbonates of the metals to form silicates, which constitute most of the important rock-forming minerals (Figure 10.22) and their weathering products, clays and soils. A great part of the Earth's crust consists of silicates. Many thousands of silicates are known, formed by the linking of $[SiO_4]^{4-}$ tetrahedra via the sharing of oxygen atoms. The classification of these compounds is of considerable importance to geologists. Before the advent of X-ray crystallography, silicates were classified according to the empirical formulae: orthosilicates, M_4SiO_4 (where M is a univalent metal), metasilicates, M_2SiO_3, pyrosilicates, $M_6Si_2O_7$, and many more such categories. After the X-ray crystallographic studies of the silicates by W. H. Bragg, W. L. Bragg, and L. Pauling in the 1920s, the minerals could be classified by structure.

Figure 10.21
An electron micrograph image of the hairs on the leaf of a stinging nettle.

Figure 10.22 Crystals of silicate minerals: aquamarine, heliodor, morganite and emerald.

The silicate structures are conveniently described in terms of the $[SiO_4]^{4-}$ unit, in which there is tetrahedral coordination of silicon by oxygen (Structure **10.4**). In minerals, these occur as discrete tetrahedra or are linked by one or more corners to give one-dimensional (chains), two-dimensional (sheets) or three-dimensional structures. Examples of all these types are illustrated in 'Silicate structures' on one of the CD-ROMs associated with this book: note particularly how the sharing of oxygen atoms between tetrahedra determines the dimensionality of the silicate structure.

10.4

STUDY NOTE

Of the structures presented on the CD-ROM, you should examine the following:

Dicrete tetrahedra: Olivine and garnet.

Chains: **Pyroxene**

Double chains: **Amphiboles** (note the hexagonal ring structures). These include the fibrous asbestos minerals.

Sheets: Biotite, a mica (note the linking of six-membered rings to generate sheets.) Similar linking is illustrated for muscovite, illite and the clay minerals montmorillonite and kaolinite.

Three-dimensional structures: Feldspar, one of the most abundant rock-forming minerals (note the formation of four-membered rings). Quartz, pure silica. (Note the formation of helical chains which link to produce a three-dimensional array.) These minerals also include the zeolites (Figure 10.23) which are used as ion exchangers, water softeners and catalysts.[*]

Figure 10.23
A sample of a zeolite, a silicate mineral.

X-ray crystallographic work on silicates has helped to explain the apparent complexity of the so-called 'silicic acids'. Orthosilicic acid, H_4SiO_4, can be made by the hydrolysis of tetrachlorosilane (silicon tetrachloride).

⬤ What is the equation for this reaction?

⬤ The equation is:

$$SiCl_4(s) + 4H_2O(l) = H_4SiO_4(aq) + 4HCl(aq) \tag{10.10}$$

The solution is gelatinous and becomes cloudy on standing. This is because the silicic acid polymerizes by splitting off water, to form Si—O—Si bridges, which results in a mixture of molecules of different chain lengths.

⬤ Form a chain molecule (diagrammatically) from four $HO-Si(OH)_2-OH$ molecules, by splitting off three (4 − 1) molecules of water. What is the repeat unit in this chain and what is the overall equation for the process?

⬤ The repeat unit is $-(H_2SiO_3)-$ (see Figure 10.24). The equation is

$$4H_4SiO_4(aq) = H_{10}Si_4O_{13}(aq) + 3H_2O \tag{10.11}$$

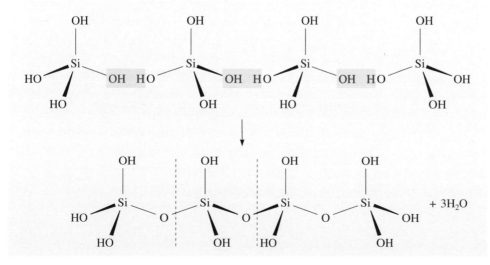

Figure 10.24
Formation of a metasilicic acid by splitting off water from orthosilicic acid. Blue dashed lines indicate the repeat unit.

[*] The structures and applications of zeolites are described in a Case Study in *Chemical Kinetics and Mechanism*.[4]

These long-chain acids are called 'metasilicic acid', and are analogous to the silicate chains found in the pyroxenes.

Ordinary 'sodium silicate' and 'calcium silicate' are, in fact, metasilicates. Sodium silicate may be made by heating sodium carbonate, Na_2CO_3, with pure sand, SiO_2, in a furnace to form a glass. Window glass (soda glass) is made by a continuous process in which sand reacts at 1 400 °C with fused carbonates of sodium, potassium and calcium (limestone and some sodium sulfate). At these temperatures the carbonates behave as a mixture of the oxide and carbon dioxide gas (for instance, $CaO + CO_2$). The liquid is stirred by the evolution of CO_2, H_2O (from the hydrated salts) and SO_3.

● Write equations for some of these reactions.

● $Na_2CO_3 + SiO_2 = Na_2SiO_3 + CO_2$ (10.12)

   $CaCO_3 = CaO + CO_2$ (10.13)

   $CaO + SiO_2 = CaSiO_3$ (10.14)

   $Na_2SO_4 + SiO_2 = Na_2SiO_3 + SO_3$ (10.15)

At these high temperatures, sodium oxide, for instance, will combine with SiO_2, opening up the Si—O—Si linkages to form two Si—O⁻ bonds (Figure 10.25); the Na^+ ions are coordinated to the O⁻ sites on the three-dimensional Si—O network (Figure 10.25c). Because of the high viscosity of the melt, glasses do not form a sufficiently regular array in order to crystallize during cooling. We can visualize a silicate glass structure as tangled and branched chains of corner-sharing $[SiO_4]^{4-}$ tetrahedra, with Na^+, K^+ and Ca^{2+} ions between them. The molten glass is drawn off and floated on pure molten tin to form sheets of glass with a flat smooth surface.

Molten SiO_2 itself crystallizes very slowly, and usually solidifes to a glass (Figure 10.25b). Fused silica is used to make special glassware because it is resistant to chemical attack, melts at a high temperature (1 500 °C) and transmits ultraviolet light.

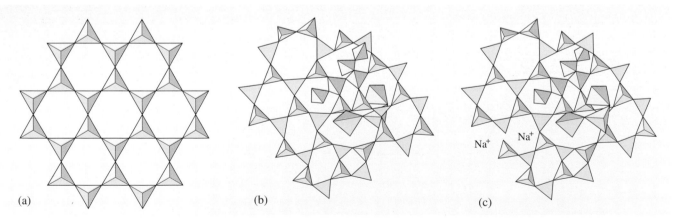

(a)   (b)   (c)

Figure 10.25 (a) Linked $[SiO_4]^{4-}$ tetrahedra in idealized crystalline SiO_2; (b) the structure of a simple SiO_2 glass, showing corner-linked, tangled $[SiO_4]^{4-}$ chains; (c) the addition of one molecule of Na_2O causes an Si—O—Si group to form two Si—O⁻ Na^+ interactions.

Pyrex glass, used in laboratories and kitchens, etc., expands less than soda glass when heated, which is why it does not crack when its temperature is suddenly changed, and has a higher melting temperature than ordinary soda glass. It is a borosilicate glass and contains less sodium and potassium than ordinary glass, and no lime at all; the triangular BO_3 groups, which are linked through their vertices, are opened up by the metal oxides, replacing the B—O—B groups with B—O⁻ bonds, so that the BO_3 groups can be linked through one or two vertices. Some silicate minerals are found as glasses, but most have had plenty of time in which to crystallize, either from magmas or after various metamorphic processes. Some of the many applications of silicates are summarized in Table 10.3.

Table 10.3 Uses of silicon–oxygen compounds

Silica gel	$SiO_2(H_2O)$	amorphous microporous structure with surface area of $700\,m^2\,g^{-1}$; a desiccant which can absorb up to half its mass of water
Vitreous silica	SiO_2	clear amorphous silica; high thermal shock resistance; very high softening temperature; transmits UV and near-IR radiation
Pyrogenic or 'fume' silica	SiO_2	made by burning $SiCl_4$ in an oxygen-rich hydrocarbon flame; used for thickening oils and greases
Exfoliated vermiculite	$(Mg,Ca)_{0.7}(Mg,Al,Fe^{III})_6$ $[(Al,Si)_8O_{20}](OH)_4.8H_2O$	a natural silicate, which on rapid heating expands to 20 times its original volume; used in packaging and thermal insulation
Soda glass	$(SiO_2)_{\approx 7}(Na_2O)_{\approx 2}$ $(CaO)_{\approx 1}(Al_2O_3)_{\approx 0.1}$	90% of all glass made has this composition, or small variants thereof; used in bottles, windows, etc.
Portland cement	$(CaO)_a(Al_2O_3)_b(SiO_2)_c$	made at high temperature from a range of materials like limestone, anhydrite ($CaSO_4$) and clays; variable composition, and made on an enormous scale for construction purposes
Soluble silicates	$(Na_2O)(SiO_2)_n$	viscous opalescent colloidal solutions; used in detergents, adhesives and high-temperature binders
Organic silicates	$(RO)_4Si$	colourless liquids, very easily hydrolysed; used as paint binders and cross-linking agents; first recognized by Mendeléev in 1860
Silicones	$(R_2SiO)_n$	a very wide range of polymers, having —Si—O—Si— chains, with organic substituents on the silicon atoms; oils, resins, rubbers, etc.

10.3.3 Halosilanes

Direct reaction of silicon with fluorine, chlorine, bromine or iodine, gives stable but moisture-sensitive compounds, SiX_4. The reaction between F_2 and Si is extremely violent, and SiF_4 is much more conveniently made by treating concentrated H_2SO_4 with a mixture of SiO_2 and sodium fluoride, NaF (Equation 10.16). The driving force for this reaction is the greater strength of the Si—F bond compared with Si—O.

$$2H_2SO_4 + 4NaF + SiO_2 = 2Na_2SO_4 + 2H_2O + SiF_4 \qquad (10.16)$$

Tetrafluorosilane, SiF_4 (silicon tetrafluoride) is a gas at room temperature. The other halosilanes are liquids with boiling temperatures increasing with molecular mass. The reaction of tetrafluorosilane with water is rather different from that of other halosilanes, in that it forms both SiF_6^{2-} ions and SiO_2; the strongly electronegative fluoride ligands bring out the maximum coordination number of six for silicon.

● Write a balanced equation for the reaction of $SiF_4(g)$ with $H_2O(l)$.

● $3SiF_4(g) + 2H_2O(l) = 4H^+(aq) + 2SiF_6^{2-}(aq) + SiO_2(s)$ (10.17)

The other halosilanes do not give SiX_6^{2-} ions; with excess water they react vigorously and completely to give SiO_2 and the hydrogen halide:

    $SiX_4(g) + 2H_2O(l) = SiO_2(s) + 4HX(g)$ (10.18)

Catenation in silicon compounds is more common in the halides than in the hydrides, in contrast to carbon chemistry. Halosilanes containing Si—Si bonds are made via reactions of the unstable silicon dihalides, SiX_2, or from metal silicides. Thus, passing chlorine gas over calcium silicide, Ca_2Si, at 150 °C gives Si_2Cl_6 and higher homologues of the alkane-like series Si_nCl_{2n+2}. The structure of polymeric $(SiCl_2)_n$ shows it to consist of infinite parallel chains.

10.3.4 Compounds of silicon with hydrogen and alkyl groups

It has been known since the pioneering development of greaseless vacuum lines by Alfred Stock in the 1920s, that magnesium silicide, Mg_2Si (made by heating Mg and Si together) reacts with dilute hydrochloric acid to give a series of compounds called **silanes** with the general formula Si_nH_{2n+2}, which are analogous to the alkanes.

● Write an equation for the formation of the lowest silane.

● $Mg_2Si(s) + 4H^+(aq) = 2Mg^{2+}(aq) + SiH_4(g)$ (10.19)

It has been shown that both straight-chain and branched-chain compounds are present, but the silanes are rather reactive, and the series has been explored only for values of n up to about 8. The lower members of the series, SiH_4, Si_2H_6 and Si_3H_8, are better made by reduction of the corresponding chlorosilanes, Si_nCl_{2n+2}, with lithium aluminium hydride, $LiAlH_4$. The silanes tend to be very reactive compounds and are spontaneously flammable in air. They have to be handled using special techniques to isolate them from contact with oxygen; this involves the use of a vacuum system or keeping them under a nitrogen atmosphere. They do not react with pure water or dilute acids, but hydrolyse rapidly in the presence of even a trace of alkali, liberating hydrogen and forming silicates.

QUESTION 10.9 ✳

-817.9 Using values of $\Delta G_f^{\ominus}$ from the *Data Book*, calculate $\Delta G_m^{\ominus}$ for the reaction of
-1467.3 (a) methane, CH_4, and (b) ethane, C_2H_6, with oxygen to give CO_2 and H_2O.
-1387.7 Calculate $\Delta G_m^{\ominus}$ values for the analogous reactions of (c) monosilane, SiH_4, and
-2551.8 (d) disilane, Si_2H_6.

The values you obtain from the calculations in Question 10.9 show you that, in oxygen, both alkanes are unstable with respect to CO_2 and H_2O, and both silanes are also unstable with respect to SiO_2 and H_2O. The silanes are spontaneously flammable in air, but the alkanes are not. Methane (natural gas) can, of course, explode in air if it is sparked.

⬤ How would you explain this difference?

⬤ The alkanes possess a *kinetic* stability to reaction with O_2 that the silanes do not. The reasons for this must lie in the *mechanism* of the reactions, and are probably concerned with the larger size of the silicon atom, which facilitates nucleophilic attack by OH^-, and its ability to expand its octet. The explosive reaction of methane when sparked is due to the production of radical species, such as $CH_3\bullet$, which initiate a chain reaction.

When the hydrogen atoms in silanes are replaced by alkyl or aryl groups, the chemical reactivity is much reduced and the thermal stability is increased. Many organosilicon compounds are made industrially by two different catalytic processes named after their inventors. In the first, the **Rochow process**, chloromethane, CH_3Cl, reacts with silicon granules in the presence of 10% copper and small amounts of oxides of metals, such as magnesium, calcium or tin at 300 °C. Under the correct conditions, the main organosilicon product is $(CH_3)_2SiCl_2$ with some $(CH_3)_3SiCl$ and CH_3SiCl_3, all volatile liquids. In the second method, the **Speier hydrosilation process**, $HSiCl_3$ reacts with an alkene, $RCH=CH_2$, by platinum-catalysed addition of Si—H across the C=C double bond, to give $RCH_2CH_2SiCl_3$. Controlled hydrolysis of organosilicon chlorides gives compounds known as **silicones** or **polysiloxanes**, compounds with a silicon–oxygen backbone (Figure 10.26) first discovered by F. S. Kipping, who proposed the name silicone in 1901.

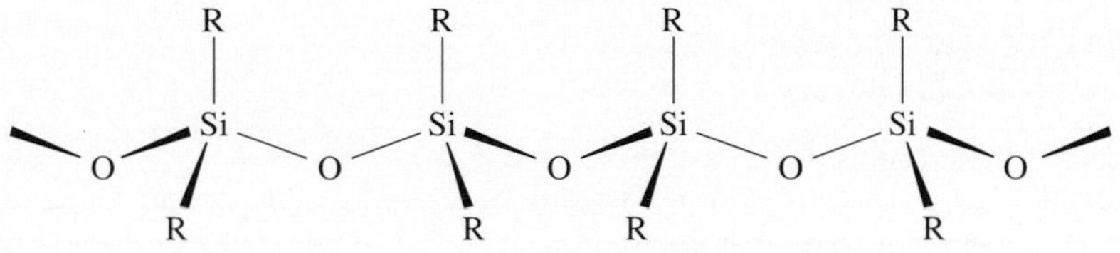

Figure 10.26
An unbranched silicone chain (R is an alkyl group, usually methyl).

Four different types of structural unit are found in silicones; they are formed by the hydrolyses shown in Equations 10.20–10.23.

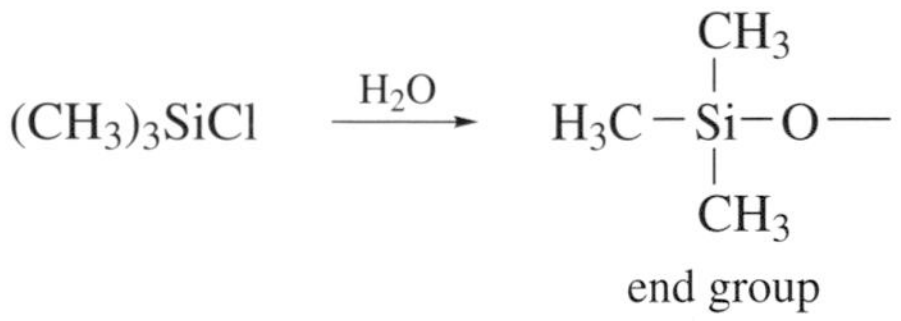

$(CH_3)_3SiCl \xrightarrow{H_2O}$

end group

(10.20)

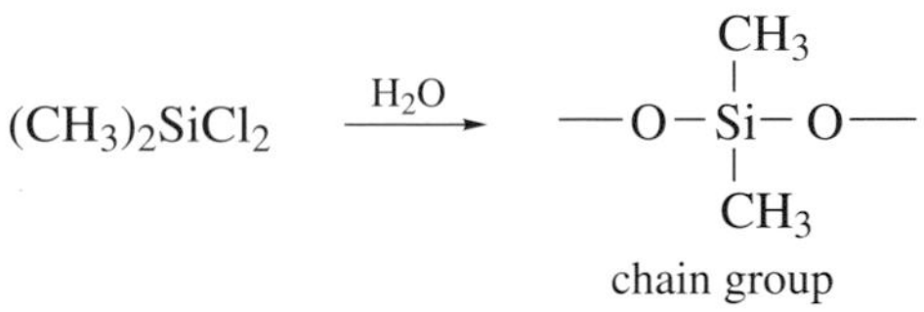

$(CH_3)_2SiCl_2 \xrightarrow{H_2O}$

chain group

(10.21)

$CH_3SiCl_3 \xrightarrow{H_2O}$

branching group

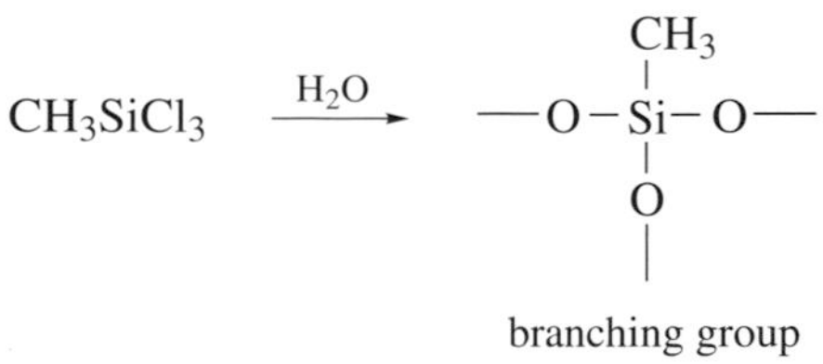

(10.22)

$$\text{SiCl}_4 \xrightarrow{\text{H}_2\text{O}} \quad \overset{\displaystyle |}{\underset{\displaystyle |}{\text{O}}} - \text{O} - \overset{\overset{\displaystyle |}{\text{O}}}{\underset{\underset{\displaystyle |}{\text{O}}}{\text{Si}}} - \text{O} - \tag{10.23}$$

branching group

The extent of polymerization and branching can be controlled by introducing end-blocking groups such as in Equation 10.20.

The important feature of the silicones is that they have a backbone containing only the very strong Si—O bonds, giving them good thermal stability. The backbone is surrounded by a sheath of methyl (or other organic) groups. The viscosity of silicones varies little over a wide temperature range. The longer the chain of the silicone, the greater its viscosity, and silicone oils can be tailor-made to a required viscosity. The oils are used for dielectric insulators, hydraulic fluids, light lubricants, car polishes, cosmetics, etc. The silicones have low surface tension and are non-toxic; they are used as antifoaming agents in textile dyeing, fermentation, sewage disposal and cooking oils, for example in making potato crisps.

When the siloxane chain is a **cross-linked polymer**, using for instance branching groups such as indicated in Equations 10.22 and 10.23, or methylene, —CH$_2$—, groups, the reaction can be used to produce rubber-like elastomers. These materials are flexible down to very low temperatures, and are used for gaskets, space-suits, soft contact lenses, and inside the body in some spare-part surgery.

Until 1981, compounds containing silicon–silicon double bonds were thought to have only a fleeting existence at room temperature. If they were made, they immediately polymerized to compounds containing silicon–silicon single bonds. Silicon–silicon double bonds are therefore *thermodynamically* unstable with respect to such polymerization. Then a way of stabilizing silicon–silicon double bonds by introducing a *kinetic* barrier was discovered. If silicon is joined to very bulky groups, polymerization is impossible and the silicon–silicon double bonds exist at ordinary temperatures. The first success was obtained by using the 2,4,6-trimethylphenyl group (known as mesityl, Mes) (**10.5**); it is bulky enough to prevent the close approach of the double bonds that is needed for polymerization to occur. The silicon–silicon double-bonded product (Mes)$_2$Si=Si(Mes)$_2$ is known as a **disilene**. The compound reacts easily with small molecules such as oxygen, which can attack the double bond, but melts without decomposition at 178 °C in the absence of air. A number of similar disilenes have now been made using different bulky groups. Success has also been achieved in stabilizing silicon–carbon double bonds (in silenes) with bulky groups to prevent the polymerization that otherwise occurs. The same trick is used to make compounds containing P=P double bonds. Steric protection with mesityl groups has also led to the preparation of the silylium ion (Mes)$_3$Si$^+$, analogous to carbocations.

The ultimate organosilicon compound is silicon carbide, SiC (Section 10.2.1). This is a very hard and strong substance, made on a huge scale as a grinding agent by the reduction of SiO$_2$ with carbon at 2 000 °C:

$$\text{SiO}_2(s) + 3\text{C}(s) = \text{SiC}(s) + 2\text{CO}(g) \tag{10.24}$$

The crystal structure of silicon carbide (Figure 10.13) is of the zinc blende type. It can be thought of as the diamond structure, in which every alternate carbon atom is replaced by a silicon atom.

10.5

Silicon carbide is also made as very strong fibres by the sequence of reactions shown in Equations 10.25 and 10.26, in which an infusible polymer containing the silicon backbone shown in Structure **10.6** is made first. This is converted by heat in an argon atmosphere into an isomer containing the silicon–carbon backbone shown in Structure **10.7**, which can be melted and spun into fibres. The fibres are heated in air at 300 °C to partly oxidize them and make them infusible, and are then heated in nitrogen to 1 300 °C to convert them into SiC fibres:

$$n(CH_3)_2SiCl_2 + 2nNa = [(CH_3)_2Si]_n + 2nNaCl \qquad (10.25)$$

$$[(CH_3)_2Si]_n \xrightarrow[\text{argon}]{400\,°C} \left(\begin{array}{c} H \\ | \\ Si-CH_2 \\ | \\ CH_3 \end{array} \right)_n \xrightarrow[\substack{\text{spin to fibres by heating} \\ \text{in air at 300 °C, and then} \\ \text{in N}_2 \text{ at 1 300 °C}}]{} n\,SiC$$

$$(10.26)$$

The fibres are used to reinforce aluminium and other materials in aircraft construction and other applications where light weight and high strength are important.

Silicon carbide is resistant to oxidation in air to about 1 600 °C because of the formation of an adherent layer of SiO_2 on its surface. Silicon nitride, Si_3N_4, made from the reaction of either silicon or a mixture of silica and carbon with a mixture of nitrogen and hydrogen at 1 300–1 600 °C, is also resistant to air oxidation, and is being tested extensively as a light-weight replacement for metal in the construction of engines for cars and jetplanes.

10.4 Germanium, tin and lead

Perhaps the most spectacular feature of these elements is the increasing tendency to form stable bivalent species, that is, the occurrence of a +2 oxidation number for the Group IV element in mononuclear compounds (see the discussion of the inert pair effect on the CD-ROM with this book). Thus, whereas all the silicon tetrahalides, SiX_4 (where X = F, Cl, Br or I) are known, the mononuclear SiX_2 species (the silylenes) are thermodynamically unstable at room temperature and have been identified as reaction intermediates. By contrast the +2 oxidation number is the most common one for lead, and all the halides PbX_2 are known and stable at room temperature. Of the PbX_4 halides, only PbF_4 is stable at room temperature; $PbCl_4$ decomposes to $PbCl_2$ and chlorine gas above 50 °C, and $PbBr_4$ and PbI_4 are unknown. In the stable dihalides, PbX_2, two of the four valence electrons of lead are exhibiting the inert pair effect. Although the inert pair of electrons is not used in bonding, its presence in the valence shell usually has an effect on the shape of the molecule. For instance, the gaseous $SnCl_2$ molecule is observed to have a bond angle of 95°. Another example is provided by the solid-state structures of SnO and PbO (litharge), which have the same crystal structure, containing square-pyramidal units of SnO_4 (or PbO_4), arranged so that the oxygens form parallel layers; the metal atoms lie alternately above and below the oxygen layer, and the non-bonded pair is assumed to lie opposite the metal atom (the O—Sn—O angle here is 75°; see Figure 10.27, overleaf). The inert pair effect is first manifested by germanium, and increases through tin to lead, where we see the dramatic change whereby oxidation number +2 becomes predominant. Clearly, we would expect only the most powerful oxidizing agents, such as fluorine (PbF_4) or oxygen (PbO_2), to be able to oxidize lead to oxidation number +4.

$$\begin{array}{c} | \quad | \\ -Si-Si- \\ | \quad | \end{array}$$

10.6

$$\begin{array}{c} | \quad | \quad | \quad | \quad | \\ -Si-C-Si-C-Si- \\ | \quad | \quad | \quad | \quad | \end{array}$$

10.7

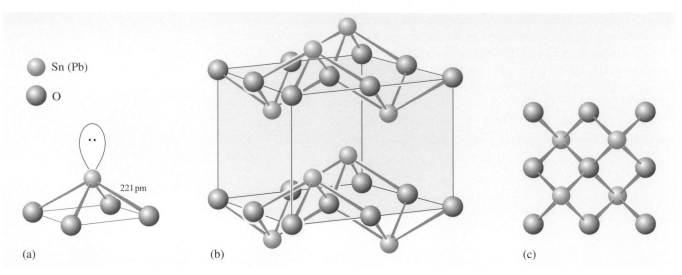

Figure 10.27 The crystal structure of SnO (and PbO), showing: (a) a square-pyramidal SnO_4 unit; (b) the linking of the square pyramids in layers; (c) a top view of a single layer.

QUESTION 10.10 ✳

The following hexahalo ions have been characterized for the first four elements of Group IV:

C: none

Si: SiF_6^{2-}

Ge: GeF_6^{2-}, $GeCl_6^{2-}$

Sn: SnF_6^{2-}, $SnCl_6^{2-}$, $SnBr_6^{2-}$, SnI_6^{2-}

What reasons can you think of to explain these differences?

10.5 Summary of Sections 10.3 and 10.4

1 Silicon is produced by carbon reduction of SiO_2. It can be purified by forming trichlorosilane, $SiHCl_3$, which is then distilled and decomposed. Single crystals of silicon are made for use in the semiconductor industry.

2 Silicon is usually four coordinate, but as a third-row element it can increase its coordination number to five and six in species such as SiF_5^- and SiF_6^{2-}.

3 The strength of the Si—O bond means that silicon forms oxygen bridges, Si—O—Si, unlike carbon, which readily forms C=O double bonds. SiO_2 occurs as the crystalline solid, quartz, with Si tetrahedrally coordinated by four oxygen atoms. SiO_2 combines with basic oxides and carbonates, and so can be said to be acidic in character.

4 Thousands of silicate structures are known. They are formed by the different arrangements of $[SiO_4]^{4-}$ tetrahedra, which link together by bridges through the apical oxygens.

5 Glass is made by treating sand (SiO_2) with metal carbonates and sodium sulfate. It consists of tangled chains of $[SiO_4]^{4-}$ tetrahedra, with Na^+, K^+ and Ca^{2+} ions in between. Pyrex glass also contains boron.

6 All the binary halides, SiX_4, of silicon are known; they are compounds that hydrolyse rather easily. The hexafluoro ions, MF_6^{2-}, are known for all the Group IV/14 elements except carbon, but the larger iodide ion is only able to form SnI_6^{2-}.

7 Silicon forms a homologous series of very reactive hydrides, Si_nH_{2n+2}, known as the silanes.

8 Hydrolysis of the organosilicon chlorides gives a group of polymeric compounds known as silicones, which contain a $(-Si-O-Si-O-)$ backbone. These compounds have been developed industrially for many uses, such as oils, polishes and rubber-like materials.

9 Recently, compounds have been made containing $Si=Si$ double bonds; polymerization is prevented by the use of bulky substituent groups.

10 The tendency to form bivalent compounds increases down the Group; the +2 oxidation state is the most common one for lead.

QUESTION 10.11

Why do you think that a homologous series of silicon hydrides, Si_nH_{2n}, analogous to the alkenes, does not exist?

QUESTION 10.12 ✳

Given that $\chi_{Si} = 1.8$ and $\chi_{Cl} = 3.0$, use Pauling's equation to predict a value for the ionic resonance energy of the Si—Cl bond, and compare it with the experimental value.

THE GROUP V/15 ELEMENTS

11

11.1 Structures and properties of the elements

Continuing our survey of the chemistry of the typical elements, we reach Group V (Figure 11.1). The first two members, nitrogen and phosphorus, are non-metals, although as we shall see they are very dissimilar in their properties. The remaining three elements are semi-metals (arsenic and antimony) and a metal (bismuth). The electronic configurations of the Group V elements are shown in Table 11.1.

Nitrogen, in the form of the dinitrogen molecule, N_2, comprises about 80% of the air around us. Despite this, nitrogen is not an abundant element overall on Earth, being found as deposits of the nitrates, KNO_3 (saltpetre) and $NaNO_3$ (Chile saltpetre).

Phosphorus occurs in various phosphate minerals, the most important belonging to the apatite family, which originate from the compressed remains of ancient organisms. Apatite has the general formula $Ca_5(PO_4)_3X$, where X can be OH^- (hydroxyapatite), F^- (fluorapatite) or Cl^- (chlorapatite).

Like carbon, phosphorus exists in allotropic forms. Highly toxic white phosphorus (also called α-phosphorus) has the tetrahedral structure shown in Figure 11.2. It is a waxy solid, which bursts into flame when in contact with air, and as a consequence is stored under water. It can be converted into the less dangerous, red, violet or black forms by heating under pressure. On heating, the P_4 molecules open out to form chains, two-dimensional layers and three-dimensional structures of covalently bonded phosphorus atoms. Figure 11.3 shows the structures of the violet and black forms.

Figure 11.1
The Group V/15 elements: nitrogen and phosphorus are non-metals, arsenic and antimony are semi-metals and bismuth is a metal.

Table 11.1 Electronic configurations of Group V/15 atoms

Atom	Electronic configuration
N	[He] $2s^2 2p^3$
P	[Ne] $3s^2 3p^3$
As	[Ar] $3d^{10} 4s^2 4p^3$
Sb	[Kr] $4d^{10} 5s^2 5p^3$
Bi	[Xe] $4f^{14} 5d^{10} 6s^2 6p^3$

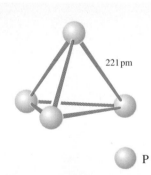

221 pm

P

Figure 11.2 The structure of white phosphorus, P_4.

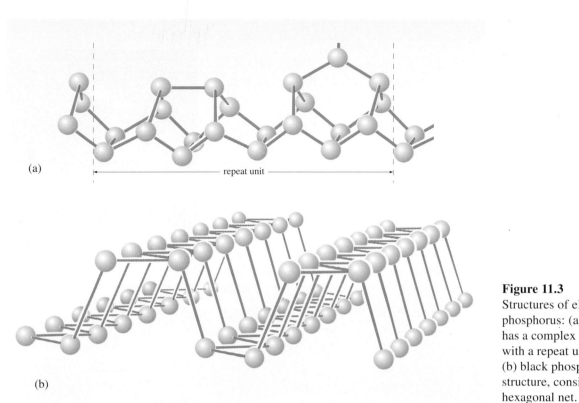

(a)

repeat unit

(b)

Figure 11.3
Structures of elemental polymeric phosphorus: (a) violet phosphorus has a complex double-layer structure with a repeat unit of 21 atoms; (b) black phosphorus has a layer structure, consisting of a puckered hexagonal net.

What do you notice about the coordination number of phosphorus in all these structures?

In each case, phosphorus forms three single bonds; that is, it has a coordination number of three.

Several of the allotropes, such as red phosphorus are **amorphous**; that is they have no long-range order of atomic packing, as is found in crystals.

Arsenic, As, antimony, Sb and bismuth, Bi have a metallic appearance but they are brittle. They are isostructural, with layer structures (Figure 11.4). The difference between the inter-layer and intra-layer distance, however, decreases from As to Bi. Hence Bi is metallic.

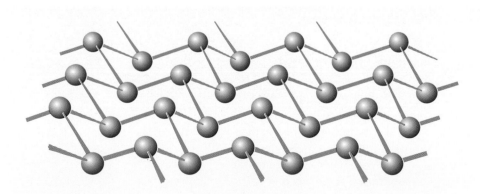

Figure 11.4
The structure of arsenic, antimony and bismuth.

11.2 Nitrogen

Nitrogen, N_2, is chemically very unreactive, in fact almost as inert as the noble gases (Group VIII). This stems from the great strength of its triple bond, which has a molar bond enthalpy of $945\,kJ\,mol^{-1}$. Gaseous nitrogen finds applications where inert atmospheres are required; these include iron and steel production and the manufacture of air-sensitive chemicals. Liquid nitrogen (T_b −196 °C) is used where extreme cold is essential (Figure 11.5), such as the preservation of blood and semen, rapid freezing of food and freeze-branding of cattle.

Nitrogen is only poorly soluble in water, although its solubility increases with increasing pressure (true of gases in general). This is a particular problem for deep-sea divers (Figure 11.6), when nitrogen from their air supply dissolves in the blood at high pressures. If the diver returns to the surface too rapidly, nitrogen comes out of solution as the pressure falls, leading to the formation of bubbles of nitrogen, particularly around the joints. This painful and often fatal condition, known as the bends, is prevented by the diver returning to the surface very slowly, or by the use of decompression chambers, or by using a breathing mixture of helium and oxygen.

Figure 11.5
Nitrogen finds many applications as a refrigerant in the liquid form.

Figure 11.6 The solubility of nitrogen in the blood at high pressure is a problem for deep-sea divers and can lead to a serious condition known as the bends.

BOX 11.1 Nature's nitrogen cycle

Cycling of nitrogen and its compounds through the environment involves a delicate balance of redox, atmospheric and biological processes. These are summarized in Figure 11.7. Plants require nitrogen, which is absorbed in the form of nitrate or ammonium ions, for the synthesis of organic compounds (amino acids and other nitrogenous materials), which are incorporated into the tissues of the plant. However, on removing a crop from a given patch of soil, some of the intrinsic nitrogen is lost. This must be made good in order to maintain the fertility of the soil.

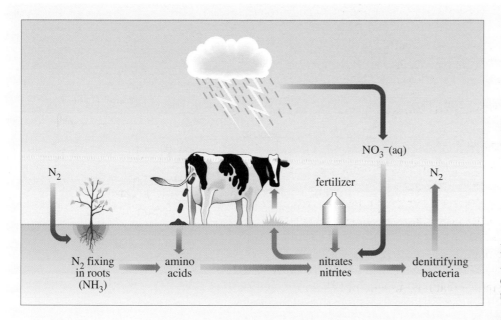

Figure 11.7
A simplified representation of the chemical processes that constitute the nitrogen cycle.

Despite a vast reservoir of nitrogen being readily available, i.e. the air, green plants are unable to use it owing to the high dissociation energy of the dinitrogen triple bond. They depend on **nitrogen fixation**, i.e. combination with hydrogen and oxygen to form ammonium compounds or nitrates. On the roots of peas, beans and other members of the legume family, there are pinkish nodules (Figure 11.8), inside which live nitrogen-fixing bacteria of the *Rhizobium* family. These bacteria fix about 10^8 tonnes of nitrogen per year worldwide, which is approximately 60% of all nitrogen fixed. This is a symbiotic relationship, with the bacteria providing the plant with nitrogen compounds, and the plant supplying nutrients to the bacteria.

To bring about cleavage of the nitrogen molecule requires a high energy input. Indeed, two molecules of the energy-transfer agent adenosine triphosphate (ATP; Section 11.4.5) are needed to bring about transfer of each electron. The reaction is catalysed by the enzyme *nitrogenase*, a complex molecule that incorporates molybdenum, iron and sulfur.

In addition there are other sources of nitrogen, the most obvious being fertilizer, either synthetic, or farmyard manure. Atmospheric lightning also causes some combination of oxygen and nitrogen, which leads to the passage of nitrogen into the soil as nitrates dissolved in rainwater. We should also include the formation of gaseous oxides of nitrogen from coal burning and internal combustion engines — a process more commonly associated with detrimental environmental effects!

The nitrogen cycle is completed by plant death, decay and bacterial denitrification, which returns nitrogen to the seas and atmosphere.

Figure 11.8
Root nodules on a pea plant containing bacteria that fix atmospheric nitrogen, and make it available to the plant.

Chemically, nitrogen is surprisingly versatile in its compounds: it can be found in oxidation numbers from −3 to +5.

● What are the oxidation numbers of the nitrogen in the nitride ion (N^{3-}), ammonia (NH_3), nitrogen fluoride (NF_3) and the nitrate ion (NO_3^-)?

● −3, −3, +3 and +5, respectively.

Like other elements in the second row, the nitrogen atom is able to form double bonds both to other nitrogen atoms and to atoms of some other elements of the row, such as boron, carbon and oxygen.

With nitrogen, we reach the first element of the second row to have a non-bonded pair of electrons in its normal valency state; this is typified by the ammonia molecule (Structure **11.1**).

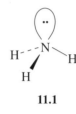

11.1

Nitrogen is also among the most electronegative elements. When discussing hydrogen bonding (Section 5.8), we described a mainly electrostatic attraction between a non-bonded pair on an electronegative element and a neighbouring hydrogen atom. This type of intermolecular attraction affects the physical properties of ammonia. Considering the boiling temperatures of the hydrides on descending Group V, we find the following values: NH_3, −33.4 °C; PH_3, −87.7 °C; AsH_3, −62.4 °C; SbH_3, −18.4 °C. The higher than expected value for ammonia is explained by the extra intermolecular forces between the ammonia molecules as a result of hydrogen bonding.

Among the relatively few chemical reactions of molecular nitrogen at room temperature is the combination with lithium to form a red ionic compound, lithium nitride, Li_3N. Many elements combine with nitrogen or ammonia on heating, to form nitrides of various types. The nitrides of the Group II metals are usually colourless (although that of magnesium is yellow), transparent and salt-like, and contain the nitride ion, N^{3-}. They are hydrolysed by water, to give the metal hydroxide and ammonia:

$$M_3N_2(s) + 6H_2O(l) = 3M(OH)_2(s) + 2NH_3(g) \qquad (11.1)$$

In contrast, the nitrides of boron and aluminium are rather different: they are refractory materials (substances with high melting temperatures) with a macromolecular structure. The B—N grouping, with (3 + 5) valence electrons, is isoelectronic with C—C. Many boron–nitrogen compounds have been made that are analogues of the corresponding organic compounds: pairs of carbon atoms are replaced by B—N groups (Figure 11.9). Thus, $B_3N_3H_6$ is a structural analogue of benzene, although its chemical properties are rather different. Boron nitride, $(BN)_x$, has a hexagonal layer structure similar to that of graphite, the difference being that in $(BN)_x$ the atoms in one layer lie directly over those in the next. Unlike graphite it is colourless and does not conduct electricity. This may be a factor of the different structure: because boron atoms lie over nitrogen atoms, the π electrons might be localized by $\overset{+}{N}-\overset{-}{B}$ interactions. If $(BN)_x$ is subjected to heat and pressure (1 800 °C and 8 500 atm), its structure changes to a cubic form analogous to diamond, known as borazon; this form is heat-resistant, and finds application as an abrasive.

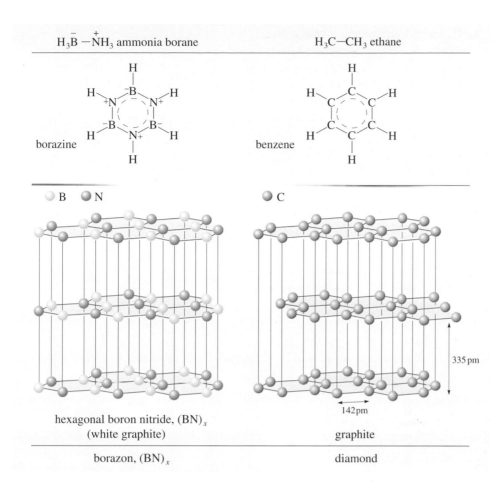

Figure 11.9
Boron–nitrogen compounds and their carbon analogues.

11.2.1 Nitrogen hydrides

Ammonia, NH_3, is a colourless gas with a very strong characteristic smell. It reacts with hydrogen chloride to form a white 'smoke', which is composed of small particles of ammonium chloride (Figure 11.10):

$$NH_3(g) + HCl(g) = NH_4Cl(s) \tag{11.2}$$

It is a reasonably strong Lewis base (electron donor), particularly towards transition metal ions, and forms an alkaline solution in water, often used in household cleaners.

$$NH_3(g) + H_2O(l) = NH_4^+(aq) + OH^-(aq) \tag{11.3}$$

Liquid ammonia (T_b −33 °C) is one of the most widely used non-aqueous solvents. Alkali metals, for example, dissolve reversibly to give blue solutions, in contrast to their more familiar reactions with water. The blue colour is thought to arise from electrons solvated by ammonia molecules. The first observation of this reaction was by Sir Humphry Davy, who in November 1807 wrote in his laboratory notebook, 'when 8 grains of potassium were heated in ammoniacal gas it assumed a beautiful metallic appearance and gradually became a pure blue colour'.

Figure 11.10 A white 'smoke' of ammonium chloride produced by the reaction of ammonia, in the flask, with hydrochloric acid on the rod.

Ammonia is produced industrially by the Haber–Bosch process (see also the *Industrial Inorganic Chemistry* Case Study) which was discovered just before World War I in Germany. It was introduced at a time when naturally occurring sodium nitrate, $NaNO_3$, from Chile was the main raw material for the synthesis of fertilizers. The available nitrate deposits were insufficient to keep pace with demand, and the Haber–Bosch process undoubtedly helped to prevent mass starvation at the time. The completion of the first functioning ammonia plants by the chemical company BASF coincided with the start of the war, and they were also used to fuel the production of explosives for the German and Austro-Hungarian armies.

The Haber–Bosch process involves direct combination of the elements:

$$N_2(g) + 3H_2(g) = 2NH_3(g); \quad \Delta H_m^{\ominus} = -92 \, kJ \, mol^{-1} \tag{11.4}$$

At room temperature the reaction is very slow, which means industrial processes must operate at high temperature (400–450 °C) and high pressure (80–350 atm) and also require the presence of an iron catalyst. The fertilizers ammonium nitrate, NH_4NO_3, and ammonium sulfate, $(NH_4)_2SO_4$ are manufactured by treating ammonia with either nitric or sulfuric acid, respectively.

Hydrazine, N_2H_4, is a fuming colourless liquid (T_m 2 °C; T_b 114 °C). In the gas phase the principal form is the conformation shown in Structure **11.2**. It is produced by the **Raschig process**, in which ammonia is oxidized by sodium hypochlorite:

$$2NH_3(aq) + OCl^-(aq) = N_2H_4(aq) + Cl^-(aq) + H_2O(l) \tag{11.5}$$

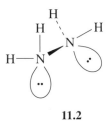

11.2

This reaction proceeds via the formation of chloramine, NH_2Cl, a highly toxic and volatile gas. For this reason, household bleach or scouring powders, which may contain the hypochlorite ion, and ammonia cleansers should never be mixed!

Hydrazine is dangerously explosive and is usually kept in aqueous solution. It burns in air with the evolution of a considerable amount of heat:

$$N_2H_4(l) + O_2(g) = N_2(g) + 2H_2O(g); \quad \Delta H_m^{\ominus} = -621 \, kJ \, mol^{-1} \tag{11.6}$$

For this reason, hydrazine finds application as a rocket fuel, along with its methyl derivatives, $H_2N-N(CH_3)_2$ and $H_2N-NH(CH_3)$. All fuels for this purpose rely on redox reactions to provide the awesome amounts of energy required for thrust to escape the Earth's gravitational field. Indeed, as rockets operate outside the Earth's atmosphere where oxygen is not available, their fuel must have an oxidant and reductant built in. One of several fuels used in the Apollo lunar missions was a 1 : 1 mixture of the methyl derivatives of hydrazine mixed with dinitrogen tetroxide, N_2O_4, which acts as the oxidizing agent; these substances ignite when mixed, producing a highly exothermic redox reaction generating hot gaseous products. Equation 11.7 shows the reaction for the dimethyl derivative:

$$H_2N-N(CH_3)_2(l) + 2N_2O_4(l) = 3N_2(g) + 4H_2O(g) + 2CO_2(g) \tag{11.7}$$

11.2.2 Nitrogen halides

NF_3 is the most stable of the nitrogen halides; it does not react with water or dilute acid. It is prepared by the direct reaction of ammonia and fluorine:

$$4NH_3(g) + 3F_2(g) = 3NH_4F(s) + NF_3(g) \tag{11.8}$$

The trichloride and tribromide are explosively unstable with respect to their constituent elements. Indeed, P. L. Dulong, who first prepared NCl_3 in 1811, lost an eye and three fingers whilst studying its chemical properties!

Until recently nitrogen triiodide (which paradoxically is the longest-known nitrogen halide) had only been prepared in combination with coordinated ammonia. Concentrated aqueous ammonia reacts with iodine to form a black crystalline product of composition $NI_3.NH_3$. When dry, this explodes with the slightest disturbance, a consequence of the strong tendency to form the dinitrogen molecule (Equation 11.9).

$$2NI_3.NH_3(s) = 3I_2(s) + N_2(g) + 2NH_3(g) \qquad (11.9)$$

However, two German chemists, Thomas Klapotke and Inis Tornieporth-Oetting, have now prepared *free* NI_3 by treating iodine monofluoride with boron nitride:

$$BN(s) + 3IF(g) = NI_3(s) + BF_3(g) \qquad (11.10)$$

Because IF disproportionates rapidly to I_2 and IF_5 at room temperature, the synthesis was carried out at $-30\,°C$. NI_3 was isolated as a deep-red solid, which was highly unstable at room temperature, decomposing with explosive violence.

11.2.3 Azides

The hydrogen azide molecule, HN_3 (**11.3**), comprises three almost co-linear nitrogen atoms with appreciably different N—N distances. It is a highly poisonous, colourless liquid, which forms an acidic solution in water:

$$HN_3(l) + H_2O(l) = H_3O^+(aq) + N_3^-(aq) \qquad (11.11)$$

Hydrogen azide is highly explosive, producing a gaseous mixture of hydrogen and nitrogen

$$2HN_3(l) = H_2(g) + 3N_2(g) \qquad (11.12)$$

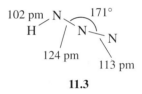

11.3

Indeed, azides in general find widespread use in the explosives industry. As a result of a serious accident in 1893, the Prussian government began an investigation into the use of lead azide, $Pb(N_3)_2$ as a shock-sensitive detonator. Lead azide is an example of a primary explosive, which will readily detonate if subjected to heat or shock. On detonation this is used to initiate a second more stable, and generally more powerful secondary explosive, e.g. TNT.

In contrast, azides are also lifesavers. One of the most common safety features in modern cars is the airbag.* If a car is involved in a collision, this inflates following the controlled explosive decomposition of sodium azide to produce a large amount of nitrogen gas:

$$2NaN_3(s) = 2Na(l) + 3N_2(g) \qquad (11.13)$$

11.2.4 Nitrogen–oxygen compounds

The nitrogen oxides, oxoions and oxoacids form a remarkable variation on the theme of σ and π bonding. All are linear or flat, with π orbitals covering the whole molecule. There are seven molecular oxides, all of which are thermodynamically unstable with respect to the formation of N_2 and O_2: their structures (showing both atom linkages and most likely resonance forms) and physical properties are summarized in Table 11.2 (overleaf). At first sight you may find the large range of nitrogen oxides fairly confusing, but by keeping track of the nitrogen oxidation numbers, their stoichiometry and chemistry become clearer.

---

* See *Chemical Kinetics and Mechanism*.[4]

Table 11.2 The oxides of nitrogen

Formula	Name (trivial name in parentheses)	Linkage/structure			Physical properties
N_2O	dinitrogen monoxide (nitrous oxide)	N—N—O, linear, $\mathbf{C}_{\infty v}$			colourless gas (T_b −89 °C)
NO	nitrogen monoxide (nitric oxide)	N—O, linear, $\mathbf{C}_{\infty v}$ N_2O_2, dimer, $\mathbf{C}_{2v}$ 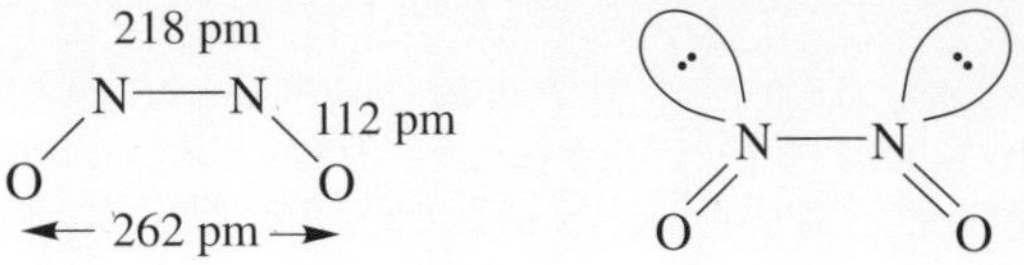			colourless paramagnetic gas (T_b −152 °C)

11.4

Formula	Name (trivial name in parentheses)	Linkage/structure			Physical properties
N_2O_3	dinitrogen trioxide	planar, $\mathbf{C}_s$			blue solid (T_m −101 °C)

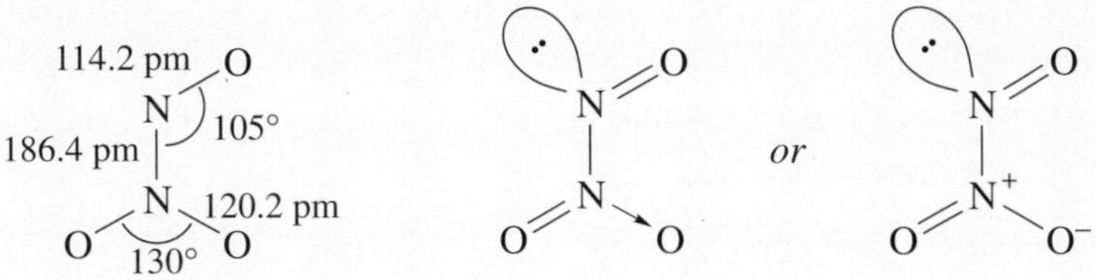

11.5

Formula	Name (trivial name in parentheses)	Linkage/structure			Physical properties
NO_2	nitrogen dioxide	bent, $\mathbf{C}_{2v}$			brown paramagnetic gas

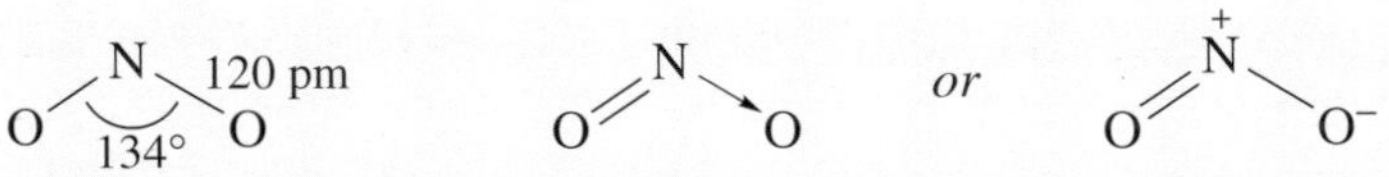

11.6

Formula	Name (trivial name in parentheses)	Linkage/structure			Physical properties
N_2O_4	dinitrogen tetroxide	planar, $\mathbf{D}_{2h}$			colourless liquid (T_m −11 °C)

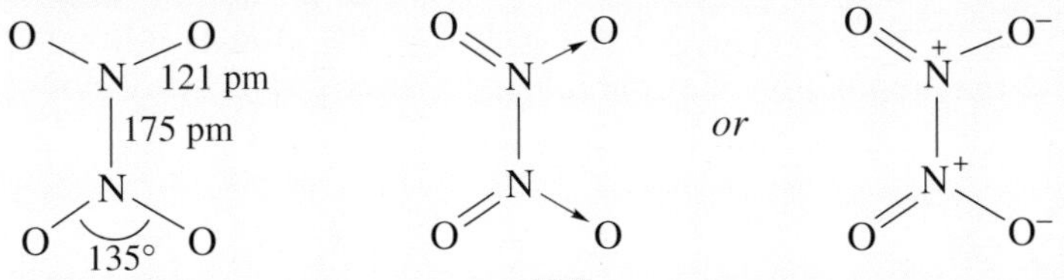

11.7

Formula	Name (trivial name in parentheses)	Linkage/structure			Physical properties
N_2O_5	dinitrogen pentoxide	ionic, $[NO_2]^+[NO_3]^-$			colourless solid (sublimes at 32 °C)
NO_3	nitrogen trioxide	trigonal planar, $\mathbf{C}_{3v}$			unstable paramagnetic radical

Dinitrogen monoxide

Dinitrogen monoxide, N_2O (nitrogen oxidation number +1), is a non-toxic, odourless and tasteless gas; it is a linear molecule with an unsymmetrical $N-N-O$ linkage. N_2O is made by *careful* thermal decomposition of ammonium nitrate at 250 °C:

$$NH_4NO_3(s) = N_2O(g) + 2H_2O(l) \tag{11.14}$$

N_2O is the only gas apart from oxygen that will relight a glowing splint, although strictly speaking this is due to the decomposition to nitrogen and oxygen that takes place above 500 °C, i.e. it is the oxygen that is responsible for rekindling the splint:

$$2N_2O(g) = 2N_2(g) + O_2(g) \tag{11.15}$$

Nitrogen monoxide

Nitrogen monoxide, NO (nitrogen oxidation number +2), is one of the most reactive of the nitrogen oxides. It is an odd-electron molecule and is therefore paramagnetic; it has a bond order of $2\frac{1}{2}$.

NO is a monomeric, colourless gas, which reacts immediately with atmospheric oxygen at room temperature to give the characteristic brown colour of gaseous NO_2; it is this reaction that is largely responsible for producing NO_2 in polluted air (the nitrogen oxides emitted by car exhausts are often given the general formula NO_x). Yet NO plays a significant role in nature (see Box 11.2). When NO is cooled to a liquid, some N_2O_2 dimers are formed; as indicated in Structure **11.4** in Table 11.2, these mainly have the *cis* structure. (The bond lengths come from X-ray diffraction studies of the solid.)

NO is prepared in the laboratory by using a mild reducing agent on a nitrogen–oxygen compound in which the nitrogen has a higher oxidation number. Thus, Equation 11.16 shows the formation of nitrogen monoxide from nitrite, NO_2^-, and iodide in acidic solution:

$$NO_2^-(aq) + I^-(aq) + 2H^+(aq) = NO(g) + \tfrac{1}{2}I_2(aq) + H_2O(l) \tag{11.16}$$

Dinitrogen trioxide

Dinitrogen trioxide, N_2O_3 (nitrogen oxidation number +3), can be isolated only at low temperatures (T_m −101 °C) as a blue solid and deep blue liquid, when stoichiometric amounts of NO and NO_2 are combined. The molecule is planar (Structure **11.5** in Table 11.2). As the temperature is raised, the liquid becomes greenish as it disproportionates to NO and the brown NO_2:

$$N_2O_3(l) = NO(g) + NO_2(g) \tag{11.17}$$

Nitrogen dioxide and dinitrogen tetroxide

Nitrogen dioxide, NO_2 (nitrogen oxidation number +4), is an odd-electron molecule (Structure **11.6** in Table 11.2) and is therefore paramagnetic. The non-bonded electron resides mainly on the nitrogen, which enables two molecules to interact to form an $N-N$ bond. Thus, when NO_2 is cooled to form a liquid, it forms the dimeric N_2O_4 (Structure **11.7** in Table 11.2).

At the boiling temperature of N_2O_4 (21.5 °C) the mixture contains about 16% NO_2:

$$2NO_2(g) = N_2O_4(l) \tag{11.18}$$

Nitrogen dioxide can be prepared by the reaction of copper metal with concentrated nitric acid (Figure 11.12):

$$Cu(s) + 4HNO_3(l) = Cu(NO_3)_2(aq) + 2H_2O(l) + 2NO_2(g) \tag{11.19}$$

BOX 11.2 Just say NO

For many years nitrogen monoxide was a molecule with an unsavoury reputation; a destroyer of ozone, a supposed carcinogen and precursor of acid rain. However, in a remarkable Cinderella story, we now realize that NO, despite its simplicity, is one of the most important functional molecules in our bodies. In fact, when its numerous biochemical roles were established, the prestigious monthly journal *Science* named NO its 'molecule of the year' for 1992, and the oxide was the subject of a BBC TV *Horizon* programme the following year.

Its major function is in the regulation of blood pressure. A blood vessel is composed of layers of muscular, elastic and fibrous tissue, and has a lining called the endothelium (Figure 11.11). Nitric oxide is released by the endothelium and enters the surrounding muscle layers. At the cellular level, NO stimulates the enzyme guanylate cyclase to synthesize cyclic guanosine monophosphate (cyclic GMP), an important messenger molecule that initiates a range of biological responses. These include muscle relaxation and hence widening of the bore of the blood vessel, thus lowering blood pressure.

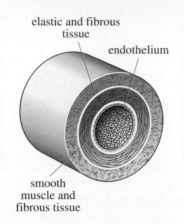

Figure 11.11
A schematic representation of a blood vessel.

In contrast to its more destructive applications, nitroglycerine is an effective drug taken by sufferers of the heart condition angina pectoris to ease the often crippling pain. Nitroglycerine and similar drugs release NO when they react with other chemicals in the bloodstream, and reduce blood pressure. Given the obvious anxiety associated with swallowing a known high explosive, the alternative name glyceryl trinitrate (GTN) is used by the pharmaceutical industry! Somewhat ironically, Alfred Nobel himself, who first prepared dynamite by mixing nitroglycerine with clay, was prescribed it by his doctor.

In males, NO is the messenger that translates sexual excitement into an erect penis. This is brought about by the dilation of blood vessels supplying the penis and relaxation of the muscles of the corpora cavernosa — the cavities that the blood fills. The latter effect, initiated by release of NO, results in increased levels of cyclic GMP. Indeed, experiments have shown that rats can be rendered impotent by administering chemicals that impede production of NO. To combat impotence (erectile dysfunction) the drug Viagra is widely prescribed. Its function is to inhibit the enzyme phosphodiesterase, which is responsible for the degradation of cyclic GMP in the corpora cavernosa.

Paradoxically, NO is also highly poisonous. It binds to haemoglobin in the blood in a similar manner to carbon monoxide (Figure 10.14, p. 138), and if inhaled reacts with oxygen and water in the tissues to form nitric acid. Indeed, during his early studies of NO, Sir Humphrey Davy suffered burning to his palate and tongue on inhaling the gas, an effect he reported as "a spasm of the epiglottis so painful as to oblige me to desist immediately".

(a)　　　　　　　　　　　　(b)

Figure 11.12 The production of NO_2 by the reaction between copper metal in a coin and concentrated nitric acid: (a) before and (b) after addition of the acid.

In addition, NO_2 is produced, together with oxygen, when nitrates of heavy metals are heated. Lead nitrate is most commonly used in this reaction, as it crystallizes with negligible water of crystallization (Figure 11.13):

$$2Pb(NO_3)_2(s) = 2PbO(s) + 4NO_2(g) + O_2(g) \qquad (11.20)$$

NO_2 and N_2O_4 are not only toxic but also corrosive, because they react with water to form nitric acid:

$$N_2O_4(g) + H_2O(l) = HNO_3(aq) + HNO_2(aq) \qquad (11.21)$$
$$3HNO_2(aq) = HNO_3(aq) + 2NO(g) + H_2O(l) \qquad (11.22)$$

Dinitrogen pentoxide

Dinitrogen pentoxide, N_2O_5 (nitrogen oxidation number +5), is the true anhydride of nitric acid; it can be made by dehydrating concentrated nitric acid with P_4O_{10} (Section 11.3.4) at low temperature:

$$4HNO_3(aq) + P_4O_{10}(s) = 2N_2O_5(s) + 4HPO_3(aq) \qquad (11.23)$$

It is a highly reactive, colourless solid, which has been shown to have an ionic structure, containing NO_2^+ and NO_3^- ions.

Nitrogen trioxide

For completeness, mention should also be made of nitrogen trioxide, NO_3: this fleeting radical species has been identified from its absorption spectrum, but has never been isolated as a pure compound.

Nitrous acid and nitrites

Nitrous acid, HNO_2, has not been isolated as a pure compound, although it is observed in equilibrium gaseous mixtures. It is a planar molecule, apparently preferring the *trans* structure shown in Structure **11.8**.

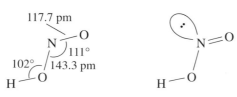

11.8

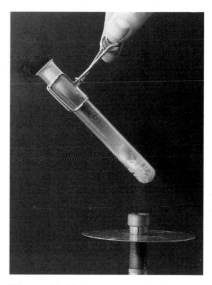

Figure 11.13
The evolution of NO_2 on heating lead nitrate.

STUDY NOTE

Nitrogen forms many oxoacids, and oxoacid salts; we consider here only the two most important.

Nitrous acid may be prepared readily by the reaction of dilute acid with metal nitrites:

$$NO_2^-(aq) + H^+(aq) = HNO_2(aq) \qquad (11.24)$$

However, at room temperature it rapidly decomposes to nitric acid and nitrogen monoxide, the latter reacting further with oxygen to yield brown fumes of nitrogen dioxide:

$$3HNO_2(aq) = HNO_3(aq) + 2NO + H_2O \qquad (11.25)$$

$$2NO(g) + O_2(g) = 2NO_2(g) \qquad (11.26)$$

Despite being slightly toxic to humans, sodium nitrite is widely used in meat processing. Nitrite forms a pink complex with haemoglobin, and inhibits the oxidation of blood, a reaction that turns meat brown. Nitrites are also responsible for the pink colour of cured bacon, ham and sausages.

Nitric acid and nitrates

Nitric acid, HNO_3, is made on a huge industrial scale by the **Ostwald process** (see also the Case Study *Industrial Inorganic Chemistry*). Ammonia is oxidized in two stages over a catalyst made from platinum metal, first to NO and then to NO_2. The NO_2 is then dissolved in water to give a concentrated aqueous solution of the acid, and the NO produced in this step is recycled back into earlier stages. The steps in the reaction can be summarized by the following equations:

$$NH_3(g) + \tfrac{5}{4}O_2(g) = NO(g) + \tfrac{3}{2}H_2O(g) \qquad (11.27)$$

$$NO(g) + \tfrac{1}{2}O_2(g) = NO_2(g) \qquad (11.28)$$

$$NO_2(g) + \tfrac{1}{3}H_2O(l) = \tfrac{2}{3}HNO_3(aq) + \tfrac{1}{3}NO(g) \qquad (11.29)$$

The anhydrous acid can be produced by distillation, and is a colourless pungent liquid. The nitric acid molecule is planar, as shown in Structure **11.9**.

11.9

In dilute aqueous solution, nitric acid behaves as a typical strong acid, being extensively dissociated into H^+ and NO_3^- ions.

QUESTION 11.1

Predict the shape of the nitrate NO_3^- by using VSEPR theory. What is its symmetry point group?

Nitric acid has many uses, but most is combined with ammonia to produce ammonium nitrate, an important fertilizer (Figure 11.14).

On gentle heating, ammonium nitrate decomposes to N_2O (Equation 11.14), but on heating above about 250 °C, an alternative explosive decomposition occurs:

$$2NH_4NO_3(s) = 2NO(g) + N_2(g) + 4H_2O(g) \qquad (11.30)$$

Figure 11.14
Ammonium nitrate is one of the most commonly used fertilizers.

Consequently, ammonium nitrate fertilizer has to be packed and handled with extreme care; in addition to being thermally unstable at high temperatures, its decomposition is catalysed by many inorganic and organic materials. Strict regulations governing the storage and transportation of the chemical were introduced, following serious explosions aboard two American ships, the *SS Grandchamp* and *SS Highflyer*, being used to transport fertilizer-grade ammonium nitrate to Europe in 1947. Today, ammonium nitrate is used extensively in industrial explosives and propellants. Commercial blasting explosive contains ammonium nitrate, fuel oil and TNT.

Nitrates of almost all metallic elements are known. They are all soluble in water, but differ in their mode of decomposition on heating. Potassium and sodium nitrates form the nitrite and oxygen gas:

$$2KNO_3(s) = 2KNO_2(s) + O_2(g) \tag{11.31}$$

Other metal nitrates tend to decompose to the metal oxide, nitrogen dioxide and oxygen (see Equation 11.20). However, because their metal oxides are unstable to heat, the nitrates of silver and mercury decompose to the metal, nitrogen dioxide and oxygen:

$$2AgNO_3(s) = 2Ag(s) + 2NO_2(g) + O_2(g) \tag{11.32}$$

$$Hg(NO_3)_2(s) = Hg(l) + 2NO_2(g) + O_2(g) \tag{11.33}$$

Both nitrates and nitrites can be reduced to ammonia in basic solution with zinc or Dervada's alloy (a combination of aluminium, zinc and copper). This is a common qualitative test for the anions, no simple precipitation from aqueous solution being possible.

$$3NO_3^-(aq) + 8Al(s) + 5OH^-(aq) + 18H_2O(l) = 3NH_3(g) + 8Al(OH)_4(aq) \tag{11.34}$$

The evolution of ammonia may be confirmed because it turns damp red litmus paper blue.

11.2.5 Summary of Sections 11.1 and 11.2

1   Nitrogen, N_2, is an unreactive, gaseous diatomic molecule with a triple bond. It has very high bond dissociation and ionization energies.

2   In combination with other elements, nitrogen is found in many oxidation numbers, from −3 to +5.

3   The Group II metal nitrides M_3N_2, are salt-like in character, but $(BN)_x$ has a macromolecular structure and is refractory in nature.

4   Ammonia, NH_3, is produced in great amounts by the Haber–Bosch process, largely to make fertilizers.

5   Hydrazine, N_2H_4, burns rapidly in air with the evolution of a considerable amount of heat. Along with its methyl derivatives, it is used as rocket fuel.

6   The trihalides of nitrogen are all known, but tend to be rather unstable.

7   The hydrogen azide molecule, HN_3, possesses three co-linear nitrogen atoms, with the hydrogen at an angle of 110°. Azides find widespread application in the explosives industry.

8   Nitrogen forms seven molecular oxides; all the molecules are either linear or planar as a result of 2p–2p π bonding. They are all thermodynamically unstable with respect to the elements.

9   Nitric acid is produced industrially by the Ostwald process. The majority of HNO_3 is used in the manufacture of fertilizers.

11.3 Phosphorus

A ruined businessman from Hamburg, Hennig Brandt, prepared phosphorus in 1669 by the rather revolting method of distilling several litres of horses' urine in the presence of charcoal. Reports vary as to his motive, but it is thought he was either searching for gold or even the 'Elixir of Eternal Life'. Subsequently, phosphorus was exhibited around the courts of Europe because of its ability to glow continuously, without heat, in the dark.

Since all living matter contains phosphate, PO_4^{3-}, in one form of another, phosphorus can in principle be obtained from any part of a plant or animal. Heating phosphate (most commonly as the calcium salt) with carbon — as a reducing agent — releases phosphorus as the volatile white phosphorus, P_4:

$$2Ca_3(PO_4)_2(s) + 10C(s) = P_4(s) + 10CO(g) + 6CaO(s) \qquad (11.35)$$

BOX 11.3 Phosphorus and the match industry

Until the discovery of phosphorus, people generally relied on sparks produced by striking together steel and flint to produce fire. The fact that white phosphorus ignites in air provided an obvious alternative. Early matches were both crude and highly dangerous. In the 1780s, the *ethereal match* was in vogue. This comprised a piece of paper with a white phosphorus tip. This was sealed in a glass tube until needed, whereupon the tube was broken and the phosphorus immediately ignited, setting fire to the paper.

The use of white phosphorus was eventually banned by international agreement, because of the terrible industrial diseases it caused. Breathing air that was contaminated with phosphorus vapour could eventually lead to tooth decay that ate its way down to the jaw bone (known as 'phossy jaw'). Because matches were first made by hand in the homes of the poor, the disease exacted a fearful toll among early workers.

Today's match industry is a much safer business, the products being wonderful examples of chemical ingenuity. These days you can buy two types of match, friction matches and safety matches. The friction or 'strike anywhere' match was invented by British chemist John Walker, and can be struck on any rough surface. The match head consists of a mixture of potassium chlorate (an oxidizing agent) and a reducing agent tetraphosphorus trisulfide, P_4S_3. The latter ignites when heated by friction and sets fire to the rest of the match.

Safety matches can only be struck on the side of the box, owing to separation of the combustible ingredients between the match head and the striking surface. They were developed in 1885 by a Swede, Johan Lundstrom, and again phosphorus is a key ingredient. The striking surface contains red phosphorus, which is converted into white phosphorus by the heat of friction. The latter spontaneously ignites in air and sets the match tip alight. The match head is a mixture of antimony sulfide, which burns at a fairly low temperature, and potassium chlorate, which supplies oxygen to promote the combustion reaction.

11.3.1 The chemistry of phosphorus

The electronic configuration of phosphorus is [Ne] $3s^2 3p^3$, and the common oxidation numbers of phosphorus combined with electronegative elements are +3 and +5. In oxidation number +3, phosphorus uses the three 3p electrons for bonding, leaving a non-bonded pair of electrons. Because phosphorus is a third-row element, it can have a coordination number greater than four, but is also found with covalent bonding arrangements involving two, three, four, five or six atoms attached to phosphorus (Figure 11.15).

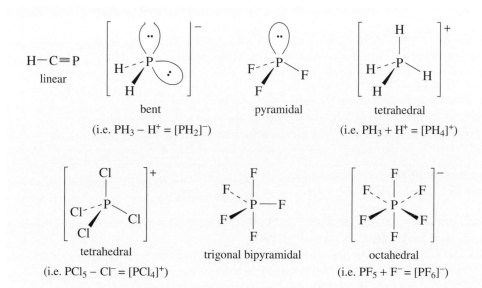

Figure 11.15
A selection of phosphorus-containing compounds, which illustrate the various coordination numbers that phosphorus can adopt in its compounds.

In the trifluoride, PF_3, the oxidation number of phosphorus is +3.

○ Account for the geometry of PF_3 depicted in Figure 11.15.

○ The three P—F single bonds and one non-bonded pair constitute four repulsion axes, leading to a pyramidal structure. The effect of the non-bonded pair will be to contract the tetrahedral angle slightly.

Phosphorus is found in oxidation number +5 in species such as $[PCl_4]^+$, PF_5 and $[PF_6]^-$.

○ How does the shape of $[PF_6]^-$ indicated in Figure 11.15 accord with that predicted from VSEPR theory?

○ There are 12 valence electrons (five from phosphorus, six from the fluorines and one negative charge). These are distributed to form six single bonds to fluorine in a regular octahedral shape. *Note the expansion of the octet, which is possible for third- and later-row elements.*

As we shall see in the following Sections, phosphorus also forms a number of **cage structures** based on expansion of the P_4 tetrahedron, containing three- or four-coordinate phosphorus. It has also been found to form ring or chain structures from linked tetrahedra, with oxygen or nitrogen substituents on phosphorus.

11.3.2 Phosphorus halides

Direct reaction of phosphorus and halogen gives the trihalide (except PF_3) if phosphorus is kept in excess. For example, phosphorus and chlorine give PCl_3, which is a volatile, easily hydrolysed, fuming liquid:

$$P_4(s) + 6Cl_2(g) = 4PCl_3(l) \tag{11.36}$$

Many other phosphorus compounds can be made from PCl_3. Industrially, it is important as a source of organophosphorus compounds for oil and fuel additives, plasticizers and insecticides.

PF_3 is a colourless, odourless gas formed by the action of a metal fluoride such as CaF_2 on PCl_3:

$$3CaF_2(s) + 2PCl_3(l) = 2PF_3(g) + 3CaCl_2(s) \tag{11.37}$$

It is poisonous because it coordinates to the haemoglobin in blood in a similar fashion to CO and NO (Figure 10.14). The trihalides are all rather volatile; even PI_3, which forms red crystals, has a melting temperature of 61 °C. All the trihalides are hydrolysed by water (although the hydrolysis of PF_3 is slow). The gaseous trihalides are pyramidal in shape.

All the PX_5 (where X is a halogen) species can be formed; for example:

$$P_4(s) + 10F_2(g) = 4PF_5(g) \tag{11.38}$$

However, although phosphorus pentachloride, PCl_5, has the trigonal bipyramidal shape in the gaseous and liquid states, at room temperature it forms off-white crystals containing the ionic species $[PCl_4]^+$ $[PCl_6]^-$. Phosphorus pentabromide forms reddish-yellow crystals, containing the ions $[PBr_4]^+$ Br^-, but dissociates in the gas phase to PBr_3 and Br_2; the pentaiodide is thought to be similar.

The diphosphorus tetrahalides, P_2X_4, are also known, as are many mixed halide compounds.

11.3.3 Phosphorus hydrides

Phosphine, PH_3, is a very reactive and poisonous gas, it tends to inflame spontaneously in air and has been described as smelling of both garlic and rotten fish! PH_3 is much less basic than NH_3, as the P—H bond is much less polar than the N—H bond. Unlike ammonia, phosphine does not form hydrogen bonds. Although **phosphonium salts**, $[PH_4]^+X^-$, are known, only PH_4I is stable at room temperature; the other phosphonium halides dissociate to PH_3 and HX.

Phosphine may be prepared by the hydrolysis of the phosphides of electropositive metals, such as calcium phosphide:

$$Ca_3P_2(s) + 6H_2O(l) = 2PH_3(g) + 3Ca(OH)_2(aq) \tag{11.39}$$

Further products of the hydrolysis of calcium phosphide are P_2H_4, P_3H_5 and other hydrides, all of which contain P—P bonds. These are very sensitive to air and also are very unstable, tending to decompose to PH_3 and phosphorus-rich polymers. The most stable is P_7H_3, which has the cage structure shown in Figure 11.16. Its stability is due to the wide separation of the hydrogen atoms, so that PH_3 is not lost readily.

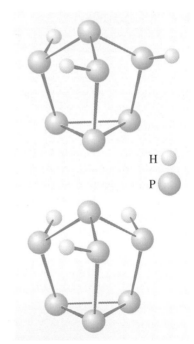

H

P

Figure 11.16
Two isomers of P_7H_3, which arise from different orientations of hydrogen atoms and non-bonded pairs of electrons on the three phosphorus atoms in the middle of the cage.

11.3.4 The oxides and sulfides of phosphorus

Phosphorus reacts readily with oxygen and sulfur. Controlled reaction of phosphorus with oxygen gives P_4O_6, a volatile molecular compound with a cage structure in which oxygen atoms are inserted into each P—P bond of P_4 (Figure 11.17a). This is commonly known as phosphor*ous* oxide (note the different spelling from the element, which is phosphor*us*). Further oxidation adds terminal oxygen atoms stepwise, giving P_4O_{10}, phosphor*ic* oxide, as the final oxidation product. P_4O_{10} has the cage structure shown in Figure 11.17b.

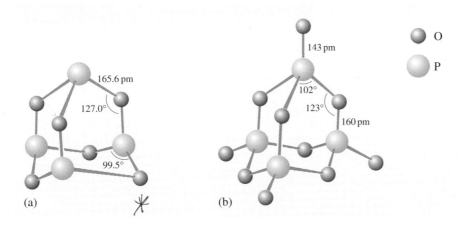

Figure 11.17
The structures of
(a) phosphorous oxide, P_4O_6, and
(b) phosphoric oxide, P_4O_{10}.

(You will also come across the names phosphorus trioxide and pentoxide, respectively, for P_4O_6 and P_4O_{10}. This stems from their empirical formulae, P_2O_3 and P_2O_5, which were known long before their structures were determined.)

● What are the oxidation numbers of phosphorus in P_4O_6 and P_4O_{10}?

○ The oxidation numbers are, respectively, +3 and +5.

● What are the coordination numbers of phosphorus in the two compounds?

○ Phosphorus in P_4O_6 and P_4O_{10} has coordination numbers of three and four, respectively.

In the chemistry of the elements of the third row of the Periodic Table, we have seen many single-bonded structures, both in the elemental forms and in the oxides, which contrast with the multiple-bonded structures formed by the second-row elements; for example, SiO_2 may be contrasted with CO_2, and P_4O_6 with N_2O_3. However, P_4O_{10} shows an important difference: each phosphorus has a terminal oxygen, and the terminal phosphorus–oxygen bond length is much shorter (143 pm) than the bridging phosphorus–oxygen bond (160 pm; see Figure 11.17); hence the terminal phosphorus–oxygen groups may be thought of as double bonded. We shall see similar behaviour in the chemistry of sulfur.

P_4O_6 and P_4O_{10} are acid anhydrides of their respective oxoacids: when P_4O_6 reacts with water, it forms phosphorous acid (Equation 11.40), whereas P_4O_{10} forms phosphoric acid (Equation 11.41):

$$P_4O_6(s) + 6H_2O(l) = 4H_3PO_3(aq) \qquad (11.40)$$

$$P_4O_{10}(s) + 6H_2O(l) = 4H_3PO_4(aq) \qquad (11.41)$$

The reaction of P_4O_{10} with water is very vigorous. The 'craving' of the phosphoric oxide for water is so great that it can be used as a **dehydrating agent**; that is, it will extract the elements of water from some compounds. An example of the use of the dehydrating property of P_4O_{10} in synthetic chemistry is the formation of a nitrile from an amide:

$$R-\overset{\displaystyle O}{\underset{\displaystyle NH_2}{C}} \quad \xrightarrow[-H_2O]{P_4O_{10}} \quad R-C\equiv N \qquad (11.42)$$

an amide → a nitrile

We have already seen how P_4O_{10} is used to dehydrate nitric acid in the preparation of dinitrogen pentoxide (Equation 11.23).

Reaction of phosphorus with a limited supply of sulfur yields P_4S_3, which, as we have already mentioned (Box 11.3), is widely used in the match industry. P_4S_3 has the cage structure shown in Figure 11.18. Further reaction with sulfur gives phosphorus sulfides with fewer P—P bonds and more P—S—P groups: the ultimate reaction product is phosphorus(V) sulfide, P_4S_{10}, which has the same structure as P_4O_{10} (Figure 11.17b); it contains P—S—P bonds and a terminal P=S bond on each phosphorus atom.

11.3.5 Phosphoric acid

Phosphoric acid is manufactured from phosphate minerals, and the pure acid forms low-melting crystals (T_m 42 °C); the structure of the H_3PO_4 molecule is shown in Structure **11.10**. Commercial phosphoric acid is 85% phosphoric acid in water; this forms a syrup, because the acid molecules are hydrogen-bonded to water molecules. The majority of phosphoric acid is used in fertilizer manufacture. It has also been widely used in metal treatment since the British patent of 1869, which was granted for the prevention of rusting in corset stays by bodily perspiration. By incorporating metal ions into the solution (typically Mn, Fe and Zn) a phosphatizing solution capable of imparting corrosion resistance to a wide variety of metal objects may be produced. Highly dilute solutions of phosphoric acid are non-toxic, and are used to impart the characteristic sour or tart taste in colas and other carbonated beverages.

11.3.6 Compounds with multiple bonds between phosphorus atoms and from phosphorus to carbon

The P_2 molecule, with a P≡P triple bond, is unstable except at high temperatures, and P_4 and other forms of phosphorus with P—P single bonds are more stable at ordinary temperatures. This behaviour is in contrast to that of nitrogen, which is stable only as N_2 molecules. The difference between the two elements is a consequence of 2p–2p π bonding being more effective than 3p–3p π bonding, and of the N—N σ bond being weaker than the P—P σ bond. The bond enthalpy terms are listed in Table 11.3.

Inspection of these data shows that:

$B(N\equiv N) > 3B(N-N)$, but

$B(P\equiv P) < 3B(P-P)$

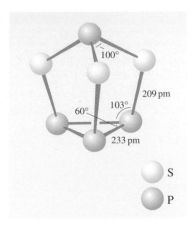

Figure 11.18
The structure of P_4S_3. This may be viewed as the P_7H_3 cage with each PH unit replaced by S. In fact the compounds P_7H_3 and P_4S_3 are isoelectronic.

$$\underset{HO}{\overset{157\ pm}{HO}}{\cdots}\overset{\displaystyle \overset{O}{\parallel}}{\underset{}{P}}\overset{152\ pm}{\underset{OH}{}}$$

11.10

Table 11.3 Molar bond enthalpy terms for some nitrogen and phosphorus bonds

Bond	B/kJ mol^{-1}
N≡N	945
N—N	158
P≡P	485
P—P	198

Early attempts to make compounds containing P=P double bonds that were stable under normal conditions all failed: only single-bonded compounds were formed. For instance, one reaction that was tried is shown in Equation 11.43. The object was the reduction of methylphosphorus dichloride with sodium:

$$2\ CH_3PCl_2\ +\ 4\ Na \begin{cases} 4\ NaCl\ +\ H_3CP{=}PCH_3 \\[1em] 4\ NaCl\ +\ 2/n\left(\!CH_3P\!\right)_n \end{cases} \qquad (11.43)$$

The compound $CH_3P{=}PCH_3$ probably forms as an intermediate, but is unstable with respect to polymerization.

● From your knowledge of silicon chemistry, how would you try to stabilize a P=P bond?

○ The bulky mesityl group stabilized the Si=Si bond, so it would seem reasonable to try the same strategy here.

In the late 1970s it was discovered that the reaction scheme in Equation 11.43 would produce the double-bonded product if the methyl group was changed for a much bulkier organic group, such as tri-*t*-butylphenyl group, [*]as shown in Structure **11.11**.

● Why should a bulky substituent prevent the formation of P—P bonded polymers?

○ As we saw earlier in the formation of Si=Si double bonds, large organic groups introduce kinetic barriers, and prevent two molecules coming close enough together to react and form a polymer.

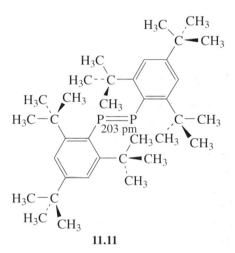

11.11

The P—P bond length in the compound shown is 203 pm, much shorter than a typical P—P bond length of 220 pm. However, the bulky groups provide no protection from attack by small molecules, and so, for instance, the compounds are air-sensitive because of reaction with oxygen.

Bulky substituent groups have also been used to stabilize compounds containing phosphorus–carbon double or triple bonds, the bonding in this case is 2p–3p. Here again, polymerization tends to occur. The linear H—C≡P molecule is stable under low pressure in the gas phase, and has been detected in interstellar space, but it polymerizes rapidly when condensed. The tertiary-butyl analogue of this compound, $(CH_3)_3CC{\equiv}P$, is stable to polymerization at room temperature, although it is more chemically reactive than the corresponding alkyne, $(CH_3)_3CC{\equiv}CH$.

11.3.7 Phosphorus–nitrogen compounds: the polyphosphazenes

● The atom combination PN is isoelectronic with which Si species?

○ SiO.

---

[*] The use of bulky substituents, such as the tertiary-butyl (*t*-butyl) group, is discussed in *Chemical Kinetics and Mechanism*.[4]

We might expect, therefore, to find phosphorus/nitrogen polymers, $(R_2PN)_n$, analogous to the silicones, $(R_2SiO)_n$, that we looked at in Section 10.3. These polymers are indeed known and are called **polyphosphazenes**; in recent years they have found various practical applications.

The parent compound of the phosphazenes is the cyclic compound, $Cl_6P_3N_3$ shown in Equation 11.44. The ring is almost planar, and although we have drawn it with alternating single and double bonds around the ring, this is just a formalism, and the experimental results show all the P—N bonds to be equal in length and rather shorter than expected for a P—N single bond. There is much debate about the bonding in these compounds because it has been found to be rather different from the delocalized pπ–pπ systems such as found in benzene. The question centres on whether or not 3d orbitals are contributing to the bonding.

$$\text{cyclic } Cl_6P_3N_3 \xrightarrow{\text{heat}} \text{high polymer} \xrightarrow{\text{NaOR}} [(RO)_2PN]_n \text{ useful high polymer}$$

(11.44)

$Cl_6P_3N_3$ is made by heating together ammonium chloride, NH_4Cl, and phosphorus pentachloride, PCl_5. It can be polymerized by heating when it is very pure. It forms a high molecular mass rubbery solid $(Cl_2PN)_n$, which is rather sensitive to water and air – not very useful! However, the P—Cl groups can be replaced by others which are more stable. Thus if the —Cl groups are replaced by —OR groups (Equation 11.44), where —OR is a mixture of —OCH_2CF_3 and —$OCH_2(CF_2)_nCF_2H$, the resulting polymers, $[(RO)_2PN]_n$, have outstanding resistance to organic solvents and retain elasticity at low temperatures.

11.3.8 Summary of Section 11.3

1   There are several allotropes of phosphorus. White phosphorus has the molecular structure P_4, whereas the red, violet and black forms contain polymeric structures.

2   Phosphorus is able to expand the octet in its bonding and, like other third-row elements, can increase its coordination number to six.

3   The phosphorus trihalides are rather volatile compounds which contain pyramidal molecules. All the phosphorus pentahalides can be formed; the fluoride has a molecular structure, whereas the chloride and bromide both have ionic structures in the solid: $[PCl_4]^+[PCl_6]^-$ and $[PBr_4]^+Br^-$, respectively.

4   Phosphine, PH_3, is the compound analogous to ammonia, but is very reactive, inflaming spontaneously in air.

5   Reaction of phosphorus with a limited supply of sulfur yields P_4S_3, an industrially important compound used in match-heads.

6   Reaction of phosphorus with a limited supply of oxygen yields P_4O_6, with its characteristic cage structure. Further oxidation adds terminal oxygen atoms to each phosphorus atom to give P_4O_{10}, a useful dehydrating agent.

7   It has proved possible to make compounds containing the P=P bond by stabilizing them with a bulky group. This prevents the close approach of molecules, which is necessary for polymerization to occur.

8   Polyphosphazenes, $[(RO)_2PN]_n$ (analogous to silicones), can be made, and have a number of practical uses.

11.4 Oxoacids

There are several oxoacids of phosphorus, some of which we have already mentioned. The three most important are phosphoric acid, H_3PO_4 (also known as orthophosphoric acid), phosphorous acid, H_3PO_3, and hypophosphorous acid, H_3PO_2. Other names for these acids are listed in Table 11.4.

● What is the oxidation number of the phosphorus in each of these three acids?

● The oxidation numbers are +5, +3 and +1, respectively.

Table 11.4 Common phosphorus oxoacids and their anions

Formula	Common name	Common anion name	IUPAC systematic nomenclature[*]	Structure
H_3PO_4	phosphoric acid or orthophosphoric acid	phosphate or orthophosphate	trihydrogen tetraoxophosphate	
H_3PO_3	phosphorous acid or phosphonic acid	phosphite or phosphonate	dihydrogen hydridotrioxo-phosphate	
H_3PO_2	hypophosphorous acid or phosphinic acid	hypophosphite or phosphinate	hydrogen dihydrido-dioxophosphate	
$H_4P_2O_7$	diphosphoric acid	diphosphate	tetrahydrogen μ-oxo-hexaoxodiphosphate	
$(HPO_3)_n$ (in the limit where $n = \infty$)	metaphosphoric acid	metaphosphate	poly[hydrogen trioxophosphate]	

[*] For an explanation see Section 11.4.2.

Some acids are **polybasic**, which means that they ionize stepwise (compare this with say hydrochloric acid, HCl, which is a *monobasic acid*). If we consider phosphoric acid, we observe that it is tribasic — it undergoes three successive ionizations:

$$H_3PO_4(aq) = H^+(aq) + H_2PO_4^-(aq) \quad \text{DIHYDROGEN PHOSPHATE} \qquad (11.45)$$

$$H_2PO_4^-(aq) = H^+(aq) + HPO_4^{2-}(aq) \quad \text{HYDROGEN PHOSPHATE} \quad \text{SALTS} \qquad (11.46)$$

$$HPO_4^{2-}(aq) = H^+(aq) + PO_4^{3-}(aq) \quad \text{PHOSPHATE} \qquad (11.47)$$

Because phosphoric acid is tribasic, it is possible to form three series of salts with a metal such as sodium: the dihydrogen phosphate, the hydrogen phosphate and the normal phosphate.

● What would you expect to be the sodium salts of phosphorous acid?

● Phosphorous acid, H_3PO_3, is dibasic and its sodium salts are NaH_2PO_3 and Na_2HPO_3. These are known as phosphites.

The structure of phosphorous acid is shown in Structure **11.12**.

A large variety of phosphorus acids are derived from 'polyacids', which contain two or more acidic phosphorus centres (these are considered in detail in Section 11.4.5).

Oxoacids by definition contain a covalent AO—H bond, which can dissociate to give a proton and an oxoanion:

$$AO-H = AO^- + H^+ \qquad (11.48)$$

There may in addition be one or more terminal oxygen atoms, so the **general formula of oxoacids** is $A(O)_t(OH)_n$, where t can equal 0. On this formulation, sulfuric acid, H_2SO_4, is written as $S(O)_2(OH)_2$ (where $t = 2$, $n = 2$), and boric acid as $B(OH)_3$ (where $t = 0$, $n = 3$).

● What are the values of t and n for phosphoric acid?

● From Table 11.4, we see that the formula of phosphoric acid can be rewritten as $P(O)(OH)_3$, giving $t = 1$ and $n = 3$.

These covalent hydroxo compounds have available a wide range of structural possibilities, which is the reason for the existence of a relatively large number of oxoacids. The variables are as follows:

1   There may be several —OH groups in the acid, each one of which can dissociate to form a proton and an oxoanion. Look again, for example, at the three successive ionizations of phosphoric acid, H_3PO_4 (Equations 11.45–11.47).

Each of the species on the left-hand side of the three equations is a different oxoacid, and each of the equilibria has a different dissociation constant. The equilibrium constant for dissociation of the first proton (stage 1), or 'first dissociation constant', is given the symbol K_1 (7.5×10^{-3} mol litre^{-1}), the second dissociation constant is K_2 (6.2×10^{-8} mol litre^{-1}), and the third dissociation constant is given the symbol K_3 (1.0×10^{-12} mol litre^{-1}).

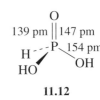

11.12

2 Oxoacids may undergo **condensation** reactions to form dimers, trimers and polymers. The term 'condensation' refers to the reaction of two or more molecules to form a larger molecule, with the elimination of a small molecule such as water. Thus, phosphoric acid, H_3PO_4, can self-condense to produce diphosphoric acid and triphosphoric acid:

diphosphoric acid

$$(11.49)$$

triphosphoric acid

$$(11.50)$$

It can also form higher polymers described by the general term 'metaphosphoric acid'.

3 The central element, A, may exist in more than one oxidation number. Remember *oxygen stabilizes high oxidation numbers*. In fact, the range of oxidation numbers found in compounds with oxygen is wider than that with any other element except fluorine. Phosphorus, for example, forms oxoacids in oxidation numbers +5, +3 and +1. Chlorine forms the oxoacids $HClO_4$, $HClO_3$, $HClO_2$ and $HClO$.

⬤ What is the oxidation number of chlorine in each of these acids?

⬤ The halogen has, respectively, oxidation numbers of +7, +5, +3 and +1.

4 A final complication, arising with sulfur, is that this Group VI element can take the place of oxygen in an oxoacid, so that there is the possibility of one or more S—S bonds, in what is then called a thioacid. We shall meet examples of this when we look at the chemistry of sulfur in detail.

11.4.1 Oxoacid formulae

The interpretation and prediction of oxoacid formulae can seem quite daunting to the beginner. This is partly because of structural complexity, but it is not helped by the irrational tradition of writing oxoacid formulae with hydrogen first, which conceals the fact that hydrogen is bonded to oxygen in many cases. For example, $(OH)_3PO$ is a more meaningful, but rarely used, formulation of phosphoric acid than H_3PO_4. Although we shall explain the new nomenclature below, you will rarely find it used for the common inorganic acids. It is further complicated by the fact that the inorganic acids and their organic derivatives also have different common names; for instance, *phosphorous* acid in inorganic chemistry becomes *phosphonic* acid for its organic derivatives! We have tried to help you through this minefield by underlining prefixes and suffixes in the next Section, and always using the name in most common use.

11.4.2 Nomenclature of oxoacids

Historically, where the central element forms oxoacids in *two* oxidation numbers, the higher state is indicated by the suffix -<u>ic</u> (as in phosphoric acid, H_3PO_4, phosphorus oxidation number +5), and the lower state by the suffix -<u>ous</u> (as in phosphorous acid, H_3PO_3, phosphorus oxidation number +3.)

If *more than two* oxidation numbers are involved, the prefixes <u>per</u>- and <u>hypo</u>- are used as well: <u>per</u>- denotes the highest oxidation number, and <u>hypo</u>- the lowest oxidation number. Thus, the oxoacids of chlorine are shown in Table 11.5.

Table 11.5 The oxoacids of chlorine

formula	$HClO_4$	$HClO_3$	$HClO_2$	$HClO$
oxidation number	+7	+5	+3	+1
name	perchlor<u>ic</u>	chlor<u>ic</u>	chlor<u>ous</u>	<u>hypo</u>chlor<u>ous</u>

Oxoanions derived from -*ic* acids are given the ending -*ate*; examples are provided by the sulfate ion, SO_4^{2-}, the phosphate ion, PO_4^{3-}, the chlorate ion, ClO_3^-, and the perchlorate ion, ClO_4^-.

Oxoanions derived from -*ous* acids are given the ending -*ite*; examples are the sulfite ion, SO_3^{2-}, the nitrite ion, NO_2^-, the hypochlorite ion, OCl^-, and the phosphite ion, PO_3^{3-}.

Condensed forms of oxoacids are also distinguished by means of prefixes. Thus, the prefix <u>ortho</u>- refers to the 'monomeric', or most highly hydroxylated, form, and the prefix <u>meta</u>- refers to the 'polymeric', or least highly hydroxylated, form.

The prefixes <u>di</u>- and <u>tri</u>- in this context are self-explanatory, and refer to 'dimers' or 'trimers'.

The oxoanions are analogously labelled.

Thus, we have orthophosphates, PO_4^{3-}, diphosphates, $P_2O_7^{4-}$, and triphosphates, $P_3O_{16}^{5-}$. (To complicate matters still further, condensed anions may also be labelled with the prefix <u>cyclo</u>- or <u>catena</u>- to distinguish cyclic from linear forms, respectively.)

Table 11.4 also contains a column with the systematic name given to the acids under the IUPAC system. Because these names have not gained wide acceptance yet, we shall only make you aware of them. The names are derived as though the acids were salts. The name is in two parts, with first the acidic hydrogens treated separately from the rest of the compound, which then is named as a coordination complex. These names are potentially very useful because they immediately tell you how many ionizable protons there are in the acid.

11.4.3 Prediction of formulae

For an element in its highest oxidation number (and sometimes others), it is possible to predict the formula of its orthoacid from coordination number considerations. The condensed oxoacid formulae are then easily derived by subtracting the appropriate number of water molecules.

Third- and fourth-row elements prefer four-coordination in their oxoanions. For example, using the oxidation number approach, we can see that when phosphorus(V) coordinates four O^{2-} ions to form a complex ion of stoichiometry PO_4, the resultant charge on this oxoanion is $(+5 - 8) = -3$. Thus, the formula of the corresponding neutral oxoacid is H_3PO_4.

● Predict the formula of the oxoacid formed by sulfur(VI).

● Sulfur(VI) coordinates four O^{2-} ions to form a complex ion of stoichiometry SO_4, with resultant charge $(+6 - 8) = -2$, so the oxoacid is H_2SO_4.

The chlorine(VII) oxoanion, with stoichiometry ClO_4, has charge $(+7 - 8) = -1$, so only one proton is required for formation of the neutral oxoacid $HClO_4$.

For oxoanions of the second period, (e.g. carbonate, $CO_3{}^{2-}$) three-coordination is the norm as it allows for the formation of a planar assembly with extensive π bonding.

The above arguments often fail to predict the formulae of oxoacids of elements in lower oxidation numbers, because the preferred coordination number is either not achieved, or is achieved only by formation of a direct link between the central atom and hydrogen, as you will see in the next Section.

QUESTION 11.2 ✳

Predict the formulae of the oxoacids of nitrogen(V) and boron(III).

11.4.4 Strengths of oxoacids

Strengths of oxoacids can be affected by many variables. For example, you may wonder how acidity depends on oxidation number, or how it varies across a row of the Periodic Table. There is also the question of the relationship between the first, second, etc., dissociation constants of polybasic acids.

Fortunately, two simple generalizations developed by Linus Pauling allow prediction of the acidities of oxoacids with a fair degree of accuracy.

Pauling's first rule relates to successive dissociation constants of a polybasic acid. From simple electrostatic considerations, it is easy to predict that successive ionization steps will take place less readily (for example, you might argue that it is easier to lose a proton from a neutral species like H_3PO_4 than from one, such as $H_2PO_4{}^-$, that already carries a negative charge). The value of the rule is that it allows us to *quantify* the relationships between successive dissociation constants.

Pauling's first rule states that successive dissociation constants K_1, K_2, K_3,…, are in the ratio $1 : 10^{-5} : 10^{-10}$…, etc. To generalize,

$$\frac{K_{n-1}}{K_n} \sim 10^5$$

This rule holds well for the common oxoacids, as Table 11.6 shows.

Table 11.6 Successive dissociation constants of H_2SO_4 and H_3PO_4

Acid	K_1/mol litre^{-1}	K_2/mol litre^{-1}	K_3/mol litre^{-1}	K_1/K_2	K_2/K_3
H_2SO_4	1.0×10^3	1.2×10^{-2}		8.0×10^4	
H_3PO_4	7.5×10^{-3}	6.2×10^{-8}	1.0×10^{-12}	1.2×10^5	6.2×10^4

Pauling's second rule relates acid strength to the number of non-hydrogenated oxygen atoms in the molecule. The more of these there are, the stronger is the acid. The first acid dissociation constant, K_1, is predicted to be of the order of 10^{5t-8} mol litre^{-1}, where t is the number of terminal (non-hydrogenated) oxygens in the acid.

If t in the general formula $AO_t(OH)_n$ is zero (that is, the acid is a *hydroxo* acid), the acid is very weak, with $K_1 \sim 10^{-8}$ mol litre^{-1}.

If $t = 1$, the acid is moderately weak ($K_1 \sim 10^{-3}$ mol litre^{-1}).

If $t = 2$ or higher, the acid is strong ($K_1 \sim 10^2$ mol litre^{-1} or above).

To a first approximation, the strength is independent of n, the number of $-OH$ groups.

Pauling's second rule can be rationalized in terms of the electron-withdrawing effect of the terminal oxygen atoms. The highly electronegative oxygen atoms withdraw electron density from the $-OH$ group, thus rendering its hydrogen more positive, and so more easily ionizable. We would expect that the more terminal oxygen atoms there are, the easier it will be to ionize the $-OH$ hydrogen and thus the stronger will be the acid.

This second rule holds fairly well for all oxoacids, although, bearing in mind the wide variation in the nature, properties and environment of the central atom, it is not surprising that the range of K_1 values for various values of t is rather wide (Table 11.7).

Table 11.7 First dissociation constants of various oxoacids.

$t = 0$		$t = 1$		$t = 2$ or more	
Formula	K_1/mol litre^{-1}	Formula	K_1/mol litre^{-1}	Formula	K_1/mol litre^{-1}
HOCl	2.9×10^{-8}	$HClO_2$	1.10×10^{-2}	$HClO_3$	large
HOBr	2.1×10^{-9}	H_3PO_4	0.75×10^{-2}	H_2SO_4	1×10^3
H_4SiO_4	1.0×10^{-10}	HNO_2	0.45×10^{-2}	$HClO_4$	large
H_3BO_3	5.5×10^{-10}				

QUESTION 11.3

Cite two acids with $t = 0$, two with $t = 1$, two with $t = 2$ and one with $t = 3$.

QUESTION 11.4

Predict the structures and order of acid strength for the chlorine oxoacids: $HClO$, $HClO_2$, $HClO_3$ and $HClO_4$.

QUESTION 11.5

Use Pauling's second rule to predict how the acidity of the highest oxoacid of an element varies across the fourth Period, from Group IV to Group VII. (You should deduce the formulae of the acids by use of the oxidation number approach, as outlined in Section 11.4.3).

QUESTION 11.6

Arrange the following oxoacids in order of acid strength using Pauling's second rule *alone*: H_5IO_6, $HReO_4$, H_3AsO_3, H_2CrO_4.

QUESTION 11.7 ✳

The acid of stoichiometry H_3PO_3 has a first dissociation constant of 1.6×10^{-2} mol litre^{-1}. Can this be explained on the basis of Pauling's second rule? What does the K_1 value suggest about the structure?

QUESTION 11.8 ✳

The acid of formula H_3PO_2 has a K_1 value of 1×10^{-2} mol litre^{-1}. Suggest a structure for this acid.

11.4.5 Condensation of oxoacids

The tendency for an oxoacid to polymerize by condensation is most marked in the less acidic (more highly hydroxylated) acids. There are many stable condensed forms of silicic and boric acid, whereas the condensed forms of phosphoric acid are unstable towards hydrolysis.

● Why do Group VII oxoacids not usually yield condensed oxoacids?

● Because they usually contain only one —OH group, so the product of condensation is an oxide. (One exception to this is periodic acid, H_5IO_6, which has five —OH groups and does form condensed oxoacids.)

Thus, dehydration of perchloric acid, $HClO_4$, gives dichlorine heptoxide, Cl_2O_7:

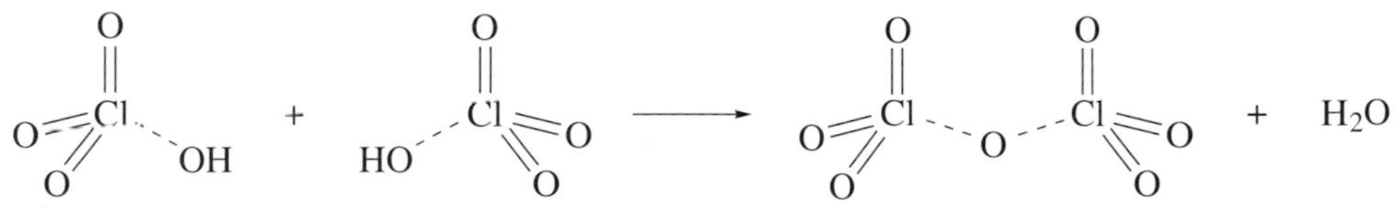

$$(11.51)$$

Condensation is most marked in structures where the charge on the uncondensed anion is high, because it is able to reduce the charge density on the anion. The dimerization of SiO_4^{4-} can be represented as follows:

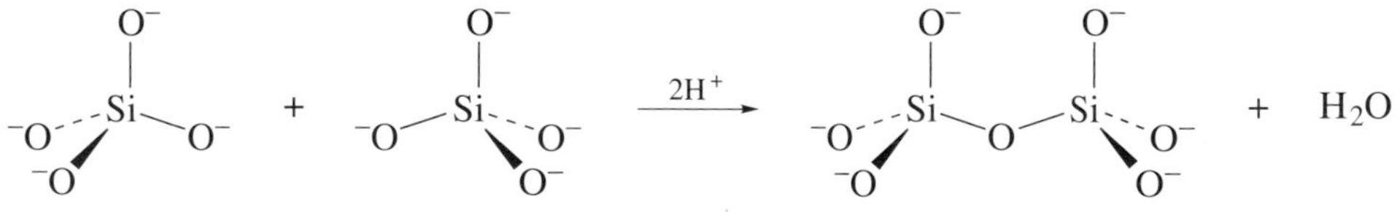

$$(11.52)$$

In the monomer there are four negative charges for four oxygen atoms; in the dimer there are six negative charges and seven oxygen atoms; in the trimer (Structure **11.13**) there are eight negative charges and ten oxygen atoms.

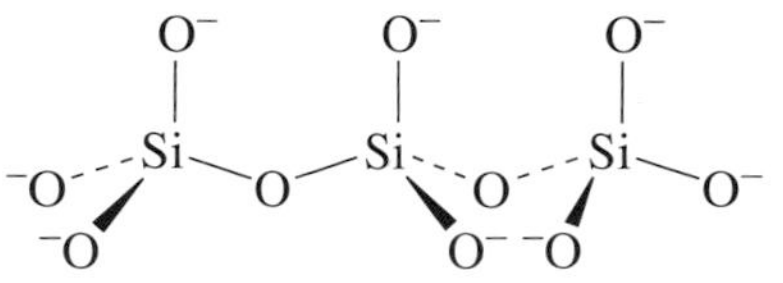

11.13

REPEATING UNIT $[H_2SiO_3]$

In the limiting case of the infinite polymer, there are two negative charges to every three oxygen atoms. In the case of protonation by two H^+ per Si, the repeat unit is $[H_2SiO_3]$ as in Figure 10.24. This is the structural unit found in the pyroxene group of minerals. As we saw in Section 10.3.2, shared SiO_4 tetrahedra can be assembled into rings, chains, double chains, sheets and three-dimensional networks to give the amazing variety of structures found in the crystalline silicate minerals.

Phosphate tetrahedra, PO_4^{3-}, can link up via oxygen-sharing to give polyphosphates in the form of chains and rings. The phosphates, unlike the silicates, contain a central atom with a valency of five. Thus, the maximum number of oxygen atoms that each tetrahedron can share is *three*, since there must always be one vertex of the tetrahedron that is a terminal oxygen, bound as P=O.

⬤ We have already met one compound where each phosphorus forms the maximum three P—O—P linkages. Which is it?

⬤ Phosphoric oxide, P_4O_{10} (Figure 11.17b).

Condensed phosphates can be formed if phosphoric oxide is treated with limited amounts of water; or by dehydrating phosphates by heating.

Two molecules of phosphoric acid can split off a molecule of water between them; the two tetrahedra share one oxygen, giving an acid of formula $H_4P_2O_7$, known as 'pyrophosphoric', or (more correctly) diphosphoric acid. In practice, this can be obtained by heating phosphoric acid ('pyro' comes from the Greek word for fire). Continuation of this process gives triphosphoric acid, $H_5P_3O_{10}$, and, eventually, chain polymers called metaphosphoric acids, which contain repeating $[HPO_3]$ units (Figure 11.19; the ring polymers shown are also called metaphosphoric acids).

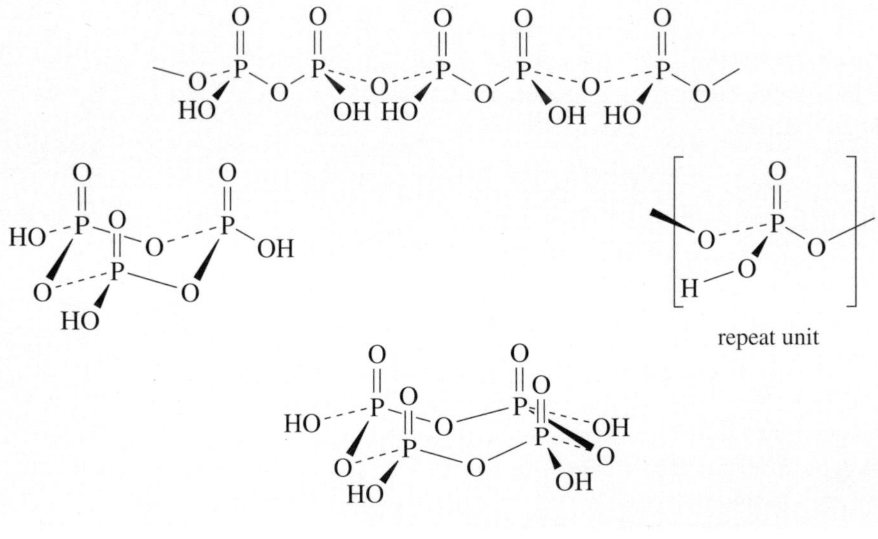

Figure 11.19 Metaphosphoric acid chains and rings.

The P—O—P link in polyphosphates is readily hydrolysed; in excess water, metaphosphates revert to orthophosphate. So, unlike the condensed silicates, polyphosphates are never found as minerals.

Sodium polyphosphates are the most well known; the general reaction for their formation is the dehydration of sodium dihydrogen phosphate by heating. The temperature of dehydration controls the nature of the product, as indicated in Figure 11.20.

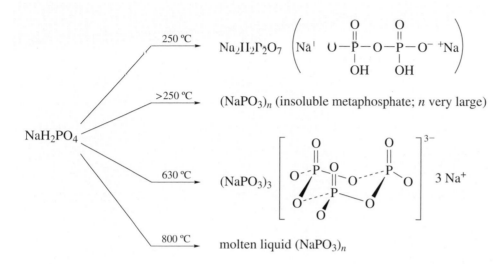

Figure 11.20 The dehydration of sodium dihydrogen phophate at various temperatures.

Partial dehydration at low temperature (*c.* 250 °C) produces disodium dihydrogen diphosphate, $Na_2H_2P_2O_7$, a substance used in baking powders as a slow-acting acid for the controlled release of CO_2 gas, which produces the aerated texture of cakes.

Long-chain polyphosphates are produced by heating NaH_2PO_4 above 250 °C.

The sodium salt $Na_3P_3O_9$, containing the cyclic $P_3O_9^{3-}$ anion, is formed when NaH_2PO_4 is heated to 600–640 °C and the melt then maintained at 500 °C.

High-temperature dehydration (800 °C) above the melting temperature of NaH_2PO_4 (628 °C) yields a liquid with very high relative molecular mass, which gives slightly soluble solids of formula $(NaPO_3)_n$:

$$nNaH_2PO_4 = (NaPO_3)_n + nH_2O \tag{11.53}$$

Heating together a 2 : 1 mixture of Na_2HPO_4 and NaH_2PO_4 at 450 °C produces a chain of only three phosphate units:

$$2Na_2HPO_4 + NaH_2PO_4 = Na_5P_3O_{10} + 2H_2O \tag{11.54}$$

The structure of the triphosphate ion, $P_3O_{10}^{5-}$, is shown in Structure **11.14**.

11.14

BOX 11.4 Adenosine triphosphate: the energy currency of life

After DNA, the polyphosphate adenosine triphosphate (**11.15**), usually abbreviated to ATP, is probably the most important phosphorus-containing molecule in the body. ATP is used to drive many biochemical reactions and cellular processes that require the input of energy; these include cell division and muscle contraction.

11.15

Energy is obtained from ATP when it is broken down to adenosine diphosphate (ADP) and free phosphate ions. Reaction 11.55 is thermodynamically favourable, having a free energy change, $\Delta G_m^{\ominus}$, of about $-40\,\text{kJ mol}^{-1}$.

Note the adenine and ribose rings are often referred to as adenosine.

$$(11.55)$$

ATP is synthesized by the reverse process, i.e. the addition of phosphate to ADP. The energy input for this reaction comes from the breakdown of organic fuel molecules, such as glucose.

You have probably noticed the negative charge on both ATP and ADP. This is conterbalanced in these, and other phosphate-containing biological molecules (e.g. DNA) by cations, usually Mg^{2+}. In fact, both ATP and DNA may be regarded as magnesium complexes.

BOX 11.5 Phosphates, detergents and the environment

Most detergents comprise long molecules with **hydrophobic** (water-hating) and **hydrophilic** (water-loving) sections as indicated in the example shown in Structure **11.16**. The basic action is for the organic hydrophobic section to bury itself in the dirt, and the hydrophilic section then allows the insoluble dirt to 'dissolve' in water.

$$CH_3(CH_2)_9 - \underset{\underset{\text{hydrophobic}}{}}{\overset{\overset{CH_3}{|}}{CH}} - \underset{\text{hydrophilic}}{SO_3^- \, Na^+}$$

11.16

Detergents are used in combination with 'builders', which soften hard water. Hard water contains significant concentrations of calcium and magnesium salts, which replace the sodium ions in the detergent molecule. The dipositive metal ions cause the long molecules to clump together and precipitate out as a scum – the familiar bathtub ring! Sodium tripolyphosphate, was for many years, added to detergents in substantial amounts to complex magnesium and calcium ions and hence remove them from solution. However, phosphates are being gradually removed from household cleaners and detergents owing to environmental concerns (Figure 11.21).

Excess phosphate in rivers and lakes, commonly arising from phosphate detergent wastes, is thought to be responsible for the occurrence of **eutrophication**. Here the ecological balance is disturbed by excessive growth of algae at certain times of year. Superficially this would seem to be a mere inconvenience, making swimming or sailing less pleasant, but it is more serious than this. As the weather gets colder, the growth ceases, and the algae sink and decay, consuming oxygen dissolved in the water. This can result in an unpleasant-smelling lake (largely due to the production of hydrogen sulfide) and, in an extreme case, the fish in the lake can die owing to oxygen starvation. Figure 11.22 shows algal growth in eutrophic water. Problems of this type have not been too serious in the UK, the main places affected being Lough Neagh in Northern Ireland, and the Norfolk Broads. In the USA there have been acute problems, particularly Lake Erie and Lake Washington near Seattle. As a consequence, most states in North America had banned the use of phosphates in detergents by 1993. Substitute builders include zeolites which are porous, non-eutrophic materials that exchange sodium ions for calcium ions and hence soften water. *

Figure 11.21
An example of an environmentally friendly phosphate-free washing powder.

Figure 11.22
An example of eutrophic water in a lake.

---

* The structures and applications of zeolites are described in a Case Study in *Chemical Kinetics and Mechanism*.[4]

11.5 Arsenic, antimony and bismuth

The halides of arsenic, antimony and bismuth illustrate the following trends down the Group: (a) increasing metallic character of the elements; (b) the inert pair effect—that is, the tendency towards an increased stability of the 'Group number minus two' oxidation number (+3); (c) a tendency to higher coordination numbers.

All three pentafluorides, AsF_5, SbF_5 and BiF_5, are known. The only other pentahalide stable at room temperature is $SbCl_5$. $AsCl_5$ is stable only at low temperatures. A demonstration of the tendency to higher coordination numbers is given by the pentafluorides. Arsenic pentafluoride is a monomeric gas. At normal temperatures, antimony pentafluoride is a viscous liquid containing six-coordinate antimony in *cis* fluorine-bridged polymers (Figure 11.23a). It crystallizes to give a structure that contains *cis*-bridged tetramers (Figure 11.23b). Bismuth pentafluoride is a white crystalline solid, in which the bismuth coordination number is six, but in this case the structure has *trans* fluorine bridges, giving infinite chains (Figure 11.24). All these fluorides are highly reactive, hydrolysing rapidly in air to give hydrofluoric acid. They have to be handled with great care!

Figure 11.23
(a) The structure of liquid SbF_5;
(b) the structure of crystalline SbF_5.

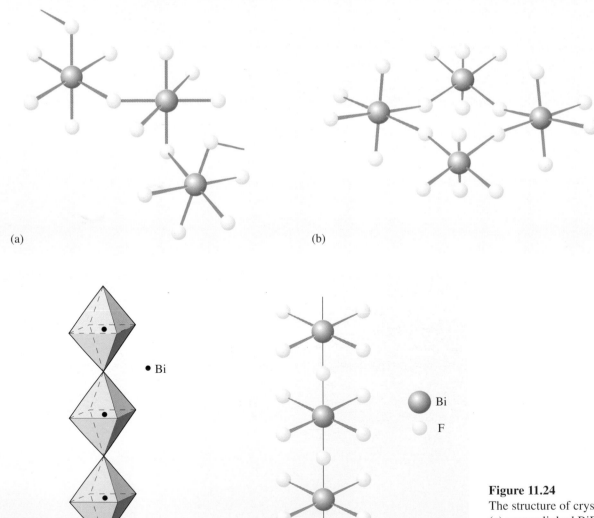

(a)   (b)

F

Sb

(a)   (b)

• Bi

Bi

F

Figure 11.24
The structure of crystalline BiF_5:
(a) corner-linked BiF_6 octahedra;
(b) structural formula.

All the trihalides, MX_3 (M = As, Sb, Bi; X = F, Cl, Br, I) are known. They are <u>all solids except for AsF_3 and $AsCl_3$</u> which are liquids at room temperature, and they are rapidly hydrolysed.

When bismuth is dissolved in molten $BiCl_3$, a black solid of formula $Bi_{24}Cl_{28}$ is obtained. This contains some very interesting ionic species: $[BiCl_5]^{2-}$, $[Bi_2Cl_8]^{2-}$ and $[Bi_9]^{5+}$, in the ratio 4 : 1 : 2. The ion $[Bi_9]^{5+}$ consists of the metal-atom cluster shown in Figure 11.25.

Arsenic(III) and antimony(III) both form oxides of formula X_4O_6. The molecules have the same structure as P_4O_6 in the gas phase. Like P_4O_6, arsenic(III) oxide is acidic; antimony(III) oxide is amphoteric. Bismuth forms a basic oxide, Bi_2O_3, which is a yellow powder at room temperature. The pentoxides of these elements are less well characterized, Bi_2O_5 being extremely unstable to loss of oxygen. Compounds containing E=E double bonds (E = As, Sb and Bi) have been synthesized. As with the corresponding phosphorus compounds, stabilization of the double bond is achieved by substitution with bulky groups, such as t-butyl.

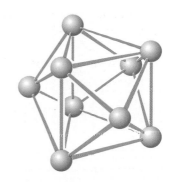

Figure 11.25
The structure of the $[Bi_9]^{5+}$ ion, in which a trigonal prism of bismuth atoms has three others outside (capping) the rectangular faces.

QUESTION 11.9

How many different types of fluorine atom are there in liquid SbF_5?

11.6 Summary of Sections 11.4 and 11.5

1 Oxoacids of elements in Groups III–VII may exist for several oxidation numbers of the central element, and as monomers (ortho oxoacids) or condensation polymers (meta oxoacids).

2 Ortho oxoacids have a general formula $AO_t(OH)_n$, which may be predicted from considerations of coordination number and oxidation number.

3 In a polybasic acid, acid strength is higher for dissociation of the first proton than for the second (and so on).

4 In a series of polybasic acids, acid strength increases as the value of t increases.

5 The relative strengths of oxoacids can be predicted from Pauling's rules:

$$\frac{K_{n-1}}{K_n} \sim 10^5 \text{ and } K_1 \sim 10^{5t-8}$$

If the number of non-hydrogenated oxygen atoms, t, is zero, the acid is weak, and $K_1 \sim 10^{-8}$ mol litre^{-1}; if $t = 1$, the acid is moderately strong, $K_1 \sim 10^{-3}$ mol litre^{-1}; if $t = 2$, the acid is strong, $K_1 \sim 10^2$ mol litre^{-1} or greater.

5 The tendency to condensation is highest with weak acids. Thus, orthosilicic acid readily converts into a meta form, which does not happen for sulfuric and perchloric acids.

6 The chemistry of arsenic, antimony and bismuth illustrates trends down a Group, such as increasing metallic character, the inert pair effect and the tendency to higher coordination numbers.

THE GROUP VI/16 ELEMENTS

12

12.1 Structures and properties of the elements

Group VI (Figure 12.1) comprises non-metals (oxygen, sulfur and selenium), a semi-metal (tellurium), and a metal (polonium). Once again you should notice an increase in metallic properties on descending a Group in the Periodic Table. The electronic configurations of the Group VI elements are shown in Table 12.1.

Oxygen is the most abundant element on our planet. It forms 46% by mass of the Earth's crust, much of it occurring in silicates; it comprises 23% of the atmosphere, where it occurs as the gaseous element, O_2; and it forms about 85% of the hydrosphere, where it is combined with hydrogen as water, H_2O.

Oxygen is generated naturally in photosynthesis, whereby plants synthesize carbohydrates, etc. from water and carbon dioxide in the atmosphere, using sunlight as the source of energy. The overall equation for the reaction is:

$$6CO_2(g) + 6H_2O(l) \longrightarrow 6O_2(g) + C_6H_{12}O_6(aq) \tag{12.1}$$

Oxygen is necessary for both plant and animal life. It is taken in during the respiration process and carbon dioxide is breathed out, so completing the cycle.

Oxygen is a colourless, odourless gas. It has three stable isotopes ^{16}O, ^{17}O and ^{18}O, of which ^{16}O is by far the most abundant (99.76%, 0.04% and 0.2%, respectively). Oxygen gas condenses to a pale blue liquid at $-180\,°C$ and atmospheric pressure.

Ozone, O_3, is a naturally occurring allotrope of oxygen. It has a characteristic strong 'metallic' smell, which the human nose can detect at levels as low as 0.01 ppm! O_3 is formed by the action of high-voltage electrical discharge, or ultraviolet radiation on oxygen; indeed, you may have smelt it around photocopiers and laser printers in offices.

- What do you predict to be the shape of O_3?

- The Lewis structure for O_3 has the two resonance forms represented by Structures **12.1** and **12.2**, and Structures **12.3** and **12.4**. The non-bonded pair on the central oxygen leads to a bent molecule.

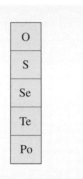

Figure 12.1
The Group VI/16 elements: oxygen, sulfur and selenium are non-metals, tellurium is a semi-metal and polonium is a metal.

Table 12.1 Electronic configurations of Group VI/16 atoms

Atom	Electronic configuration
O	[He] $2s^2\,2p^4$
S	[Ne] $3s^2\,3p^4$
Se	[Ar] $3d^{10}\,4s^2\,4p^4$
Te	[Kr] $4d^{10}\,5s^2\,5p^4$
Po	[Xe] $4f^{14}\,5d^{10}\,6s^2\,6p^4$

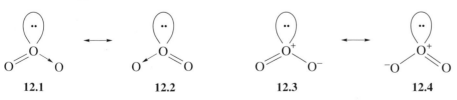

| 12.1 | 12.2 | 12.3 | 12.4 |

Note that we cannot describe O_3 as in Structure **12.5**, because this would involve expanding the octet of a second-row element. Microwave spectroscopy of O_3 indicates a bond angle of 116.8° and an O—O bond distance of 127.8 pm: this compares with the O—O distance of 120.7 pm in doubly bonded O_2 and 149 pm in the singly bonded peroxide ion, O_2^{2-}.

Because of its dramatic associations with volcanic eruptions, sulfur has been recognized since antiquity; the first reference to the element as 'brimstone' was in the Book of Genesis. However, in the 18th century, sulfur was one of the first substances to be firmly established as an element in the modern sense, by the work of the French chemist Antoine Lavoisier. Sulfur is found in sulfide ores combined with metals, such as galena, PbS, iron pyrites, FeS_2 (known as 'fools gold'), cinnabar HgS, and zinc blende, ZnS. Removal of sulfur from these ores is a necessary accompaniment to the metal extraction process. There are widespread deposits of elemental sulfur located in areas of high volcanic activity or found in association with petroleum. The latter sedimentary sulfur deposits result from the reducing action of bacteria on sulfate during rock formation. Sulfate deposits form a large proportion of world reserves of sulfur but, as we (unlike bacteria!) are not yet able to convert sulfate efficiently into sulfur these are not commercially useful.

Petroleum-linked sulfur deposits often occur at some depth below the surface, and it was not until the beginning of the twentieth century, when the elegant **Frasch** recovery method was developed, that they become commercially important. This process is covered in detail in the Case Study *Industrial Inorganic Chemistry*.

BOX 12.1 Blackpowder and the firework industry

'By the flash and combustion of fires, and by the horror of sounds, wonders can be wrought, and at any distance that we wish, so that a man can hardly protect himself or endure it.'

The above quote was attributed to Roger Bacon (Figure 12.2), an English monk who introduced blackpowder (later known as gunpowder) to Europe in the 13th century. The explosive material consists of sulfur and charcoal (a fuel), mixed with potassium nitrate (an oxidizer).

A simplified equation for the deflagration (a self-propagating burning surface reaction) of gunpowder is:

$$2KNO_3 + S + 3C = K_2S + 3CO_2 + N_2 \tag{12.2}$$

The basic mixture was, in fact, originally discovered by eighth century Chinese alchemists. However, Bacon experimented with the relative proportions and undertook a detailed scholarly investigation into the explosive. Bacon incurred the wrath of the Church, and suffered imprisonment for his efforts, but managed to preserve his recipe for blackpowder in the form of an anagram (Figure 12.3).

As a propellant for ammunition, blackpowder strongly influenced military tactics in the 13th century. Its first recorded use in battle in the West was during Edward III's triumph at the Battle of Crecy (1346); however, the victory was largely attributed to the skill of the longbow men, rather than bombardment from field artillery.

Figure 12.2
Roger Bacon, who introduced blackpowder to Europe in the 13th century.

Blackpowder has been used as a major constituent of fireworks since the early firecrackers made by the Chinese in about AD 1000. A schematic representation of a simple 'banger' is shown in Figure 12.4. The blue touchpaper is impregnated with potassium nitrate, which smoulders when lit. This reaches the sulfurless priming powder, which is a mixture of potassium nitrate and charcoal. This then ignites a delay fuse, which burns noisily until the end is reached. Smoke, flame and hot particles are then showered onto the main blackpowder filling. Being in a confined space, this promptly explodes.

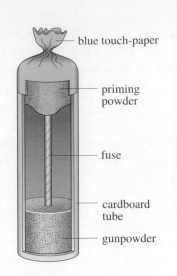

Figure 12.4
A schematic diagram of a 'banger' firework.

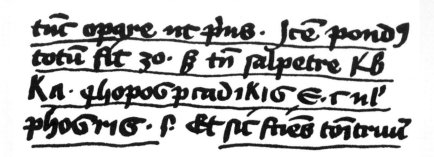

Figure 12.3 Roger Bacon preserved his recipe for gunpowder as an anagram; you can just about make out the word saltpetre.

Sulfur is a bright yellow solid. The form stable at room temperature is called **orthorhombic sulfur** because of the crystalline structure adopted (Figure 12.5). The unit cell contains crown-shaped S_8 molecules (Figure 12.6), stacked in a complex array. The only difference between orthorhombic sulfur and the high-temperature (above the transition temperature of 96 °C) monoclinic modification, with which it may readily be interconverted, is in the stacking pattern of the crowns. At room temperature the monoclinic form slowly changes into the orthorhombic form.

It is possible to make modifications that do not contain octasulfur rings — ring sizes from S_5 to S_{20} are known — but all revert in time to the more stable S_8 form. Why?

In terms of the bonding around a sulfur atom in a chain of sulfur atoms, two of the six valence electrons are used to make bonding pairs with the adjacent sulfurs, leaving two non-bonded pairs of electrons.

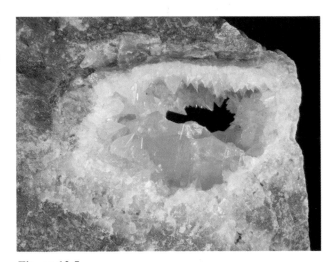

Figure 12.5
Orthorhombic sulfur crystals and crystalline calcite, found in a cavity in Sicilian limestone.

○ How will the electron pairs arrange themselves?

○ The electron pairs will want to be as far apart as possible and to give minimum interference with electrons on the neighbouring sulfur atoms. With effectively four pairs of electrons, an approximately tetrahedral arrangement will be preferred.

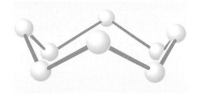

Figure 12.6
Crown-shaped S_8 molecule.

Look at the chain of four sulfur atoms in Figure 12.7 taken from an S_8 ring. We find that the S—S—S bond angle is 108°, and the **dihedral angle** (the angle between the two planes) is 98°. In this arrangement the non-bonded pairs are as far away from each other as possible. All other forms of sulfur have less ideal dihedral angles: in S_6 the bond angles are 102° and the dihedral angles are 74°; in the more-strained structure of S_7, the bond angles range from 101° to 107°, with the dihedral angle varying between almost zero and 108°.

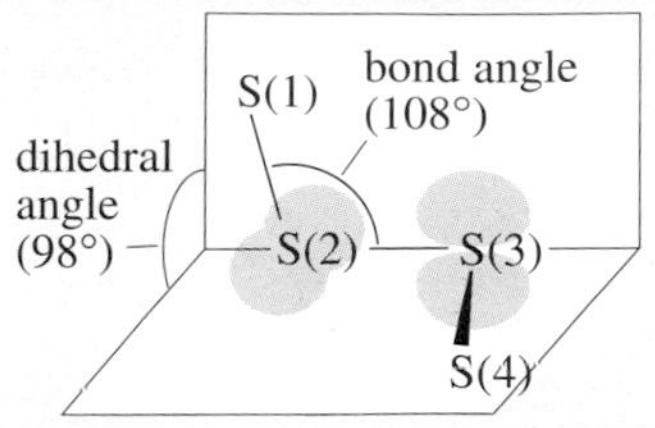

Orthorhombic sulfur melts to a yellow liquid at 112 °C. As the temperature of the liquid is raised above about 160 °C, its colour changes to orange, brown and almost black as the S_8 rings open to form **diradicals** (that is, radical species in which there are two atoms that have an unpaired electron), •S—(S_6)—S•. These join together to form chains of various lengths. The darkening is due to the presence of the unpaired electrons on the chain ends. The average chain length at 170 °C is about 10^6 sulfur atoms, decreasing markedly at higher temperatures. When liquid sulfur at 160 °C is cooled rapidly by pouring into cold water, a dark rubbery polymeric solid called **plastic sulfur** results (Figure 12.8). This can be stretched to about 20 times its original length, with a strength comparable with that of Nylon. Unfortunately, the conversion into brittle orthorhombic sulfur takes less than a month at normal room temperatures, ruling out potential applications.

The vapour of sulfur, at temperatures below 500 °C and at atmospheric pressure, consists mainly of S_8 rings, but, as the temperature is increased and the pressure reduced, the dark violet paramagnetic substance S_2 becomes the major constituent.

An important use of sulfur is in the **vulcanization process**. This involves heating rubber with sulfur to induce cross-linkage between the long hydrocarbon chains of which rubber is composed, and results in a product with superior mechanical properties. However, the major uses of sulfur are in the heavy chemical industry. Nearly 90% of all the sulfur used goes to the production of sulfuric acid.

Selenium is classified as a non-metal and tellurium as a semi-metal (Figure 12.1). They both have spiral polymeric structures, although it is possible to form red allotropic forms of selenium, containing Se_8 molecules, by rapid quenching of the melt; there is no analogous molecular form of tellurium.

Polonium is a radioactive metal, the only one to have a primitive cubic structure.

- What is the coordination around each polonium atom in the solid?

- In a primitive cubic structure each polonium atom is surrounded by six others at the corners of an octahedron.

Figure 12.7
Angles involved in a sequence of four S atoms in the S_8 molecule.

Figure 12.8
Molten sulfur being poured into cold water. ⌨

BOX 12.2 Io: a sulfur-rich moon of Jupiter

Sulfur is not very abundant in the Solar System, but tends to concentrate (in association with iron) in the cores of the inner planets. One interesting exception is Io (Figure 12.9a), one of the moons of Jupiter whose composition was determined following the visits of the NASA *Voyager* spacecraft in 1979, and the *Galileo* probe, which arrived at Jupiter in December 1995, following a six-year journey from Earth. Io, the most active volcanic body in the Solar System, is incredibly sulfur rich; the surface is littered with sulfur volcanoes which spew plumes of sulfur and sulfur dioxide (Figure 12.9b). Its thin atmosphere is mostly SO_2, and there is speculation that there are many new and exotic sulfur compounds on Io, waiting to be discovered. The surface of Io is continually changing due to the volcanic activity, indeed dramatic differences were observed by the 1995 *Galileo* mission since NASA had last paid a visit.

(a)

Figure 12.9
(a) An image from the *Galileo* probe of Io, the innermost large satellite of Jupiter. (b) Io is the most active world in the Solar System: this image taken by *Voyager 1* shows volcanic eruptions on the surface of the moon.

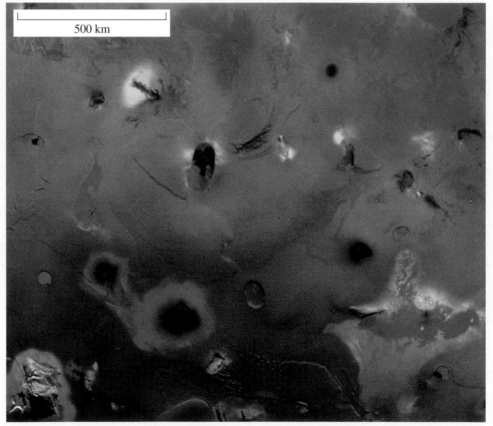

500 km

(b)

12.2 Oxygen

The bond dissociation energy of the oxygen molecule, O_2, is quite high (498 kJ mol^{-1}; cf. N_2, 945 kJ mol^{-1}), but oxygen is nevertheless a very reactive gas, which oxidizes many of the elements directly (for example, hydrogen is oxidized to water).

● Why is the bond dissociation energy high for the O_2 molecule?

◐ O_2 has a double bond, owing to both σ and π overlap of the 2p orbitals. ✗

When an oxidation reaction gives out a lot of heat and proceeds spontaneously, we call this **combustion**. Oxygen willingly supports combustion partly because the molecule has two unpaired electrons. If you look at the molecular orbital energy level diagram for dioxygen (Figure 12.10) you will see two unpaired electrons in the degenerate 1πg (2pπ$_g$) orbitals. Oxygen is a diradical, and as such, is very reactive. In the combustion process a flame or spark initiates the reaction by breaking some bonds to form free atoms or radicals, these start *chain reactions* in which radicals (or atoms) regenerate radicals (or atoms) and so on. The reaction gives out heat and light, and can be explosive, because, as well as being exothermic, it goes very fast. This provides a mechanism for the strong O=O bond to react, whereby radicals interact with the unpaired electrons to open up the double bond, which then allows the weak O—O single bond to react.

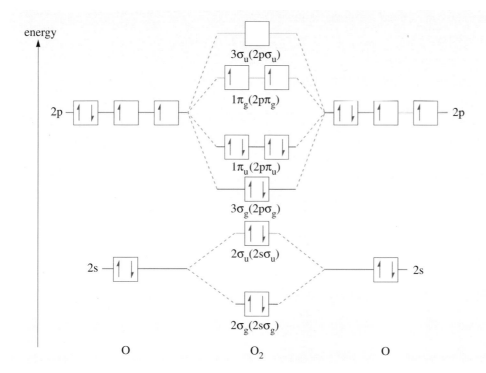

Figure 12.10
The molecular orbital energy-level diagram for O_2.

● What is the magnetism of the O_2 molecule?

◐ O_2 is paramagnetic; it has two unpaired electrons in the 2pπ$_g$ level. Liquid oxygen is attracted towards the poles of a magnet.

12.2.1 Peroxides

Hydrogen peroxide, H_2O_2, is the other hydride of oxygen, apart from water, and its structure is shown in Figure 12.11.

Peroxide salts, such as Na_2O_2 and BaO_2, are formed when alkali metals and alkaline earth metals are heated in air. The action of dilute acids on these salts produces hydrogen peroxide; for example:

$$BaO_2(s) + 2H^+(aq) + SO_4^{2-}(aq) = BaSO_4(s) + H_2O_2(aq) \qquad (12.3)$$

This was first demonstrated in 1818 by J. L. Thenard, who is credited with the discovery of hydrogen peroxide. The reaction in Equation 12.3 is a particularly useful method of preparation of H_2O_2 because, unusually for the salt of a Group I or II metal, barium sulfate is extremely insoluble, which shifts the equilibrium to the right.

Hydrogen peroxide is thermodynamically unstable with respect to disproportionation (Equation 12.4). However, kinetic factors dictate that the decomposition is negligibly slow for the pure material, although numerous catalysts are known, including metal surfaces, manganese dioxide and blood. This is shown to dramatic effect in Figure 12.12.

$$H_2O_2(l) = H_2O(l) + \tfrac{1}{2}O_2(g); \quad \Delta G_m^{\ominus} = -116.7 \text{ kJ mol}^{-1} \qquad (12.4)$$

Hydrogen peroxide is a strong oxidizing agent in either acidic or basic solution, but towards very strong oxidizing agents such as MnO_4^- it will behave as a reducing agent. In acidic solution, hydrogen peroxide can oxidize iodide to iodine:

$$H_2O_2(l) + 2I^-(aq) + 2H^+(aq) = I_2(s) + 2H_2O(l) \qquad (12.5)$$

and reduce the permanganate ion to manganese(II) ions

$$MnO_4^-(aq) + 2\tfrac{1}{2}H_2O_2(l) + 3H^+(aq) = Mn^{2+}(aq) + 4H_2O(l) + 2\tfrac{1}{2}O_2(g) \qquad (12.6)$$

Under alkaline conditions, hydrogen peroxide is a very efficient bleaching agent and finds widespread application in the textile and paper industries. The latter application has increased dramatically in recent years, where environmental considerations have led to the replacement of the classical multistage bleaching process using chlorine and chlorine-containing bleaching agents.

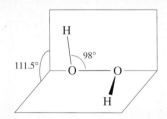

Figure 12.11
The most stable conformation of the hydrogen peroxide molecule, H_2O_2, in the gas phase. The dihedral angle is 111.5°.

Figure 12.12
A dramatic demonstration of the decomposition of hydrogen peroxide using manganese dioxide as a catalyst.

12.2.2 Oxides and the Periodic Table

Except for the lighter noble gases, all the elements form oxides. Table 12.2 shows the s- and p-Block elements, indicating the acidity, basicity or amphoteric nature of the oxides. In the amphoteric section the elements underlined are amphoteric in their lower oxidation states and acidic in their higher oxidation states.

Table12.2 Oxides of the typical elements

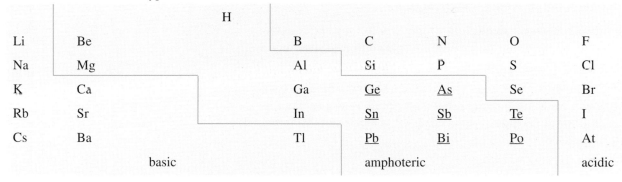

Certain oxides are covalent but are neither acidic nor alkaline; examples we have seen so far include CO and N_2O. Such oxides are called **neutral oxides**.

Some oxides, when melted at very high temperatures and then rapidly quenched, can form glasses, notably B_2O_3 and SiO_2. Al_2O_3 is refractory (a substance with a high melting temperature), with an extended structure, which is largely covalent and single bonded.

Oxygen resembles fluorine in many of its properties, for example in its electronegativity and its ability to stabilize high oxidation numbers. *The highest oxidation numbers of the elements are found in their oxides and fluorides.*

12.2.3 Summary of Section 12.2

1 Oxygen is a reactive, gaseous, diatomic molecule constituting approximately 23% of the air we breathe.

2 The oxygen molecule has a strong double bond owing to both σ and π overlap of the 2p orbitals.

3 Oxygen is paramagnetic with two unpaired electrons in the 2pπ antibonding molecular orbitals.

4 Ozone, O_3, an allotropic form of oxygen, is a bent triatomic molecule.

5 Hydrogen peroxide, H_2O_2, may behave as an oxidizing agent or a reducing agent. It forms metal salts containing the peroxide ion, O_2^{2-}.

6 Apart from the lighter noble gases, all elements form oxides. Generally speaking, basic oxides (such as Na_2O) are formed by the typical metal on the left-hand side of the Periodic Table, and acidic oxides (such as SO_3) by the non-metals on the right. Amphoteric oxides (such as Al_2O_3) fall in the middle. A few oxides, such as CO and N_2O, do not fall into these categories, and are said to be neutral.

7 Oxygen, like fluorine, tends to stabilize higher oxidation numbers.

12.3 Sulfur

In its lower oxidation states, sulfur resembles oxygen in its chemical behaviour, but differs from oxygen in that higher oxidation numbers such as +4 and +6 are also available. A brief examination of the compounds of sulfur with hydrogen, halogens and oxygen should serve to illustrate the similarities and differences.

12.3.1 Sulfur hydrides (sulfanes)

Higher oxidation numbers are stabilized, in general, by oxygen and fluorine; however, compounds of hydrogen with sulfur are limited to those in which sulfur is bivalent.

The most important hydride, H_2S, is an important constituent of many types of natural gas. H_2S, with its characteristic 'rotten egg' smell, is as poisonous as hydrogen cyanide (fatal at 100 ppm!), and because it is much more common, presents a greater menace.

BOX 12.3 Hydrogen sulfide: the key to life in a darkened world

Whereas hydrogen sulfide is known to be highly toxic to us, certain exotic sea creatures actually *thrive* in an environment of H_2S. These include large tubeworms, which live on the sea floor near hydrothermal vents that spew out superheated water saturated with H_2S and heavy-metal sulfides (Figure 12.13). In contrast to the Earth's surface, where life relies on the light-driven process of photosynthesis (Equation 12.1), these tubeworms contain bacteria that produce energy by the oxidation of hydrogen sulfide to the sulfate ion (Equation 12.7). This energy is then used to convert dissolved carbon dioxide into the complex carbon-containing molecules in the tubeworm's structure.

$$H_2S(aq) + 4H_2O = SO_4^{2-}(aq) + 10H^+(aq) + 8e^- \qquad (12.7)$$

An alternative host for these bacteria are shrimps which tend to dominate around vents in the Atlantic Ocean.

On the sea bed, at a depth of several thousand metres, signs of life are usually sparse. However, around hydrothermal vents, complex ecosystems have developed, and around 500 new species have been discovered, with hydrogen-sulfide-consuming bacteria forming the first link in the food chain.

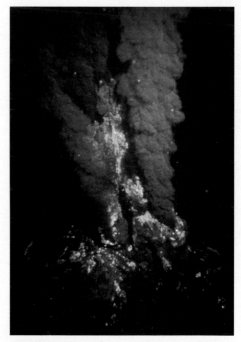

Figure 12.13
A 'black smoker': hydrogen sulfides and heavy-metal sulfides being spewed from a vent in the sea floor. These can rise to heights of 15 metres.

● Account for the fact that H_2S is a gas ($T_b - 60\,°C$), whereas its oxygen analogue, H_2O, is a liquid at room temperature

● This is a consequence of the stronger hydrogen-bonding between molecules in H_2O.

Like water, the H_2S molecule is V-shaped. The bond angle in H_2S (**12.6**) is found to be close to 90°, which can be taken to indicate that sulfur makes use of almost pure p orbitals in its bonds to hydrogen. The strength of the S—H bond is lower than that of the O—H bond, as a comparison of bond enthalpy terms shows: $B(O-H)$, $464\,kJ\,mol^{-1}$; $B(S-H)$, $364\,kJ\,mol^{-1}$.

H_2S can be prepared by treating a metal sulfide with a dilute acid:

$$FeS(s) + 2H^+(aq) = Fe^{2+}(aq) + H_2S(g) \qquad (12.8)$$

In nature, it is produced by bacteria, present in rotting vegetation, and bogs (as you can tell from the smell!)

If acid is added to polysulfides (see Section 12.3.2), a mixture of acids containing S—S bonds results: H_2S_2, H_2S_3, H_2S_4, etc. As these are analogues of carbon and silicon hydrides, they are called **sulfanes**. Sulfur has the highest tendency after carbon to form bonds to itself, that is, to **catenate**.

12.3.2 Sulfides

Most metals react directly with sulfur to form sulfides, which in the main are very insoluble in water, although the sulfides of metals from Groups I and II are ionic, and *are* soluble. However, like oxides, they are salts of a weak acid, and are therefore extensively hydrolysed in solution:

$$S^{2-}(aq) + H_2O(l) = OH^-(aq) + HS^-(aq) \qquad (12.9)$$

Solutions of Groups I and II sulfides can dissolve sulfur to form anions containing sulfur chains of type S_n^{2-} (n up to 6), which are known as **polysulfides**, for example calcium pentasulfide, CaS_5. The disulfide ion, $^-S-S^-$ is found naturally in the metal ore, iron pyrites, FeS_2.

The polysulfide anion S_3^- is the source of the rich blue colour in the highly prized mineral *lapis lazuli*. This rare stone, found mainly in Asia can be refined to produce the pigment *ultramarine* (an example of an artefact decorated with this pigment is shown in Figure 12.14). Lapis lazuli is based on the mineral sodalite (a zeolite of formula $Na_8[Al_6Si_6O_{24}]Cl_2$) where S_3^- ions substitute partly or completely for chloride ions. Owing to its expense, ultramarine was only found in manuscripts and paintings of the wealthy until 1828, when a cheap process to produce synthetic ultramarine from the clay, kaolin, was discovered. The manufacture of a cheap blue pigment played an important role in the rise of the Impressionist movement of painters, whose work frequently used large amounts of synthetic ultramarine.

Figure 12.14
A decorated Russian lapis lazuli carving.

12.3.3 Sulfur halides

Sulfur uses a wide range of oxidation numbers in its compounds with halogens, particularly with fluorine. The sulfur halides (also known as *halosulfanes*) that have been characterized are listed in Table 12.3.

12.6

Table12.3 Sulfur halides

	+1	+2	+4	+5	+6
			Oxidation number		
	S_2F_2	$[SF_2]^\dagger$	SF_4	S_2F_{10}	SF_6
$S_nCl_2^*$	S_2Cl_2	SCl_2	SCl_4		
$S_nBr_2^*$	S_2Br_2				
	S_2I_2	~~S_2I_2~~			

*Terminal sulfur atoms in the chain have oxidation number +1; others have oxidation number 0.

†SF_2 rapidly disproportionates to sulfur and SF_4. However, we include it because its structure and dipole moment are known from spectroscopic studies on the gas at low pressure.

The product of the direct combination of sulfur with fluorine is the octahedral sulfur hexafluoride, SF_6 (**12.7**). Fluorine is the only halogen able to oxidize sulfur to the +6 state. Sulfur hexafluoride is chemically unreactive. It is both the most inert sulfur compound and the most inert covalent fluoride. This is in contrast to the lower fluorides and the other sulfur halides, which are reactive and, in particular, are readily hydrolysed. The reluctance of SF_6 to take part in chemical reaction is purely kinetic; reactions such as the hydrolysis

$$SF_6(g) + 3H_2O(l) = SO_3(g) + 6HF; \quad \Delta G_m^\ominus = -518.5\,kJ\,mol^{-1} \tag{12.10}$$

are highly favoured thermodynamically. The lack of a low-energy pathway for these reactions may be due to the fact that the central atom has achieved its maximum coordination number (it is said to be **coordinatively saturated**).

As a result of its inertness and high stability, SF_6 finds application as an insulating gas in high-voltage electrical systems. At a pressure of 2–3 bar it will prevent discharge across a 1.0–1.4 MV potential difference between electrodes 50 mm apart. Other applications include filling the soles of sports shoes, and because the velocity of sound is low in SF_6, it is used to fill the space between the glass sheets in double-glazing units found in airport concourses. However, great care must be exercised when using SF_6, as it has proved an extremely efficient greenhouse gas, exhibiting a global warming potential about 23 900 times greater than that of carbon dioxide. Indeed, at the Kyoto Summit on Climate Change in 1997, SF_6 was added to the list of gases whose use is to be carefully controlled.

The lower fluorides, S_2F_2 and SF_4, disproportionate readily into sulfur and SF_6. The difluoro compound, S_2F_2, demonstrates unusual structural isomerism in that it can exist in two forms: disulfur difluoride, F—S—S—F (**12.8**) and thiothionyl fluoride, S=SF_2 (**12.9**).

SF_4 with its characteristic shape (Structure **12.10**), is a useful selective fluorinating agent, which is capable, for example, of replacing the C=O group in organic compounds by CF_2 without affecting other functional groups. SF_2 is unusual in that it exists both as the monomer and as a dimer (Structures **12.11** and **12.12**, respectively).

12.7

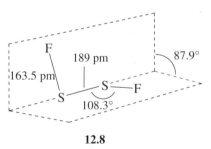

12.8

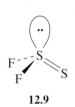

12.9

12.10

12.11

12.12

The normal product of the combination of chlorine or bromine with sulfur is the so-called monohalide S_2X_2 (where X is Cl or Br). S_2Cl_2 is formed by bubbling chlorine gas through molten sulfur, and is a toxic yellow liquid with a foul odour. The shape of the molecule is the same as that of S_2F_2. Chlorine does not stabilize oxidation number +6 in sulfur; nor indeed is it very successful at stabilizing the +4 state (SCl_4 is stable only at low temperatures, decomposing at $-31\,°C$ into SCl_2 and chlorine). However, unlike bromine, it does stabilize the +2 state, and some dichloride, SCl_2, a cherry-red liquid, can be extracted from the reaction mixture if S_2Cl_2 is treated with a large excess of chlorine in the presence of a trace of catalyst. In accordance with VSEPR theory, the SCl_2 molecule has a V-shaped structure. SCl_2 is used in the production of several sulfur-containing compounds, including 'mustard gas' (Structure **12.13**), a chemical weapon, widely used in World War I, which causes debilitating burns to the skin and eyes.

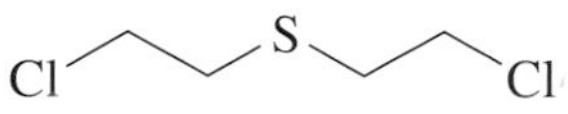

12.13

Bromine forms only disulfur dibromide, S_2Br_2, and the bromosulfanes, S_nBr_2. Chloro- and bromosulfanes are most conveniently made by treating sulfanes with the monohalide. The terminal hydrogens are replaced in this reaction by $-SCl$ or $-SBr$ groups, increasing the sulfur chain length by two. Chloro- and bromosulfanes with up to eight sulfur atoms in the chain have been characterized. Here again sulfur is demonstrating its propensity for catenation.

S_2I_2 is unstable, but can be prepared by shaking aqueous KI with S_2Cl_2 in carbon disulfide: it decomposes to S_8 and other even-numbered ring allotropes.

12.3.4 Oxides of sulfur

The sulfur–oxygen bond is an important feature of sulfur chemistry, just as the silicon–oxygen bond is of silicon chemistry. The best-known simple oxides are the dioxide, SO_2, and the trioxide, SO_3 (although highly reactive lower oxides SO and S_2O have been described). The bonding in SO_2 and SO_3 is usually described in terms of S=O double bonds, VSEPR theory predicting the shapes shown in Structures **12.14–12.17**.

12.14

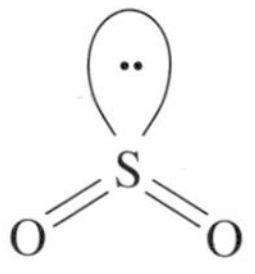

Sulfur dioxide

SO_2, with its characteristic choking smell, is well known as the product of combustion of sulfur or sulfides in air or oxygen. SO_2 is readily soluble in water, dissolving to form sulfurous acid (an acid that has never been isolated as a pure compound) and the sulfite ion:

12.15

$$H_2O(l) + SO_2(g) = 2H^+(aq) + SO_3^{2-}(aq) \qquad (12.11)$$

As a consequence, sulfur dioxide in the presence of water is an effective reducing agent, supplying electrons to an oxidizing agent, via the sulfite ion:

$$SO_3^{2-}(aq) + H_2O(l) = SO_4^{2-}(aq) + 2H^+(aq) + 2e^- \qquad (12.12)$$

This can easily be demonstrated visually by bubbling sulfur dioxide into an aqueous solution of acidified potassium dichromate. There is a rapid colour change from orange to green as the chromium(VI) is reduced to chromium(III):

$$3SO_2(g) + Cr_2O_7^{2-}(aq) + 2H^+(aq) = 3SO_4^{2-}(aq) + 2Cr^{3+}(aq) + H_2O(l) \qquad (12.13)$$

The reducing properties of sulfur dioxide are put to good use commercially in bleaching natural fibres (straw, wool and newsprint). Sulfur dioxide has long been used as an antiseptic and antioxidant in the food industry; wine casks have been fumigated with SO_2 for thousands of years. The home winemaker of today, who adds a sodium metabisulfite tablet to the wash water, is following a very ancient practice! The sodium metabisulfite dissolves to release SO_2, which sterilizes the equipment.

Sulfur trioxide

The reaction of SO_2 with oxygen in the gas phase

$$SO_2(g) + \tfrac{1}{2}O_2(g) = SO_3(g); \quad \Delta G_m^\ominus = -71.0 \, kJ \, mol^{-1} \tag{12.14}$$

or in aqueous solution, is thermodynamically favoured but kinetically hindered. From one point of view, the unfavourable kinetics are a good thing; SO_2 oxidation is not a reaction that we would wish to promote in the air of our cities! On the other hand, Reaction 12.14 is vitally important to the chemical industry. Sulfuric acid, which is the product of the reaction of SO_3 with water, is the most heavily used industrial chemical (see Section 12.3.5, and the Case Study *Industrial Inorganic Chemistry*).

Sulfur trioxide is monomeric in the gas phase (Structures **12.16** and **12.17**). However, at room temperature it exists as a white solid, which can occur in several polymeric forms. There is a low-melting trimer, S_3O_9 (Figure 12.15), which adopts an ice-like ring structure, as well as an involatile asbestos-like chain structure in which chains of SO_4 tetrahedra are linked through apical oxygens.

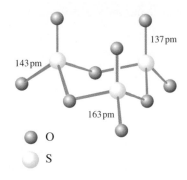

Figure 12.15
The ice-like form of the solid trimer of SO_3.

As well as being a strongly acidic oxide, sulfur trioxide is also quite a strong oxidizing agent. It is used directly in the detergent industry as a sulfonating agent. The SO_3^- group thus introduced confers water solubility on long-chain hydrocarbon detergent molecules.

12.3.5 Oxoacids of sulfur

We considered oxoacids in some detail in Section 11.4. This Section will now give us the opportunity to revise those rules and apply them to the sulfur oxoacids. With sulfur, remember, we have the added complication that an oxygen atom can be substituted by sulfur, giving a range of **thioacids**, for example thiosulfuric acid (**12.18**), dithionic acid (**12.19**) and polythionic acid (**12.20**).

Table 12.4 shows the rich variety of oxoanions of sulfur that are possible. In fact, few of these oxoacids are known in the undissociated state. Apart from some of the polythionic acids, no oxoacid in which sulfur has an oxidation number <6 is stable, even in concentrated aqueous solution, and only sulfuric acid and its 'dimer' disulfuric acid exist as free acids at room temperature. The derivative, obtained by replacing one of the —OH groups of sulfuric acid by an —SH group, thiosulfuric acid, is stable only below 0 °C.

Table 12.4 Oxoacids and oxoanions of sulfur[*]

Oxidation number	Oxoacid/oxoanion		

+6   sulfuric   thiosulfuric   disulfuric

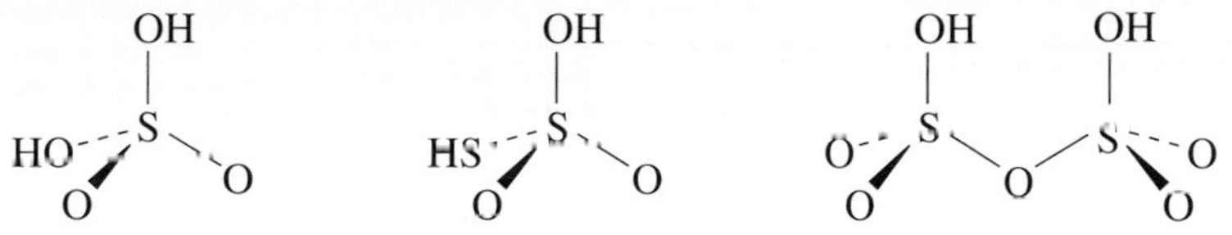

+5   dithionate   thionate (tri-, tetra-, penta-, hexa-)

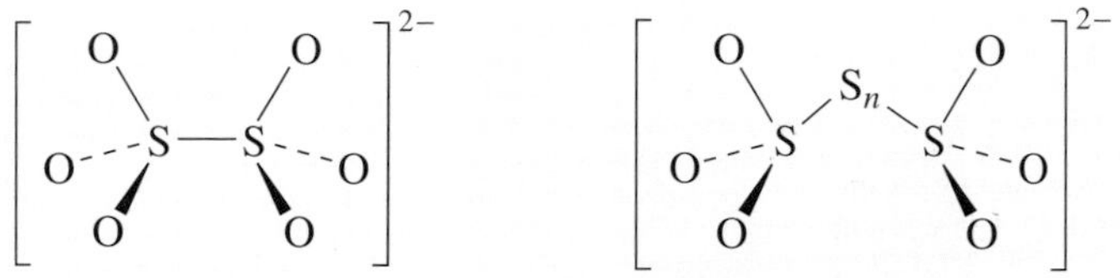

+4   disulfite   sulfite   hydrogen sulfite

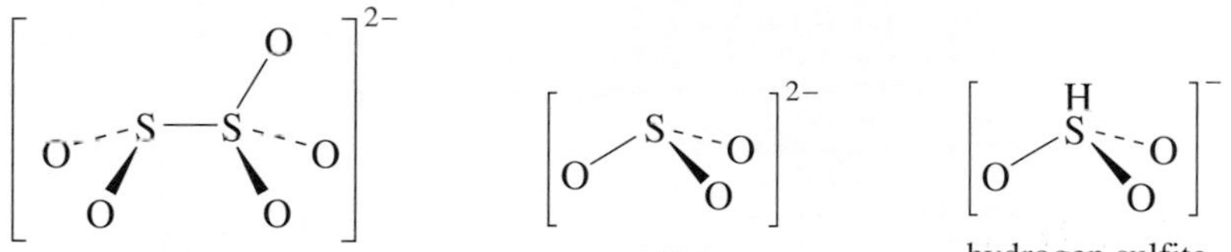

+3   dithionite

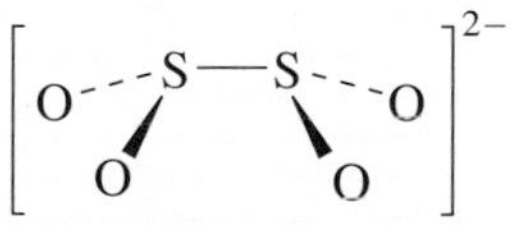

Sulfuric acid is manufactured commercially by the **contact process** (see also the Case Study *Industrial Inorganic Chemistry*). Sulfur dioxide mixed with air is passed over a catalyst (usually vanadium pentoxide, V_2O_5) whereupon it is oxidized to sulfur trioxide:

$$2SO_2(g) + O_2(g) = 2SO_3(g) \tag{12.15}$$

Because sulfur trioxide cannot be satisfactorily absorbed by water, it is treated with 98% concentrated sulfuric acid to form a fuming liquid called *oleum*, $H_2S_2O_7$, which is then carefully diluted with water to give sulfuric acid:

$$SO_3(g) + H_2SO_4(l) = H_2S_2O_7(l) \tag{12.16}$$
$$H_2S_2O_7(l) + H_2O(l) = 2H_2SO_4(l) \tag{12.17}$$

The 19th century claim that there is 'no better barometer to show the state of an industrial nation than the consumption of sulfuric acid per head of population' still holds true today. The chemical finds numerous applications, including the production of fertilizers, petrochemicals, dyestuffs and detergents.

Pure sulfuric acid is a colourless oily liquid, boiling at 317 °C, which, because of its high affinity for water, can dehydrate many organic compounds (recall the similar property of phosphoric oxide). Carbohydrates, for example, can be degraded to elemental carbon (Figure 12.16):

$$H_2SO_4(l) + C_6H_{12}O_6(s) = 6C(s) + H_2SO_4.6H_2O(l) \tag{12.18}$$

(a)

(b)

Figure 12.16
The dehydration of sucrose to elemental carbon by concentrated sulfuric acid (a) before and (b) after addition of acid.

Sulfuric acid, a very strong dibasic acid, forms two ions, the hydrogen sulfate (HSO_4^-) and sulfate (SO_4^{2-}) ions. Most of the salts of sulfuric acid (sulfates and hydrogen sulfates) are water-soluble, with the notable exceptions of $SrSO_4$, $BaSO_4$ and $PbSO_4$; $CaSO_4$ is slightly soluble in water.

12.3.6 Sulfur–carbon and sulfur–nitrogen compounds

A vast number of compounds are known in which sulfur forms single bonds with carbon of types such as RSH, RSR or RSSR, where R is an organic group. The more volatile of these compounds have powerful smells: hence the use of butanethiol, C_4H_9SH, to give an odour to natural gas, so that leaks are easily noticed.

Stable compounds with carbon–sulfur multiple bonds are somewhat more common than compounds with either carbon–silicon or carbon–phosphorus double bonds. The reaction of sulfur vapour with carbon at high temperatures gives carbon disulfide, S=C=S, which is isostructural with CO_2. CS_2 is a smelly liquid at room temperature and a valuable solvent. Compounds of type $R_2C=S$, which are equivalent to ketones such as acetone, $(CH_3)_2C=O$, are readily made, but many are unstable and polymerize to cyclic trimers with carbon–sulfur single bonds.

In biological systems, sulfur is found as a constituent of a number of carbon-based organic molecules. Most sulfur assimilated by organisms is incorporated into the two amino acids cysteine and methionine, having been reduced to the −2 oxidation state. Structure **12.21** shows the formation of a disulfide bridge between two cysteine molecules; these units are essential features of the structural protein keratin, which forms the main component of hair and nails.

$$\begin{array}{c} H_2N-CH-COOH \\ | \\ CH_2 \\ | \\ S \\ | \\ S \\ | \\ CH_2 \\ | \\ H_2N-CH-COOH \end{array}$$

12.21

The chemistry of sulfur–nitrogen compounds is rich and varied. Tetrasulfur tetranitride, S_4N_4, with its closed, basket-like shape (Figure 12.17a) is prepared by passing NH_3 into a warm solution of S_2Cl_2 in benzene:

$$6S_2Cl_2 + 16NH_3 = S_4N_4 + 8S + 12NH_4Cl \qquad (12.20)$$

The crystals formed are yellow when cold, becoming orange at room temperature and red when heated. They are kinetically stable in air, but can decompose explosively to the elements if struck or heated. The S—N bond lengths are all the same, but quite short compared with the sum of the covalent radii (162 pm compared with 178 pm). It is difficult to describe the bonding in classical terms, but some possible resonance structures are shown in Figure 12.17b.

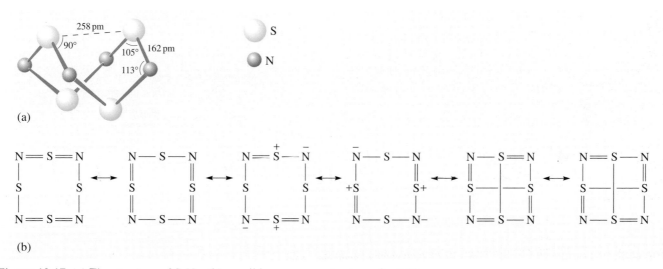

Figure 12.17 (a) The structure of S_4N_4; (b) possible resonance structures for S_4N_4.

If S_4N_4 is heated and passed over silver wool at low pressure, an unstable dimer, S_2N_2, is formed, which has square-planar geometry (Structure **12.22**). The instability of S_2N_2 is demonstrated by striking or warming, but if left at room temperature it spontaneously polymerizes to give $(SN)_x$, as shown in Figure 12.18. Poly(sulfur nitride), $(SN)_x$, first synthesized in 1910, is a bronze-coloured material with remarkable physical properties. It is a metallic conductor at room temperature with conductivity being greatest along the chain, where π orbitals on sulfur and nitrogen overlap to form a conduction band. However, if cooled below 0.3 K, it loses all resistance to the flow of electric current, i.e. it becomes a superconductor.

12.22

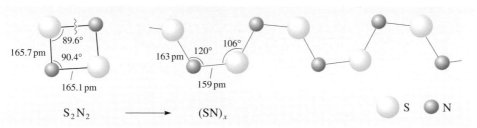

Figure 12.18 The structure of $(SN)_x$ and its relation to S_2N_2.

12.4 Selenium, tellurium and polonium

We have already noted that the properties of the elements in Group VI show the characteristic trends that we have come to expect on descending a Group. The elements become more metallic in character: oxygen is a covalently bonded gaseous diatomic molecule; sulfur is a solid containing S_8 molecules and is an insulator; selenium (non-metal) and tellurium (semi-metal) are semiconductors with polymeric structures; polonium is a metal. The compounds of selenium, tellurium and pollonium also illustrate the inert pair effect and a tendency to higher coordination numbers.

BOX 12.4 Selenium in photocopying

One of the most important applications of selenium lies in the photocopying industry. The process of xerography (derived from the Greek, '*dry writing*') first patented by American scientist Chester Carlson in 1942 (Figure 12.19a and b) involves a number of steps leading to the production of a copied document. However, crucial to the process is an electrostatically charged photoreceptor, which is exposed to a light and dark image pattern; in the light areas the photoreceptor becomes discharged due to the photoelectrons produced. To develop the image, black toner particles are used which are attracted to the charged areas on the plate, and which correspond to the dark areas of the original pattern. The toner can then be transferred to paper and fixed, usually by heating, to make a final copy. In Carlson's original patent, anthracene, sulfur and mixtures of sulfur and selenium were used as photoconductors. However, these proved very limited, particularly in the visible region of the spectrum. In the modern xerographic process, the photoreceptor typically consists of amorphous selenium deposited on an aluminium drum. Frequently selenium is alloyed with tellurium which ensures sensitivity at all visible wavelengths.

(a)

(b)

Figure 12.19 (a) An extract from the 1942 patent filed by Chester Carlson for electrophotography. (b) This special commemorative stamp was issued in honour of Chester Carlson.

BOX 12.5 Selenium and diet

Despite being highly toxic, selenium is known to be essential for human life, being a constituent of the enzyme *glutathionine peroxidase*, which catalyses reactions resulting in the removal of toxic peroxides from the body. The daily requirement of selenium is around 70 mg; major sources include meat and fish, dairy foods and brazil nuts. Selenium levels in food are related to the concentration in the soil where it is grown. In parts of China, Keshan disease caused by selenium deficiency was once endemic. This tends to affect young children in whom it causes heart failure, although regular dietary supplements of sodium selenite, Na_2SeO_3, have virtually eradicated the problem (Figure 12.20).

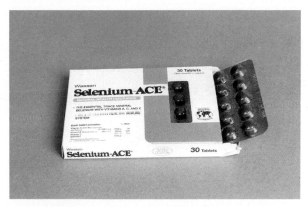

Figure 12.20
A commercial selenium food supplement.

Moving down Group VI, we see, as we have seen for Groups IV and V, a decreasing ability to form double bonds. Take the compounds with carbon as an example. Carbon dioxide is a gas at room temperature, which exists only in the form of O=C=O molecules. In contrast, carbon disulfide and carbon diselenide are liquids at room temperature, which contain S=C=S and Se=C=Se molecules, respectively. However, under high pressure, and on heating, CS_2 gives a black solid polymer containing both double and single bonds; with CSe_2, this change occurs slowly at room temperature and pressure; Te=C=Te is unknown.

The halides formed by selenium, tellurium and polonium are summarized in Table 12.5.

Table 12.5 The halides of selenium, tellurium and polonium

		Oxidation number		
<1	+1	+2	+4	+6
	Se_2F_2	SeF_2*	SeF_4	SeF_6
	Se_2Cl_2	$SeCl_2$*	$(SeCl_4)_4$	
	Se_2Br_2	$SeBr_2$*	$(SeBr_4)_4$	
		~~TeF_4~~		TeF_6
Te_2Cl		$TeCl_2$	$(TeCl_4)_4$	
Te_3Cl_2			TeF_4	
Te_2Br		$TeBr_2$*	$(TeBr_4)_4$	
Te_2I	Te_4I_4		$(TeI_4)_4$	
			PoF_4†	
		$PoCl_2$	$PoCl_4$	
		$PoBr_2$	$PoBr_4$	
		PoI_2	PoI_4	

*Unstable at room temperature.

†Existence uncertain.

It should be remembered that polonium is a rather rare element, which is difficult to handle because of its radioactivity; because of this its chemistry has not been as fully explored as that of selenium and tellurium.

Table 12.5 shows evidence for the trends that we have come to expect. *The highest oxidation number (+6) is stabilized by fluorine only in the cases of selenium and tellurium; PoF$_6$ is not known.* There are no low oxidation number fluorides, apart from F—Se—Se—F and Se=SeF$_2$, and these can only be trapped at low temperatures. Selenium forms no iodides, but the more electropositive tellurium and polonium do.

We see the inert pair effect increasing down the Group as the stability of oxidation number +4 increases. All four tetrahalides exist for tellurium and polonium, but not for selenium.

The tendency towards higher coordination number down the Group can be seen in SeF$_4$ and TeF$_4$. Thus, SeF$_4$ is a colourless volatile liquid, which gives monomeric four-coordinate molecules in the gas phase (like SF$_4$), whereas TeF$_4$ forms colourless crystals containing chains of *cis*-linked five-coordinate square-pyramidal TeF$_5$ units (Figure 12.21).

The oxides in which the element is in oxidation number +6, MO$_3$, are known for Se and Te. The structure of SeO$_3$ contains cyclic tetramers in which the selenium is four coordinate, and the α-TeO$_3$ structure contains TeO$_6$ octahedra sharing all vertices to give a three-dimensional lattice; *again we see the trend to higher coordination number on descending the Group.* All three dioxides, MO$_2$, are known; use Question 12.5 to try to predict the changes in structure that you might expect to see as you go down Group VI. This will be further discussed in Section 13.2.2.

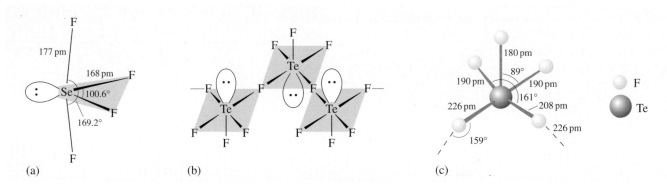

Figure 12.21 The structures of (a) gaseous SeF$_4$ and (b) solid TeF$_4$. (c) The geometry around a TeF$_5$ group in solid TeF$_4$.

12.5 Summary of Sections 12.3 and 12.4

1   Sulfur forms hydrides (sulfanes) only in oxidation number +2 (or lower). These have the general formula H_2S_n (where n represents the number of sulfur atoms in the chain).

2   The higher sulfanes are unstable with respect to decomposition into H_2S and solid sulfur.

3   The sulfanes are weak acids, liberated by acidification of their salts (sulfides or polysulfides).

4   Sulfur can occur in many oxidation numbers in its halogen compounds. The highest oxidation number achieved is graded in order of electronegativity (or oxidizing power) of the halogen: for bromosulfanes it is +2, for chlorosulfanes +4, for fluorosulfanes +6 (there are no known iodosulfanes).

5   Halosulfanes (apart from SF_6, which is kinetically inert) are readily hydrolysed, and fume in moist air owing to formation of HX.

6   The important oxides of sulfur are the dioxide, SO_2, and trioxide, SO_3. As with fluorine, use of an electronegative ligand favours higher oxidation numbers.

7   Sulfur dioxide is a gas at room temperature and is a reducing agent.

8   Sulfur trioxide is a white solid at room temperature, occuring in several polymeric forms. SO_3 is a fairly strong oxidizing agent.

9   Sulfur exhibits a rich variety of oxoacids and thioacids. Sulfuric acid, prepared industrially on a massive scale by the contact process, is a strong dibasic acid. It is both an oxidizing and effective dehydrating agent, and forms sulfate and hydrogen sulfate salts.

10   The chemistry of selenium, tellurium and polonium illustrates trends down a Group such as increased metallic character, increased inert pair effect and increased tendency to higher coordination numbers.

QUESTION 12.1

Predict the formula of the oxoacid of sulfur in which sulfur is in its highest oxidation number. What is the formula of the condensed acid made from two molecules of this acid?

QUESTION 12.2

Using Pauling's rules, predict the strength of sulfuric acid, H_2SO_4. If its first dissociation constant is 1×10^3 mol litre^{-1}, what value would you predict for the second dissociation constant?

QUESTION 12.3

S_4N_4 decomposes explosively. Suggest the likely decomposition route, and therefore what drives this reaction.

QUESTION 12.4

Suggest some resonance structures to describe the bonding for the S_2N_2 molecule.

QUESTION 12.5

How might you expect the structures of the Group VI oxides, MO_2, to vary down the Group?

THE TYPICAL ELEMENTS: A SUMMARY OF TRENDS IN THE PERIODIC TABLE

13

You have now completed the survey of the p-Block elements, as laid out in the mini-Periodic Table of Figure 13.1. In this last Section we summarize some of the trends in chemical properties that occur both across the Periods and down the Groups of Figure 13.1.

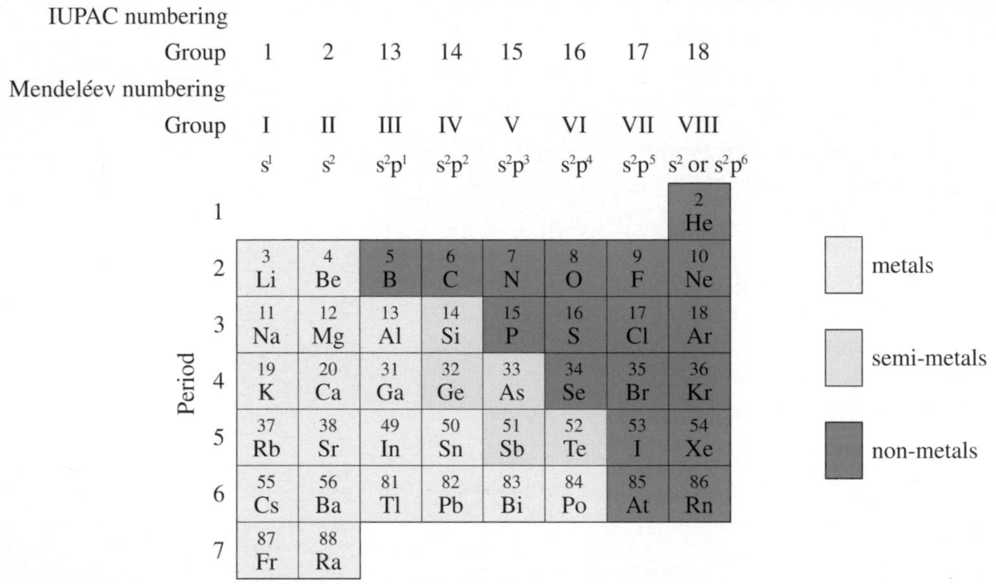

IUPAC numbering

| Group | 1 | 2 | 13 | 14 | 15 | 16 | 17 | 18 |

Mendeléev numbering

| Group | I | II | III | IV | V | VI | VII | VIII |
| | s^1 | s^2 | s^2p^1 | s^2p^2 | s^2p^3 | s^2p^4 | s^2p^5 | s^2 or s^2p^6 |

Period

1								2 He
2	3 Li	4 Be	5 B	6 C	7 N	8 O	9 F	10 Ne
3	11 Na	12 Mg	13 Al	14 Si	15 P	16 S	17 Cl	18 Ar
4	19 K	20 Ca	31 Ga	32 Ge	33 As	34 Se	35 Br	36 Kr
5	37 Rb	38 Sr	49 In	50 Sn	51 Sb	52 Te	53 I	54 Xe
6	55 Cs	56 Ba	81 Tl	82 Pb	83 Bi	84 Po	85 At	86 Rn
7	87 Fr	88 Ra						

metals

semi-metals

non-metals

Figure 13.1
A periodic arrangement of the typical or main-Group elements (note that the helium atom has the outer electronic configuration s^2 rather than s^2p^6).

13.1 Trends across the Periods

As noted earlier, the first ionization energies and electronegativities of the typical elements show an overall increase across the rows of Figure 13.1, whereas the single-bond covalent radii show a decrease.

13.1.1 Metals, semi-metals and non-metals

According to the simplest theory of metallic bonding, electrons hold a metal together through interaction with the residual positive ion cores. The tendency is for metals to give way first to semi-metals, and then to non-metals, across a row of Figure 13.1.

How is this tendency manifest in the structure of the following elements of Period 3: sodium, silicon and sulfur?

Sodium is a metal with the body-centred structure, silicon is a semi-metal, and sulfur has a molecular covalent structure of S_8 rings. These structures are shown in Figure 13.2.

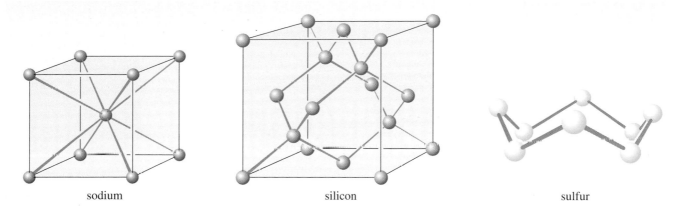

sodium   silicon   sulfur

Figure 13.2 The structures of elements sodium, silicon and sulfur.

Sodium, silicon and sulfur are in the same row of the Periodic Table. The approximate electrical conductivities of sodium, silicon and sulfur at 0 °C are $10^7\,S\,m^{-1}$, $10\,S\,m^{-1}$ and $10^{-15}\,S\,m^{-1}$, respectively. Consistent with its high metallic conductivity and other metallic properties, sodium has a characteristic body-centred cubic metallic structure. According to the simplest theory, this consists of Na^+ cores and an electron gas. At first sight, the increase in nuclear charge at silicon has confined the valence electrons to two-electron covalent bonds in a diamond-type structure. However, the confinement is not so great that thermal energies at normal temperatures cannot detach *some* of these bonding electrons and transfer them to an electron gas (simple theory) or to a conduction band (band theory). Silicon is therefore a semi-metal and a semiconductor. By contrast, at sulfur, confinement of the valence electrons to S—S covalent bonds or non-bonded pairs is complete, and the structure consists of discrete S_8 ring molecules, between which van der Waals forces operate.

13.1.2 The structures of halides, hydrides and oxides

Halides of the Group I elements Li, Na, K and Rb, involving large differences in electronegativity, have the structure of NaCl.

- How does electronegativity control the structure of the halides in moving across the rows of the Periodic Table?

- In moving across the rows, the electronegativity of the elements increases, and there is a corresponding decrease in the electronegativity difference between the two kinds of atom in their halides and oxides. Consequently, the ionic character of the compounds decreases. Structurally, this is apparent in the way that three-dimensional ionic structures give way first to either layer, chain or macromolecular structures, and then to discrete covalent molecules.

To take the specific example of chlorides in the third Period, sodium chloride, which is typically ionic, gives way to $MgCl_2$, which adopts a $CdCl_2$ layer structure (Figure 13.3, overleaf), and then to $AlCl_3$ with the layer structure described in Figure 9.20.

The loss of ionic character between NaCl and $AlCl_3$ is also apparent in the conductivity of the molten salts: molten NaCl and $MgCl_2$ are excellent conductors of electricity; molten $AlCl_3$ is a very poor conductor. Beyond aluminium, there are chlorides with discrete covalent molecules. Thus, $SiCl_4$ and SCl_2 are non-conducting volatile liquids containing tetrahedral and V-shaped molecules, respectively, and Cl_2 is a gas.

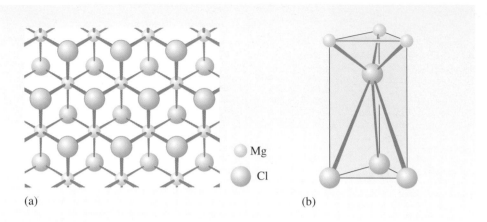

(a) Mg Cl (b)

Figure 13.3 The $CdCl_2$ layer structure of $MgCl_2$. In (a) one layer is viewed from above; it has three decks. The top and bottom decks consist of chlorines; these have been distinguished in the Figure by giving the top-deck atoms a heavy border; the middle deck consists of octahedrally coordinated magnesiums. In (b) the environment of each chlorine is shown. It is a distorted trigonal prism: there are three magnesiums on one side, but on the other side there are three chlorines substantially farther away in an adjacent layer. This is not what would be expected of a collection of ions.

How do trends of this kind reveal themselves in the structures of the normal oxides of sodium, silicon and chlorine?

Na_2O has an ionic-type antifluorite structure, SiO_2 has a macromolecular structure in which silicon and oxygen attain noble gas structures by forming covalent bonds with each other, and the oxides of chlorine (Cl_2O, ClO_2, Cl_2O_7) consist of discrete covalent molecules.

13.1.3 Trends in the formulae of oxides and hydrides

It is worth recalling the trends in the formulae of the highest normal oxides and of the highest hydrides, which were so successfully exploited by Mendeléev when he put together his Periodic Table. The elements usually exercise their highest valency and highest oxidation number in the highest normal oxide. In most cases, both are equal to the number of outer electrons, which is taken to be zero for the noble gases.

What are the formulae of the highest normal oxides of the elements of the third Period?

Na_2O, MgO, Al_2O_3, SiO_2, P_2O_5, SO_3, Cl_2O_7 and no oxide at argon; the highest valencies and highest oxidation numbers run from one at sodium to seven at chlorine, equal to the number of outer electrons.

If we allow expansion of the octet, this valency pattern occurs because in its highest oxide, each element uses all its outer electrons in forming bonds. Thus, with six outer electrons, sulfur can form six electron-pair covalent bonds, and an oxide SO_3 results. In the case of the hydrides the pattern is different. There is no expansion of the octet, and the elements attain noble gas configurations in the hydrides. This gives formulae AH_x, where x takes values 1, 2, 3, 4, 3, 2, 1, 0 across a row.

13.1.4 Acid–base properties of oxides and hydrides

There is a tendency for normal oxides, especially the highest normal oxides, to change from basic, through amphoteric to acidic from left to right across a Period of Figure 13.1. Thus, at the start of the third Period, Na_2O is basic, and at the end, Cl_2O_7 is acidic (Section 12.2.2).

This is apparent from the reactions with water, when basic and acidic solutions are formed, respectively:

$$Na_2O(s) + H_2O(l) = 2Na^+(aq) + 2OH^-(aq) \tag{13.1}$$

$$Cl_2O_7(l) + H_2O(l) = 2H^+(aq) + 2ClO_4^-(aq) \tag{13.2}$$

In between these two oxides comes Al_2O_3, which has negligible solubility in pure water, but reveals amphoteric character by dissolving in both acids and bases:

$$Al_2O_3(s) + 6H^+(aq) = 2Al^{3+}(aq) + 3H_2O(l) \tag{13.3}$$

$$Al_2O_3(s) + 2OH^-(aq) + 3H_2O(l) = 2[Al(OH)_4]^-(aq) \tag{13.4}$$

A related trend is apparent in the reactions of the hydrides with water. At the beginning of the Periods, the Group I and Group II hydrides give hydrogen gas and basic solutions with water, for example

$$NaH(s) + H_2O(l) = Na^+(aq) + OH^-(aq) + H_2(g) \tag{13.5}$$

These basic properties contrast with the acidity of the hydrides of Group VII elements; for example

$$HCl(g) + H_2O(l) = H_3O^+(aq) + Cl^-(aq) \tag{13.6}$$

13.2 Trends down the Groups

On descending a Group of typical elements, there are *overall* decreases in first ionization energy and electronegativity, and *overall* increases in covalent and ionic radii. The influence of these properties is seen in metallic/non-metallic behaviour, structure and bonding patterns within the Groups.

13.2.1 Metals, semi-metals and non-metals

At the top of the Groups, the outer electrons are closer to the nucleus, and ionization is difficult. Electrons are therefore less likely to be lost to become part of an electron gas that can engender metallic bonding. The increasing distance of the outer electrons from the nucleus therefore accounts for the tendency for non-metals to give way first to semi-metals, and then to metals, down a Group of Figure 13.1. Group IV provides a good example: carbon is a non-metal, silicon and germanium are semi-metals and semiconductors, and tin and lead are metals.

13.2.2 Structure and bonding in halides and oxides

Within the Groups in Figure 13.1, the electronegativity is highest at the top, and it is here where it is usually closest to the high electronegativities of oxygen and the halogens. On descending a Group, electronegativity decreases, so electronegativity differences in oxides and halides tend to increase. Oxides and halides should therefore become more ionic.

In the Group II chlorides [*] this trend is marked first by a progression from the chain structure of $BeCl_2$ (Figure 13.4) to the layer structure of $MgCl_2$. Neither are what we would expect of a collection of ions. Dichloride structures more consistent with an ionic model are observed only for calcium, strontium, barium and radium: $CaCl_2$ has the rutile structure (Figure 13.5), $SrCl_2$, the fluorite structure (Figure 13.6), and, in $BaCl_2$ and $RaCl_2$, each metal atom is surrounded by nine chlorides (Figure 13.7), and each chloride by either four or five metal atoms.

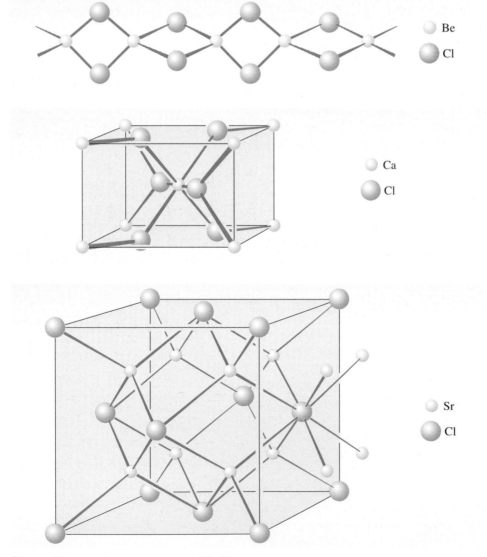

Be
Cl

Figure 13.4
The chain structure of solid $BeCl_2$.

Ca
Cl

Figure 13.5
The rutile structure of $CaCl_2$.

Sr
Cl

Figure 13.6 The fluorite structure of $SrCl_2$.

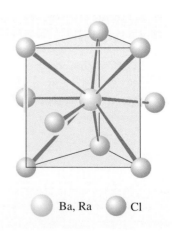

Ba, Ra   Cl

Figure 13.7
The structure of $BaCl_2$ and $RaCl_2$.

[*] The chemistry of the Group II elements is covered in detail in *Metals and Chemical Change*.[3]

⬤ What are the coordination number of calcium in $CaCl_2$, and of strontium in $SrCl_2$?

◐ Six and eight, respectively; in Figure 13.5, calcium is in approximately octahedral coordination, and in $SrCl_2$, strontium has the fluorite structure of Figure 13.6.

Thus, along with the structural evidence of increasing ionic character, we see the influence of the increase in ionic radii down the Group: the coordination number increases from four in $BeCl_2$ to six in $MgCl_2$ and $CaCl_2$, through eight in $SrCl_2$ to nine in $BaCl_2$ and $RaCl_2$.

A similar trend in a very different form is revealed in the Group IV dioxides.

⬤ What forms are taken by the dioxides of the Group IV elements C, Si, Ge, Sn and Pb? How do these structures reflect trends in bonding down the Group?

◐ Carbon dioxide is a gas consisting of linear covalent molecules, $O=C=O$, whereas SiO_2 and GeO_2 form quartz-like structures, consistent with the presence of Si—O and Ge—O covalent bonds; SnO_2 and PbO_2, however, have the rutile structure. Here, there is a progression from a discrete molecular covalent substance, through macromolecules to ionic structures. At the same time, the coordination number increases from two in CO_2, to four in SiO_2 and GeO_2, to six in SnO_2 and PbO_2. The progression is partly explained by the observation that second-row elements have a greater tendency than later-row elements to form multiple-bonded compounds. Because carbon forms double bonds in its dioxide, the coordination number is lower than in the singly bonded SiO_2 and GeO_2.

In the case of the Group VI dioxides, the trend is recognizably similar. There is a transition from covalent molecules at O_3 and SO_2, through macromolecular chain structures in SeO_2 and TeO_2 (Figure 13.8) to a fluorite, ionic-type structure in PoO_2. The coordination number increases in the sequence two, two, three, four, eight, but the drift away from π bonding is in this case more gradual, and is completed only at tellurium.

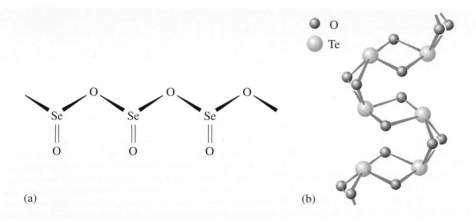

(a)   (b)

Figure 13.8 The macromolecular chain structures of (a) SeO_2 and (b) tellurite, the naturally occurring form of TeO_2. In both structures, the Group VI element has been drawn in a tetravalent state.

13.2.3 Acid–base properties of oxides

Normal oxides of a particular formula type tend to become less acidic, and more basic in descending a Group. Thus, in Group II, BeO is amphoteric and has very low solubility in water, but dissolves in both acids and bases. The other Group II monoxides, however, are basic, and yield basic solutions when added to water; for example

$$BaO(s) + H_2O(l) = Ba^{2+}(aq) + 2OH^-(aq) \qquad (13.7)$$

The full change from acidic to basic is best seen in Group III. The oxide B_2O_3 is acidic, Al_2O_3 and Ga_2O_3 are amphoteric, and In_2O_3 and Tl_2O_3 are basic.

13.2.4 Special differences between the second and subsequent Periods

In Figure 13.1, the second Period is the first to contain an element from each Group of the typical elements, and the differences between each of its elements, and the one beneath it in the Table, are greater than in any subsequent pair of Periods. One of these differences is the greater preference for multiply bonded structures in the chemistry of second-row elements. This is especially obvious in the forms of the elements of Groups IV, V and VI. Thus, nitrogen and oxygen are gases, but phosphorus and sulfur assume solid, singly bonded forms at room temperature.

⬤ To what extent is this trend apparent in the known forms of the elements carbon and silicon?

◗ Carbon, in the form of diamond, and silicon share the same single-bonded structure, but carbon is also found as graphite (Figure 10.2) and fullerenes (Figure 10.4), which consist of shared π-bonded rings.

Although molecules containing silicon–silicon and phosphorus–phosphorus multiple bonds are known, they are thermodynamically unstable, and exist only when bulky groups attached to the silicon or phosphorus atoms provide a kinetic barrier to conversion into single-bonded polymers.

Two distinct factors tend to make π-bonded structures more stable in the second Period than the third. First, the strength of π bonding seems to decrease down a Group; secondly, in Groups V and VI, alternative single-bonded structures are disadvantaged by a weakness of the N—N and O—O single bonds, which is attributed to the repulsion between non-bonding electrons over the necessarily short internuclear distances.

Third-row typical elements in Groups V, VI and VII show another difference from second-row elements, as they can exist in higher oxidation numbers. For example, the highest fluorides of oxygen and sulfur are OF_2 and SF_6; the highest normal oxides of fluorine and chlorine are F_2O and Cl_2O_7. One possible explanation is the expansion of the octet in the third and later Periods through the use of d orbitals of the same principal quantum number as the s and p valence electrons, a possibility that is not open to second-row elements. This may allow higher coordination numbers to be reached by second-row elements in both oxidation reactions and the formation of complex ions. Thus, potassium fluoride and silicon tetrafluoride will react to form K_2SiF_6, which contains the octahedral complex ion, $[SiF_6]^{2-}$. The silicon octet in SiF_4 has been expanded by the donation of non-bonded pairs by two fluoride anions, giving silicon a coordination number of six. As this possibility is not open to carbon, CF_4 does not react with KF, and K_2CF_6 has not been prepared. The ability of silicon to expand its coordination number provides a possible reason why the hydrolysis of $SiCl_4$ is fast but that of CCl_4 is immeasurably slow; as CCl_4 is less willing to allow entry of an incoming ligand into its coordination sphere, there is a kinetic energy barrier to the thermodynamically favourable hydrolysis.

13.2.5 The inert pair effect

If one ignores catenated compounds (those in which atoms of the element in question are bound to each other), the typical elements tend to form oxidation numbers that differ by multiples of two. Thus, sulfur exists in oxidation numbers -2, 0, $+2$, $+4$ and $+6$; as usual, the highest oxidation number is equal to the number of outer electrons. The so-called inert pair effect is most clearly seen among the elements of Periods 4, 5 and 6 in Groups III, IV and V (Figure 13.9): from Period 3 to Period 6, the lower oxidation number, which is two less than the number of outer electrons, tends to increase in stability relative to the higher oxidation number.

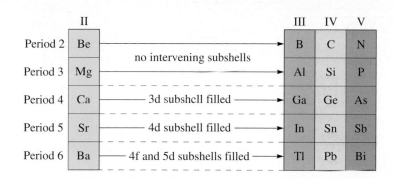

Figure 13.9
The Groups and elements in which the so-called inert pair effect is seen.

Which elements in Figure 13.9 show the most marked inert pair effect, and which of their oxidation numbers does the effect stabilize?

Thallium(I), lead(II) and bismuth(III); the effect is most marked in Period 6, and the oxidation number stabilized is two less than the Mendeléev Group number.

Chemical evidence proves the stability of these oxidation numbers: the higher thallium(III), lead(IV) and bismuth(V) are all powerful oxidizing agents. Bi_2O_5, for example, loses oxygen very easily and forms Bi_2O_3. Figure 13.9 shows that the phenomenon occurs among those elements that immediately follow the filling of f and/or d subshells in the Periodic Table. It is most marked in Period 6, where the filling of the 4f *and* 5d subshells in the 24 preceding elements will be associated with a significant increase in nuclear charge. This increased nuclear charge will tend to bind the valence electrons in thallium, lead and bismuth more tightly and resist their removal for ion or bond formation in higher oxidation numbers. In particular, when the outer p electrons have been lost, the ions Tl^+, Pb^{2+} and Bi^{3+} are left with the configuration $[Xe]4f^{14}5d^{10}6s^2$. The unusually high nuclear charge resists removal of the outer $6s^2$ electrons for bond formation in higher oxidation numbers. It is because of this resistance of the outer s^2 pair for bonding that the phenomenon is called the inert pair effect.

QUESTION 13.1

By referring to Figure 13.1 only, state which element of the following pairs has (a) the higher first ionization energy, (b) the higher electronegativity, (c) the larger covalent radius: (i) aluminium and sulfur; (ii) oxygen and tellurium; (iii) chlorine and germanium.

QUESTION 13.2

Consider the compounds IBr, $CaCl_2$ and $CaMg_2$. One is ionic, one is covalent and one is metallic. Identify which is which, match each compound to one of the three descriptions (i)–(iii) below, and state which one contains discrete molecules, with the given formula, in the solid state.

(i) A white solid that melts at 782 °C. It is a poor conductor of electricity in the solid state, but a good one when molten or when dissolved in water.

(ii) A brown–black solid that melts at 41 °C to give a liquid with low electrical conductivity.

(iii) A silvery-looking solid that melts at 720 °C. It is an excellent conductor of electricity in both solid and liquid states.

QUESTION 13.3

Substance A is a typical element and is classified as a semi-metal. It forms two normal oxides, AO_2 and AO_3, and the fluorides AF_4, A_2F_{10} and AF_6. Identify A. What will be the formula of its hydride?

QUESTION 13.4

Consider the three oxides PbO, BaO and SeO_3. One is acidic, one is amphoteric and one is basic. Identify which is which.

QUESTION 13.5

Samples of the oxides GeO_2, SnO_2 and PbO_2 in crucibles are all placed in a sand bath at a temperature of 650 °C. One of the oxides decomposes. Which one is it, and what will be the decomposition reaction?

LEARNING OUTCOMES

Now that you have completed *Elements of the p Block*, you should be able to do the following things:

1 Recognize valid definitions of, and use in a correct context, the terms, concepts and principles in the following Table. (All Questions)

List of scientific terms, concepts and principles introduced in *Elements of the p Block*

Term	Page number	Term	Page number	Term	Page number
acid anhydride	79	eutrophication	187	oxoacid general formula	178
acidic oxide	79	fractional distillation	87	Pauling electronegativity scale	32
activated charcoal	133	Frasch process	191	Pauling's first rule	181
amorphous substance	157	fullerenes	133	Pauling's second rule	182
amphibole	147	greenhouse effect	141	pH	26
anisotropic structure	132	Grotthuss mechanism	49	phosphonium salts	172
bond enthalpy term	30	hydrogen bond	54	pK_a	26
Brønsted–Lowry theory	23	hydrogenation	43	plastic sulfur	193
buckminsterfullerene	133	hydrophilic substance	187	polar molecule	34
cage structures	171	hydrophobic substance	187	polarizability	38
carbonyl compound	138	induced dipole–induced dipole interaction	37	polybasic acid	178
catenation	199	inert pair effect	125	polyphosphazenes	176
combustion	195	interstitial carbides	138	polysulfides	199
condensation of oxoacids	179	ionic resonance energy	33	pyroxenes	147
conjugate acid	23	isostructural form	129	Raschig process	162
conjugate base	23	isotropic structure	132	Rochow process	151
contact process	203	Lewis acid	29	silanes	150
coordinative saturation	200	Lewis base	29	silicon wafer	143
cross-linked polymer	152	Mond process	138	silicones (polysiloxanes)	151
dehydrating reagent	174	nanotubes	134	silylenes	143
dihedral angle	193	neutral oxide	197	Speier hydrosilation process	151
dipole	34	nitrogen fixation	159	strong acid	22
dipole moment	34	orthorhombic sulfur	192	sulfanes	199
diradical	193	Ostwald process	168	thioacids	202
disilene	152	oxidation number	18	vulcanization process	193
dry ice	139	oxidation state	18	weak acid	22
energies of hydration	71				

2   Assign oxidation numbers to elements in simple compounds or ions, and use them to identify elements that are oxidized or reduced. (Questions 2.1 and 2.2)

3   Use oxidation numbers to balance redox equations in which the reactants and products are known. (Questions 2.3 and 2.4)

4   Write Brønsted–Lowry acid–base reactions and identify the conjugate acid/base pairs. (Question 3.1)

5   Given the relative strengths of Brønsted–Lowry acids, predict whether a reaction will occur between one acid and the conjugate base of another. (Question 3.2)

6   Relate values of $[H^+]$, pH, K_a and pK_a, and perform simple calculations involving these quantities. (Question 3.3)

7   Identify Lewis acids and Lewis bases in chemical reactions. (Question 4.1)

8   Draw structural formulae in which a dative bond A→B is represented $\overset{+}{A}-\overset{-}{B}$. (Question 4.2)

9   Understand the relationship between bond enthalpy terms and the enthalpies of atomization of gaseous covalent molecules, and calculate one from the other. (Question 4.3)

10 Understand the relationship between bond enthalpy terms and electronegativity differences embodied in Pauling's electronegativity equation, and calculate one from the other. (Question 4.4)

11 Relate the dipole moments of molecules to molecular structure, electronegativity differences, and the influence of non-bonded pairs (Question 4.5)

12 Assess the relative polarizabilities of two simple molecules, and use your conclusion to decide which will have the higher boiling temperature. (Question 4.6)

13 Recall important items of information about hydrogen and its compounds, and combine them with others to gain new insights about the chemistry of the element. (Questions 5.1, 5.2, 5.3 and 5.4)

14 Recall important items of information and generalizations about the halogens and their compounds, and combine them with others to gain new insights into the chemistry of the elements. (Questions 6.1–6.10)

15 Relate the acidity of an oxoacid to the number of terminal oxygens in the undissociated molecule. (Question 11.3–11.8 and 12.2)

16 Recall important items of information about the noble gases and their compounds, and combine them with others to gain new insights into the chemistry of the elements. (Questions 7.1–7.5)

17 Provide a molecular orbital treatment of the bonding in XeF_2 and related systems, including a partial orbital energy-level diagram. (Question 7.5)

18 Use your understanding of the trends in the Periodic Table to make predictions about the chemistry of the elements of Groups III–VI. (Question 8.1, 8.2 and 13.1–13.5)

19 Apply skills and understanding to the chemistry of the elements of Groups III–VI. (Questions 9.1–9.6, 10.2–10.10, 10.12, 11.1, 12.4 and 12.5)

20 Recall important items of information about the Group III elements, and combine them to gain new insights about the chemistry of those elements. (Questions 9.3–9.5 and 9.7)

21 Recall important items of information about the Group IV elements, and combine them to gain new insights about the chemistry of those elements. (Questions 10.1, 10.5, 10.11 and 13.5)

22 Predict the formulae of ortho oxoacids from considerations of coordination number and oxidation number. (Questions 11.2, 11.4 and 12.1)

23 Recall important items of information about the Group V elements, and combine them to gain new insights about the chemistry of those elements. (Questions 11.3 and 11.9)

24 Use Pauling's second rule to predict (approximately) the first dissociation constant of an acid, from a consideration of the value of t in the general formula $AO_t(OH)_n$. (Questions 11.4–11.7 and 12.2)

25 Use Pauling's first rule to predict (approximately) the dissociation constants for each step in the dissociation of polybasic acids. (Questions 11.5, 11.8 and 12.2)

26 Derive structural information for an unknown oxoacid, given the value of t in the general formula $AO_t(OH)_n$. (Question 11.7)

27 Recall important items of information about the Group VI elements, and combine them to gain new insights about the chemistry of those elements. (Questions 12.1 and 12.3–12.5)

QUESTIONS: ANSWERS AND COMMENTS

QUESTION 2.1 (Learning Outcome 2)

(i) zero (rule 2); (ii) −2 (rule 1); (iii) +4 (rules 4 and 7); (iv) +6 (rules 4 and 7); (v) −2 (rules 5 and 7); (vi) +2 (rules 6 and 7); (vii) +6 (rules 4, 5 and 7); (viii) +4 (rules 4 and 7); (ix) +6 (rules 3 and 7); (x) +6 (rules 4, 5 and 7).
The substances are grouped by oxidation number of sulfur in the following table:

Sulfur oxidation number				
−2	0	+2	+4	+6
$H_2S(g)$	S(s)	$SCl_2(l)$	$SO_2(g)$	$SO_3(s)$
$S^{2-}(aq)$			$SO_3^{2-}(aq)$	$H_2SO_4(l)$
				$HSO_4^-(aq)$
				$SF_6(g)$

QUESTION 2.2 (Learning Outcome 2)

(i) Carbon is oxidized from oxidation number zero to oxidation number +4; chlorine is reduced from oxidation number zero to oxidation number −1.

(ii) Nitrogen is reduced (zero to −3); hydrogen is oxidized (zero to +1).

(iii) This is not a redox reaction: according to rules 4, 5 and 7, the oxidation numbers of hydrogen and oxygen are +1 and −2, respectively, on both sides of the equation.

(iv) Sulfur is oxidized and sulfur is reduced: the sulfur in H_2S is oxidized to elemental sulfur (−2 to zero) and the sulfur in SO_2 is reduced to elemental sulfur (+4 to zero). Oxygen remains in oxidation number −2 on both sides of the equation.

QUESTION 2.3 (Learning Outcome 3)

Using the rules in Box 2.1, rule 1 tells us that in $Br^-(aq)$, the oxidation number of bromine is −1; rule 2 tells us that in $Br_2(g)$ it is zero. Rules 4 and 7 imply that in SO_2 and SO_4^{2-}, the oxidation numbers of sulfur are +4 and +6, respectively. Thus the unbalanced equation takes the form:

$$SO_2(g) + \tfrac{1}{2}Br_2(g) = Br^-(aq) + SO_4^{2-}(aq)$$
$$+4 0 -1 +6$$

Here, sulfur is oxidized (+4 to +6) and bromine is reduced (zero to −1). The sulfur and bromine are not present in the right proportions because the total change in oxidation number is +1. It can be made zero by doubling the bromine on each side so that the total change in bromine oxidation number becomes −2 to balance the sulfur change of +2:

$$SO_2(g) + Br_2(g) = 2Br^-(aq) + SO_4^{2-}(aq)$$

As there are four oxygens on the right and only two on the left, we add $2H_2O(l)$ to the left-hand side:

$$SO_2(g) + Br_2(g) + 2H_2O(l) = 2Br^-(aq) + SO_4^{2-}(aq)$$

As there are no hydrogens on the right and four on the left, we now add $4H^+(aq)$ to the right-hand side:

$$SO_2(g) + Br_2(g) + 2H_2O(l) = 2Br^-(aq) + 4H^+(aq) + SO_4^{2-}(aq) \qquad (Q.1)$$

This balances the equation in terms of atoms; if we have done the job properly it will also be balanced with respect to charge. This is so: the charges add up to zero on each side.

QUESTION 2.4 (Learning Outcome 3)

Rules 1, 2, 4 and 7 tell us that the oxidation numbers of bromine in Br^-, Br_2 and BrO_3^- are −1, zero and +5, respectively. Thus, in the unbalanced equation:

$$BrO_3^-(aq) + Br^-(aq) = Br_2(aq)$$
$$+5 -1 0$$

the bromine in bromate is reduced to Br_2 (+5 to zero) and the bromine in bromide is oxidized to Br_2 (−1 to zero). The bromate and bromide are not present in the right proportions because the total change in oxidation number is −4. It can be made zero if $5Br^-(aq)$ react with each BrO_3^- (aq), because then the increase of +5 for the five bromides balances the change of −5 for the bromine of the single bromate:

$$BrO_3^-(aq) + 5Br^-(aq) = 3Br_2(aq)$$

With three oxygens on the left and none on the right, we must add $3H_2O(l)$ to the right-hand side. The absence of hydrogen on the left then means that $6H^+(aq)$ must be added to the left-hand side:

$$BrO_3^-(aq) + 5Br^-(aq) + 6H^+(aq) = 3Br_2(aq) + 3H_2O(l) \qquad (Q.2)$$

The equation is now balanced with respect to atoms. That it has been correctly balanced is made apparent by checking the charges: these add up to the same figure (zero) on each side.

QUESTION 3.1 (Learning Outcome 4)

(a) Removal of a proton from the acids HI, $HClO_4$ and H_2O gives the conjugate bases I^-, ClO_4^- and OH^-. (b) Addition of a proton to the bases Br^-, NO_3^- and NH_3 gives the conjugate acids HBr, HNO_3 and NH_4^+. (c) Removal of a proton from HSO_4^- gives the conjugate base SO_4^{2-}; addition of a proton gives the conjugate acid H_2SO_4.

QUESTION 3.2 (Learning Outcome 5)

The acid lies above the base in Figure 3.3 for combinations (a) and (c), so these are the cases where a reaction occurs. For combination (b) no reaction occurs.

$$\text{(a) } HClO_4(aq) + F^-(aq) = HF(aq) + ClO_4^-(aq) \qquad (Q.3)$$

The acids are $HClO_4$ and HF, and the respective conjugate bases are ClO_4^- and F^-.

(c) Assuming sodium hydride to be a source of hydride ions:

$$H_2O(l) + H^-(aq) = H_2(g) + OH^-(aq) \qquad (Q.4)$$

The acids are H_2O and H_2, and the respective conjugate bases are OH^- and H^-.

QUESTION 3.3 (*Learning Outcome 6*)

From Equation 3.13,

$$pH = -\log[H^+(aq)/mol\,litre^{-1}]$$

$$pH = -\log(4 \times 10^{-3}\,mol\,litre^{-1}/mol\,litre^{-1})$$

$$pH = 2.40$$

QUESTION 4.1 (*Learning Outcome 7*)

1 $[Cu(H_2O)_4]^{2+}$; 2 Cu^{2+}; 3 H_2O; 4 Cl^-; 5 $[CuCl_4]^{2-}$. Note how the ligands H_2O and Cl^- contain the electronegative atoms O and Cl, which have attained noble gas configurations with non-bonding pairs through which to act as a Lewis base.

QUESTION 4.2 (*Learning Outcome 8*)

(a) The alternative structural formula is Structure **Q.1**: each Al←Cl bond in Structure **4.7** (repeated below) has been replaced by Al$^-$—Cl$^+$. In this new representation the bridge bonds all become identical, so this fits the experimental data better than Structure **4.7**. To make the arrow representation fit the observed bond lengths, it is necessary to use two equivalent resonance structures, **4.7** and **Q.2**.

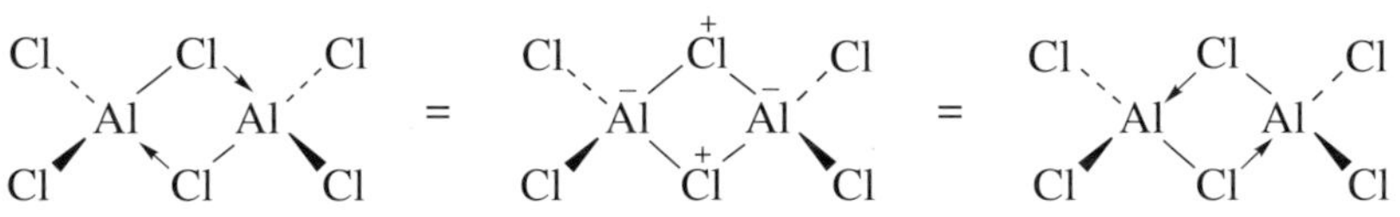

4.7   **Q.1**   **Q.2**

(b) The alternative version is Structure **Q.3**:

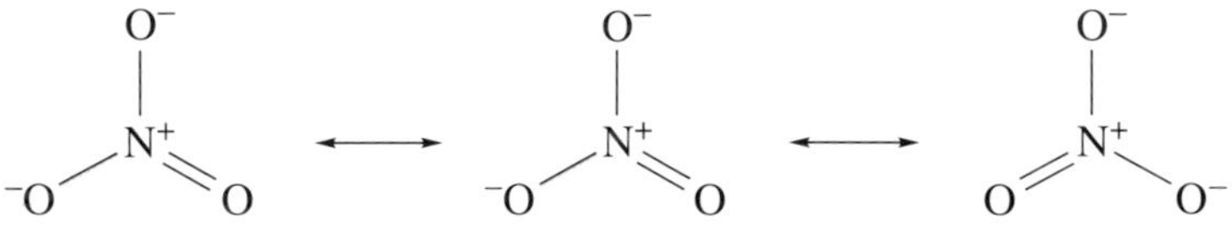

Q.3

Each N→O bond has been replaced by N$^+$—O$^-$, and as the nitrogen atom forms two dative bonds, a charge of 2+ must be added to the single negative charge on the nitrogens in Structure **4.8**. This gives a resulting charge of +1 on nitrogen in the new version, but note that the total charge on the ion is −1 as before.

QUESTION 4.3 (*Learning Outcome 9*)

$B(N-H) = 391\,kJ\,mol^{-1}$ and $B(N-N) = 158\,kJ\,mol^{-1}$; the unnecessary information is $\Delta H_f^{\ominus}(N_2H_4, l)$, because bond enthalpy terms are calculated from the standard enthalpies of atomization of *gaseous* molecules.

The atomization process for $NH_3(g)$ is as follows:

$$NH_3(g) = N(g) + 3H(g)$$

Hence

$$\Delta H_m^{\ominus} = \Delta H_f^{\ominus}(N,\ g) + 3\Delta H_f^{\ominus}(H,\ g) - \Delta H_f^{\ominus}(NH_3,\ g)$$

$$= [472.7 + (3 \times 218.0) - (-46.1)]\,kJ\,mol^{-1}$$

$$= 1\,172.8\,kJ\,mol^{-1}$$

Three N—H bonds are broken in this process, which allows us to calculate B(N—H):

$3B$(N—H) = 1 172.8 kJ mol^{-1}

B(N—H) = 391 kJ mol^{-1}

The atomization process for N_2H_4(g) is as follows:

N_2H_4(g) = 2N(g) + 4H(g)

Hence

$\Delta H_m^{\ominus} = 2\Delta H_f^{\ominus}$ (N, g) + $4\Delta H_f^{\ominus}$ (H, g) − $\Delta H_f^{\ominus}$ (N_2H_4, g)

$\Delta H_m^{\ominus}$ = [(2 × 472.7) + (4 × 218.0) − (95.4)] kJ mol^{-1}

= 1 722.0 kJ mol^{-1}

Four N—H bonds and one N—N bond are broken in this process so,

$4B$(N—H) + B(N—N) = 1 722 kJ mol^{-1}

Assuming that B(N—H) is the same as in NH_3(g):

(4 × 391) kJ mol^{-1} + B(N—N) = 1 722 kJ mol^{-1}

B(N—N) = (1 722 − 1 564) kJ mol^{-1}

= 158 kJ mol^{-1}

QUESTION 4.4 (Learning Outcome 10)

The arithmetic mean of B(N—N) and B(H—H) is 297 kJ mol^{-1}, and B(N—H) exceeds this by 94 kJ mol^{-1}. Thus,

$(\chi_N - \chi_H)^2 = 94/96.5$

$(\chi_N - \chi_H) = 1.0$

This value agrees closely with the value from Figure 4.4 of 0.9.

QUESTION 4.5 (Learning Outcome 11)

The shapes predicted by VSEPR theory for CO_2(g) and PF_5(g) are linear and regular trigonal bipyramidal, respectively. In these cases, the bond dipoles cancel out (Structures **Q.4** and **Q.5**).

O=C=O

Q.4

Q.5

This is not so in the case of the bent molecule SO_2(g) and the square-pyramidal BrF_5(g) (Structures **Q.6** and **Q.7**). (In these cases, the dipole moments are 1.62 D and 1.51 D, respectively.)

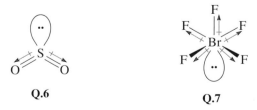

Q.6   **Q.7**

(b) Both molecules have a similar V-shaped structure (**Q.8** and **Q.9**), so differences in shape cannot account for the difference in dipole moment. Oxygen is more electronegative than hydrogen, so in H_2O, the bond and non-bonding-pair dipoles reinforce each other (Structure **Q.8**); fluorine is more electronegative than oxygen, so in F_2O, the two kinds of dipole are opposed (Structure **Q.9**).

Q.8

Q.9

QUESTION 4.6 (*Learning Outcome 12*)

(a) Krypton atoms contain more electrons than argon atoms; its outer electrons are further from the nucleus than those of argon, and have a lower ionization energy. Thus, the krypton atom is the more polarizable, so krypton (120 K) has a higher boiling temperature than argon (87 K).

(b) SiF_4 (Structure **Q.10**) will have a low polarizability because the molecule is a regular tetrahedron like CF_4, and the exterior consists of small fluorine atoms with tightly constrained electrons. Hence, its boiling temperature (187 K) is much lower than that of SCl_2 (332 K).

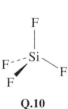

Q.10

(c) SnH_4 has a low polarizability for the same reason that SiF_4 does: like fluorine atoms, hydrogen atoms are small with a tightly constrained electron. Thus, the boiling temperature of SnH_4 (220 K) is less than that of BCl_3 (286 K).

(d) In the series GeH_4, AsH_3 and H_2Se, the shapes of the molecules (Structures **Q.11–Q.13**) and the number of hydrogen atoms are such that the exterior of the molecule becomes more polarizable. Thus, the boiling temperatures increase in the following order: GeH_4 (185 K) < AsH_3 (211 K) < H_2Se (232 K).

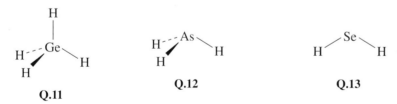

Q.11

Q.12

Q.13

(e) SF_6 is regular octahedral with an outer sheath of fluorine atoms, which have low polarizability. Its boiling temperature (209 K) is much lower than that of Br_2 (333 K).

QUESTION 5.1 (*Learning Outcome 13*)

(i)   First there is an industrial method in which methane reacts with steam; for example

$$CH_4(g) + H_2O(g) \xrightarrow{\text{Ni catalyst}} CO(g) + 3H_2(g) \qquad (Q.5)$$

(ii)   Secondly, water is rendered conducting by the addition of a little sulfuric acid:

$$H_2SO_4(aq) = 2H^+(aq) + SO_4^{2-}(aq)$$

Electrolysis then yields hydrogen at the negative electrode:

$$2H^+(aq) + 2e^- = H_2(g)$$

(iii) Finally, there is the method of Figure 5.1, but with magnesium replacing zinc, and dilute sulfuric acid replacing dilute hydrochloric acid:

$$Mg(s) + 2H^+(aq) = Mg^{2+}(aq) + H_2(g)$$

QUESTION 5.2 (*Learning Outcome 13*)

$$\Delta H_m^{\ominus} = \tfrac{1}{2}D(\text{Cl}-\text{Cl}) - E(\text{Cl}) = -227\,\text{kJ mol}^{-1}$$

This is $372\,\text{kJ mol}^{-1}$ more negative than the corresponding value of $145\,\text{kJ mol}^{-1}$ given for hydrogen. In a Born–Haber cycle of the type shown in Figure 5.5, but where MgX_2 replaces NaX, this difference is decisive in making $\Delta H_f^{\ominus}(\text{MgCl}_2, \text{s})$ more negative than $\Delta H_f^{\ominus}(\text{MgH}_2, \text{s})$. This is so despite differences in the lattice energies of $MgH_2(\text{s})$ and $MgCl_2(\text{s})$. Thus, $MgH_2(\text{s})$ is much less stable with respect to its elements than $MgCl_2(\text{s})$.

QUESTION 5.3 (*Learning Outcome 13*)

Tellurium is in Group VI/16, so there are $(8 - 6) = 2$ hydrogen atoms in the hydride: the gas is hydrogen telluride, H_2Te. There are four tetrahedrally disposed repulsion axes around each tellurium, two of them being occupied by non-bonding pairs. Thus, the molecule should be V-shaped with an angle less than the regular tetrahedral one of 109°. Structure **Q.14** gives the experimental geometry, which agrees with these predictions. Note that the bond angle observed experimentally is *considerably* less than 109°.

Q.14

QUESTION 5.4 (*Learning Outcome 13*)

A structural formula for the chain is given in Figure Q.1; the hydrogen bonds are shown as purple dashed lines. From Figure 5.22, their length is 154 pm. The C=O bonds and C—O⁻ bonds must involve oxygen atoms covalently bonded just to carbon, and the C=O bond will be the shorter of the two. This justifies the distribution of bonds in Figure 5.22. The chains consist of HCO_3^- ions linked end to end by hydrogen bonds.

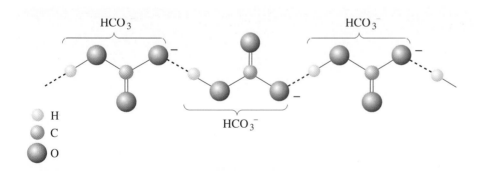

Figure Q.1 A structural formula for the HCO_3^- chain in the compound $NaHCO_3$.

QUESTION 6.1 (*Learning Outcome 14*)

As the oxidizing strength of the halogens decreases down the Group, bromine is a stronger oxidizing agent than iodine, but less strong than chlorine. Thus, bromine will oxidize iodide, but not chlorine: only in (ii) is there a reaction:

$$\text{Br}_2(\text{l}) + 2\text{I}^-(\text{aq}) = 2\text{Br}^-(\text{aq}) + \text{I}_2(\text{aq}) \qquad \text{(Q.6)}$$

QUESTION 6.2 (Learning Outcome 14)

The explanation is analogous to that used to account for the low value of $B(F-F)$: the oxygen atom is small and divalent, and the repulsion between non-bonding electrons over the short internuclear distance weakens the bond (Figure Q.2a). In Group IV/14, the atoms are tetravalent and there are no non-bonding electrons (Figure Q.2b), so this effect is absent. It is the weakness of the O—O bond which explains why elemental oxygen does not contain single bonds, in contrast to sulfur which does (see Section 4.2). Oxygen forms O=O molecules and is a gas; sulfur forms S_8 rings composed of S—S bonds and is a solid at room temperature.

QUESTION 6.3 (Learning Outcome 14)

The relative molecular mass of HCl is 36.5, so HCl is denser than air. This is consistent with the way in which HCl is collected in Figure 6.6: by upward displacement of air. Contrast this with the opposite method used for hydrogen in Figure 5.1.

QUESTION 6.4 (Learning Outcome 14)

The reaction of metals or metal oxides with hydrogen halides is one of the best-known ways of making metal halides. Beryllium in Group II forms dihalides, so the reactions are:

$$Be(s) + 2HF(g) = BeF_2(s) + H_2(g) \tag{Q.7}$$
$$BeO(s) + 2HF(g) = BeF_2(s) + H_2O(g) \tag{Q.8}$$

QUESTION 6.5 (Learning Outcome 14)

Fluorine is the best halogen for bringing out high oxidation numbers, so the two halides will be fluorides. The highest oxidation number is usually equal to the Mendeléev Group number, so the formulae will be SeF_6 and BiF_5, respectively. At room temperature, SeF_6 is a colourless gas like SF_6, and BiF_5 is a colourless solid, which melts at 155 °C.

QUESTION 6.6 (Learning Outcome 14)

In the case of copper, Equation 6.27 becomes:

$$\Delta H_f^{\ominus}(CuX_2,s) = \Delta H_f^{\ominus}(Cu^{2+},g) + 2\Delta H_f^{\ominus}(X^-,g) + L\,(CuX_2,s) \tag{Q.9}$$

When we compare $CuBr_2$ and CuI_2, the values of $\Delta H_f^{\ominus}(Cu^{2+},\,g)$ on the right-hand side are identical, and Table 6.4 shows that the values of $\Delta H_f^{\ominus}(Br^-,\,g)$ and $\Delta H_f^{\ominus}(I^-,\,g)$ differ by only 24 kJ mol^{-1}. The difference must arise mainly because of the difference in $L(CuX_2,\,s)$: $L(CuBr_2,\,s)$ is substantially more negative than $L(CuI_2,\,s)$ because $r(Br^-)$ is considerably smaller than $r(I^-)$; see Equation 6.28. This is why formation of the dibromide is favourable but formation of the diiodide isn't.

QUESTION 6.7 (Learning Outcome 14)

(a) $3Br_2(aq) + 6OH^-(aq) = BrO_3^-(aq) + 5Br^-(aq) + 3H_2O(l)$ (Q.10)

Bromine disproportionates to bromate and bromide whether the alkali is hot or cold. The reaction is analogous to the chlorine reaction in hot alkali (Equation 6.36). The bromate ion will be pyramidal like chlorate (Structure **6.5**).

(b) $ClF_5(g) + AsF_5(g) = [ClF_4]^+[AsF_6]^-(s)$ (Q.11)

AsF_5 is a powerful fluoride ion acceptor (Equations 6.63 and 6.64). According to VSEPR theory, $[ClF_4]^+$ will have five repulsion axes; there is one non-bonding pair, which occupies an equatorial position (Figure Q.3).

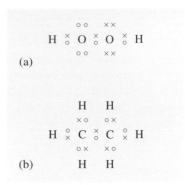

Figure Q.2
(a) There are non-bonding electrons on divalent oxygen atoms forming an O—O bond; (b) there are no such non-bonding electrons on the tetravalent carbon atoms that form a C—C bond.

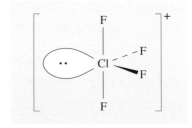

Figure Q.3
The shape of the ion $[ClF_4]^+$ according to VSEPR theory.

(c) $CsCl(s) + ICl(l) = Cs^+[ICl_2]^-(s)$ (Q.12)

This reaction is analogous to the formation of I_3^- (Equation 6.61). ICl accepts a Cl^- ion from CsCl at its less electronegative iodine end. Like I_3^-, ICl_2^- will be linear.

QUESTION 6.8 (Learning Outcome 14)

The halogen Group number is VII/17, so the required oxidation number is +7. The only binary compounds in this state are Cl_2O_7 and IF_7: Br_2O_7, I_2O_7, ClF_7 and BrF_7 have not been made, and fluorine does not occur in positive oxidation numbers. Chlorine, bromine and iodine all occur in the +7 oxidation number as perhalates and perhalic acids such as KXO_4 and HXO_4.

QUESTION 6.9 (Learning Outcome 14)

The anhydride would be Cl_2O_5; this is the formula obtained by doubling the formula of the acid, $HClO_3$, and then abstracting H_2O from the result. It is also the normal oxide in which chlorine has the same oxidation number as in the acid (+5). The anhydride of $HClO_4$ is made by displacing $HClO_4$ from $KClO_4$ with concentrated H_2SO_4 and then dehydrating the acid with P_2O_5. But the action of the acid on $KClO_3$ gives not $HClO_3$, but ClO_2 and $KClO_4$ (Equation 6.47), the products that would result from a *disproportionation* of chlorate. This suggests that $HClO_3$, and especially its unknown anhydride Cl_2O_5, tend to disproportionate into ClO_2 and ClO_4^-, and that this is why Cl_2O_5 is hard to make. Certainly, attempts to concentrate $HClO_3$ by evaporation lead eventually to this kind of disproportionation; for example

$$3HClO_3(aq) = HClO_4(aq) + 2ClO_2(g) + H_2O(l)$$ (Q.13)

QUESTION 7.1 (Learning Outcome 16)

Figure Q.4 shows how the two successive distillations are coupled. The boiling temperatures are O_2 (90 K), Kr (120 K), Xe (165 K); thus, oxygen is the most volatile, followed by krypton and then xenon. In the first column, the liquid is separated into a gaseous oxygen fraction, and a liquid Kr/Xe fraction. The latter passes to a second column where it is separated into a gaseous krypton fraction and a liquid xenon fraction.

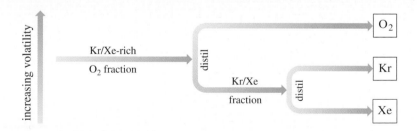

Figure Q.4 A scheme for the separation of oxygen, krypton and xenon by the fractional distillation of the liquid mixture.

QUESTION 7.2 (*Learning Outcome 16*)

Figure Q.5 shows the required cycle. It shows that for the O_2/PtF_6 reaction

$$\Delta H_m^{\ominus} = I(O_2) - E(PtF_6) + L(O_2^+\,[PtF_6]^-,\,s) \tag{Q.14}$$

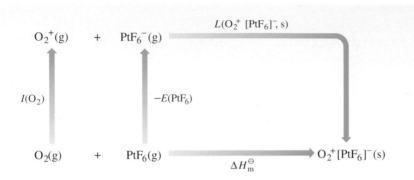

Figure Q.5
A thermodynamic cycle constructed around the reaction in which O_2^+ $[PtF_6]^-$ is formed from O_2 and PtF_6.

The balance of terms on the right-hand side of this equation must make $\Delta H_m^{\ominus}$ favourable to the reaction. When O_2 is replaced by Xe, the similarity in ionization energies leaves the first term virtually unchanged. The second term, $-E(PtF_6)$ is unaffected by replacement and is completely unchanged. Thus any significant changes in $\Delta H_m^{\ominus}$ will be caused by those in the third term, the lattice energy of the solid.

From Equation 6.28, this lattice energy is given approximately by

$$L = -\frac{2W}{r_+ + r_-}$$

There are reasons for thinking that when $O_2^+[PtF_6]^-$ is changed to $Xe^+[PtF_6]^-$, the change in L will be small. The change is caused by replacement of $r(O_2^+)$ by $r(Xe^+)$. Although the xenon atom is bigger than the oxygen atom, there are two oxygens in O_2^+ and not one, so the radii of O_2^+ and Xe^+ may well be similar. In any case, the unchanged negative ion, $[PtF_6]^-$ is very big, so r_- will be extremely large. This means that it will dominate the denominator in the equation, and a small alteration in r_+ will have little effect on L. We conclude that replacement of O_2 by Xe in Figure Q.5 will indeed have little effect on $\Delta H_m^{\ominus}$.

QUESTION 7.3 (*Learning Outcome 16*)

(i) $XeF_4(g) + 2H_2(g) = Xe(g) + 4HF(g)$ (Q.15)

(ii) $2KrF_2(g) + 2H_2O(l) = 2Kr(g) + 4HF(aq) + O_2(g)$ (Q.16)

(iii) $XeF_6(g) + AsF_5(g) = [XeF_5]^+ [AsF_6]^-(s)$ (Q.17)

(iv) $3Mg(s) + XeO_3(s) = 3MgO(s) + Xe(g)$ (Q.18)

(v) $Xe(g) + 3KrF_2(g) = XeF_6(g) + 3Kr(g)$ (Q.19)

The fluorides and oxides of krypton and xenon are strong oxidizing agents. In Reactions Q.15 and Q.18, they fulfil this role with obvious and strong reducing agents. In Reaction Q.16, KrF_2 oxidizes water in the same way as XeF_2 (Equation 7.7). Reaction Q.19 marks the fact that KrF_2 is a stronger oxidizing or fluorinating agent than fluorine; it will match the best that fluorine can do, and yield the highest-known fluoride of xenon. Only Reaction Q.17 is not a redox reaction: the strong fluoride acceptor, AsF_5, removes a fluoride from XeF_6.

QUESTION 7.4 (Learning Outcome 16)

The predicted shapes are shown in Figure Q.6; $XeOF_4$ is a square pyramid with oxygen at the peak. Four of the eight xenon electrons are taken up by the Xe—F bonds, and two by the Xe=O bond, leaving one non-bonding pair. Thus, there are six repulsion axes. As double bonds contain more electrons than single bonds, they are likely to exercise stronger repulsions, so the minimization of non-bonding pair–double bond repulsions may be regarded as a priority. They take up opposite positions in the octahedral distribution of repulsion axes. This predicted shape is confirmed experimentally.

With XeO_3F_2, six xenon electrons are assigned to three Xe=O bonds and two to two Xe—F bonds, leaving no non-bonding pairs. There are five repulsion axes, and the three oxygen atoms occupy the equatorial positions in the trigonal bipyramid to minimize double bond–double bond repulsions. This too matches the shape found by experiment.

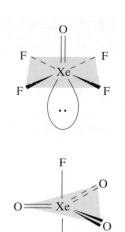

Figure Q.6
The shape of the molecules $XeOF_4$ and XeO_3F_2 according to VSEPR theory.

QUESTION 7.5 (Learning Outcomes 16 and 17)

(a) The molecular orbital diagram populated for the case of the HF_2 molecule, is shown in Figure Q.7. The ionization energy of fluorine is larger than that of hydrogen, so the two fluorine $2p_z$ atomic orbitals lie below the hydrogen 1s. The three atomic orbitals generate three molecular orbitals. The hydrogen orbital is used to give the lower bonding orbital with matching signs in the overlap regions, and the upper, antibonding orbital with mismatched signs. This leaves the third non-bonding orbital to be formed from just the two fluorine orbitals; as these two orbitals have opposite orientations along the z axis in the bonding and antibonding combinations, they have the same orientations in the non-bonding orbital. In the HF_2^- ion there will be an additional electron in the non-bonding orbital.

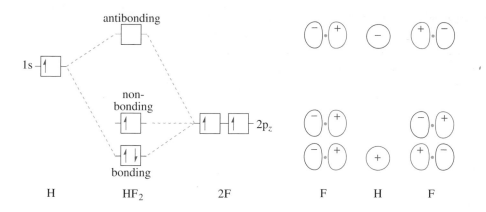

Figure Q.7
Partial energy-level diagram and molecular orbitals for the HF_2 molecule; the HF_2^- ion would have one more electron in the non-bonding orbital.

(b) With the HF_2 molecule, there are three electrons to feed into the diagram — one from each of the three orbitals. Two occupy the bonding orbital, and one the non-bonding one. Following the argument used for XeF_2, because the overall bond order of one is spread over two H—F distances, the H—F bond order is one-half, or at least much less than one. When we turn to the HF_2^- ion, there is one extra electron, but the bond order is unchanged because this extra electron goes into the *non-bonding* orbital, which is then full. Thus, once again, the bond order is much less than one. This is quite consistent with the very substantial lengthening of the bond by some 20% relative to the H—F single bond in HF(g).

QUESTION 8.1 (Learning Outcome 18)

In white phosphorus, and in N≡N molecules, each atom forms three bonds to other atoms of the same sort. Thus, each Group V atom is trivalent, and by exercising this valency it acquires an octet of electrons. In phosphorus there are three separate single bonds; in nitrogen, however, the atoms express the trivalency in multiple bonds. These preferences exemplify a characteristic difference between second- and third-row elements.

QUESTION 8.2 (Learning Outcome 18)

(i) The outer electronic configuration of the sulfur atom is $3s^2 3p^4$ (Figure Q.8). To form six S—F single electron-pair bonds, six unpaired electrons in six separate orbitals must be generated. This can be achieved by promoting an electron from each of the pairs in the 3s and 3p orbitals to separate 3d orbitals (Figure Q.9). Thus, two 3d orbitals are involved in this bonding scheme.

3s   3p   3d   3s   3p   3d

Figure Q.8 The outer electronic configuration of sulfur.   **Figure Q.9** Promotion of electrons to 3d orbitals.

(ii) Each of the six fluorine 2p orbitals involved in the bonding scheme contains one electron. When these six are added to the six sulphur electrons, the total is twelve. These twelve electrons exactly fill the four bonding orbitals and the two non-bonding ones. The total bond order is therefore four, distributed over six S—F bonds, which gives an average bond order of two-thirds.

QUESTION 9.1 (Learning Outcome 19)

The halides, BX_3, are stabilized by π bonding in which non-bonding halogen electrons are delocalized into the $2p_z$ orbital on boron (Figure 6.3). In BH_3 this stabilization is impossible because hydrogen has no non-bonding electrons.

QUESTION 9.2 (Learning Outcome 19)

The short B(1)–B(3) distance is almost identical with that in B_2Cl_4 (Structure **Q.15**), which suggests that this is a B—B single bond. The structure therefore contains one B—B single bond, six B—H terminal single bonds and four B—H—B three-centre bridging bonds. Each of these eleven bonds takes two valence electrons, making 22 in all. The four boron atoms have three valence electrons each, and the ten hydrogens each have one. Again, this comes to 22; in other words, the bonds account for all the valence electrons.

Cl⟍
⟍B—B
Cl⟋   ⟍Cl

Cl

Cl

Q.15

QUESTION 9.3 (Learning Outcomes 19 and 20)

(a) The B_2O_3 structure is shown in Figure Q.10. For the sake of clarity it has been simplified by making the B—O—B sequences linear: this gives the structure the form of a planar sheet. However, the B—O—B angle is not 180°, and this destroys the sheet, making the structure a complex three-dimensional one. Nevertheless, Figure Q.10 shows correctly how the BO_3 triangles are joined. The oxygen atoms occur at the vertices, and each vertex is shared by two triangles. Therefore, each central boron is entitled to one half of each of its three surrounding oxygens. This gives the formula $BO_{1.5}$ or B_2O_3, as required.

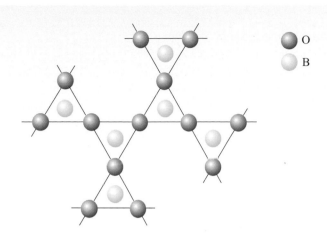

Figure Q.10
Simplified structure of B_2O_3.

(b) In the chain anion, only two of the three vertices in each triangle are shared with another triangle: the third is occupied by an unshared, terminal oxygen with a negative charge. Here then, each boron is assigned one half of each of two surrounding oxygens, and the whole of the third oxygen, along with its negative charge. The empirical formula of the anion is therefore BO_2^-, and the calcium salt of this anion must be $Ca(BO_2)_2$ or CaB_2O_4. Hence the proportion of the constituents of the mixture was $1 : 1$.

QUESTION 9.4 (*Learning Outcomes 19 and 20*)

We have argued that aluminium dissolves in alkali as well as in acid because the protective oxide is amphoteric; that is, it is soluble in both media. As gallium oxide is amphoteric, and indium oxide is basic, this suggests that only gallium should dissolve in alkali. This is correct; the reaction corresponds to that of aluminium (Reaction 9.15):

$$2Ga(s) + 6H_2O(l) + 2OH^-(aq) = 2[Ga(OH)_4]^-(aq) + 3H_2(g)$$

QUESTION 9.5 (*Learning Outcomes 19 and 20*)

We have noted the very marked tendency of aluminium to form the oxide, Al_2O_3. This suggests that the sulfate of aluminium should revert to the oxide at a lower temperature than does Na_2SO_4. This is correct: $Al_2(SO_4)_3$ decomposes at about 600 °C, whereas Na_2SO_4 remains undecomposed at 1 000 °C. Our conclusion can also be supported by lattice energy arguments: aluminium compounds tend to decompose more readily to the oxide because the small sizes and high charges of the Al^{3+} and O^{2-} ions give Al_2O_3 a large, negative lattice energy.

QUESTION 9.6 (*Learning Outcome 19*)

Two types of explanation have been used for such observations. First, one can argue that the electronegativity difference between aluminium and the halogen decreases from AlF_3 to AlI_3, so covalent character will increase; secondly, in AlI_3, covalency is likely to be introduced by the polarization of the very large I^- ion by the small, highly charged Al^{3+} ion. This effect will diminish as the anion becomes smaller, and less polarizable, in the sequence I^-, Br^-, Cl^- and F^-.

QUESTION 9.7 (*Learning Outcome 20*)

Lead(II) in Group IV and bismuth(III) in Group V; we saw that the stability of the oxidation number that is two less than the Group number increases down Groups III, IV and V. Thus, in Groups IV and V, lead and bismuth are the elements in which it should be most prominent. Note that lead(II) and bismuth (III) are the oxidation numbers that are associated with the Tl^+ electronic configuration, $[Xe]4f^{14}5d^{10}6s^2$, which contains an inert pair of electrons in the 6s subshell. The conclusion is correct: lead(IV) and bismuth(V) are even less stable with respect to lead(II) and bismuth(III) than thallium(III) is with respect to thallium(I).

QUESTION 10.1 (*Learning Outcome 21*)

A metal complex, or coordination compound, consists of a metal atom or ion bonded to several neutral or anionic ligands. Carbon monoxide, CO, is a neutral unidentate ligand, which forms a four-coordinate complex with Ni; it coordinates through the carbon atom and the complex, $Ni(CO)_4$, is tetrahedral.

QUESTION 10.2 (*Learning Outcome 19*)

CO has a permanent dipole moment. The dipole changes on stretching, and so CO will also absorb in the infrared region.

QUESTION 10.3 (*Learning Outcome 19*)

The central carbon atom has four valence electrons, which together with two electrons from each oxygen, makes eight in all. These are distributed in two double bonds to the oxygens; the bond order for each bond is therefore 2. VSEPR theory predicts a linear shape for two repulsion axes.

QUESTION 10.4 (*Learning Outcome 19*)

A linear molecule has $3n - 5$ normal modes. For CO_2 this will be four — two stretching modes and two degenerate bending modes. CO_2 is a linear molecule with $D_{\infty h}$ symmetry, so it has a centre of symmetry and the mutual exclusion rule will apply. The symmetric stretch, v_1, will be Raman-active, and the bend, v_2, and the antisymmetric stretch, v_3, will be infrared-active.

QUESTION 10.5 (*Learning Outcomes 19 and 21*)

CO has a triple bond, whereas the carbon–oxygen bonds in CO_2 are double bonds. In the harmonic oscillator model

$$v = (1/2\pi)\sqrt{(k/\mu)}$$

we would expect the force constant, k, to be greater in CO, and therefore the C—O stretch should occur at a higher frequency than in CO_2. In practice, the stretching frequency of CO is $2\,140\ cm^{-1}$ and the two stretching frequencies of CO_2 are found at $2\,349\ cm^{-1}$ (antisymmetric stretch, infrared) and $1\,330\ cm^{-1}$ (symmetric stretch, Raman); the large difference between the antisymmetric and symmetric stretch frequencies for CO_2 is due to the coupling of the two vibrations. The effect of increasing the isotope mass (and therefore μ) is to decrease the stretching frequency, so v_1 and v_2 for ^{13}CO should be at lower frequencies than the corresponding ^{12}CO vibrations.

QUESTION 10.6 (Learning Outcome 19)

According to the Ellingham diagram, reduction of SiO_2 with C is not possible below 1 950 K.

$$\tfrac{1}{2}SiO_2(s) + C(s) = CO(g) + \tfrac{1}{2}Si(s)$$

$$\Delta H^{\ominus}_m = \Delta H^{\ominus}_f(CO, g) + \tfrac{1}{2}\Delta H^{\ominus}_f(Si, s) - \tfrac{1}{2}\Delta H^{\ominus}_f(SiO_2, s) - \Delta H^{\ominus}_f(C, s)$$

$$= \{-110.5 + 0 - (\tfrac{1}{2} \times -910.9) - 0\} = +344.9 \text{ kJ mol}^{-1}$$

$$\Delta S^{\ominus}_m = \Delta S^{\ominus}_f(CO, g) + \tfrac{1}{2}\Delta S^{\ominus}_f(Si, s) - \tfrac{1}{2}\Delta S^{\ominus}_f(SiO_2, s) - \Delta S^{\ominus}_f(C, s)$$

$$= \{197.7 + (\tfrac{1}{2} \times 18.8) - (\tfrac{1}{2} \times 41.8) - 5.7\} = +180.5 \text{ J K}^{-1} \text{ mol}^{-1}$$

Using the relationship $\Delta G^{\ominus}_m = \Delta H^{\ominus}_m - T\Delta S^{\ominus}_m$, when $\Delta G^{\ominus}_m = 0$ then

$$T = \Delta H^{\ominus}_m / \Delta S^{\ominus}_m$$

$$= (344.9 \times 10^3/180.5) \text{ K}$$

$$= 1910.8 \text{ K}$$

The above calculation assumes that $\Delta H^{\ominus}_m$ and $\Delta S^{\ominus}_m$ do not change with T, and that none of the elements or compounds undergoes a phase change. In fact, silicon does melt within the range of our plot (T_m 1 683 K); $\Delta H_{fus}(Si)$, however, is small and there is very little change in slope at this point on the Ellingham diagram.

98% pure silicon is prepared commercially by the reduction of SiO_2 with carbon in an electric furnace, using a slight excess of SiO_2 to prevent formation of silicon carbide, SiC.

QUESTION 10.7 (Learning Outcome 19)

In SiF_5^-, Si has four electrons and the five fluorines provide five electrons. Together with the single charge on the ion, there is a total of 10 electrons, or five bonding pairs: these are used to make five single bonds to the fluorines, and so the shape of SiF_5^- expected from VSEPR theory is trigonal bipyramidal.

In SiF_6^{2-}, the six fluorine ligands together provide six electrons; the anion possesses a charge of −2, so there are 12 electrons in total, equivalent to six electron pairs. The six fluorines around the silicon form an octahedral structure.

QUESTION 10.8 (Learning Outcome 19)

$$CCl_4(l) + 2H_2O(l) = CO_2(g) + 4HCl(g)$$

$$\Delta G^{\ominus}_m = \Delta G^{\ominus}_f(CO_2, g) + 4\Delta G^{\ominus}_f(HCl, g) - \Delta G^{\ominus}_f(CCl_4, l) - 2\Delta G^{\ominus}_f(H_2O, l)$$

$$= \{-394.4 + (4 \times -95.3) - (-65.2) - (2 \times -237.1)\} \text{ kJ mol}^{-1}$$

$$= -236.2 \text{ kJ mol}^{-1}$$

$$SiCl_4(l) + 2H_2O(l) = SiO_2(s) + 4HCl(g)$$

$$\Delta G^{\ominus}_m = \Delta G^{\ominus}_f(SiO_2, s) + 4\Delta G^{\ominus}_f(HCl, g) - \Delta G^{\ominus}_f(SiCl_4, l) - 2\Delta G^{\ominus}_f(H_2O, l)$$

$$= \{-856.6 + (4 \times -95.3) - (-619.8) - (2 \times -237.1)\} \text{ kJ mol}^{-1}$$

$$= -143.8 \text{ kJ mol}^{-1}$$

So the hydrolysis of CCl_4 is thermodynamically *more* favourable than that of $SiCl_4$. (However, the equilibrium constant for the $SiCl_4$ hydrolysis is 1.1×10^{25} mol^2 l^{-2} at 298.15 K, so that can hardly be said to be unfavourable). This result will be very surprising for those of you who have worked with silicon tetrachloride, since you will know that $SiCl_4$ is a fuming liquid that hydrolyses very easily, whereas CCl_4 is a common solvent that is insoluble in water and does not react with it. Why is this? CCl_4 must be *kinetically stable* to reaction with water.

QUESTION 10.9 (Learning Outcome 19)

Using values from the CD-ROM:

(a) $CH_4(g) + 2O_2(g) = CO_2(g) + 2H_2O(l)$

$\Delta G_m^{\ominus} = \Delta G_f^{\ominus}(CO_2, g) + 2\Delta G_f^{\ominus}(H_2O, l) - \Delta G_f^{\ominus}(CH_4, g) - 2\Delta G_f^{\ominus}(O_2, g)$

$= \{(-394.4) + (2 \times -237.1) - (-50.7) - 0\}\, kJ\, mol^{-1}$

$= -817.9\, kJ\, mol^{-1}$

(b) $C_2H_6(g) + \frac{7}{2}O_2(g) = 2CO_2(g) + 3H_2O(l)$

$\Delta G_m^{\ominus} = 2\Delta G_f^{\ominus}(CO_2, g) + 3\Delta G_f^{\ominus}(H_2O, l) - \Delta G_f^{\ominus}(C_2H_6, g) - \frac{7}{2}\Delta G_f^{\ominus}(O_2, g)$

$= \{(2 \times -394.4) + (3 \times -237.1) - (-32.8) - 0\}\, kJ\, mol^{-1}$

$= -1\,467.3\, kJ\, mol^{-1}$

(c) $SiH_4(g) + 2O_2(g) = SiO_2(s) + 2H_2O(l)$

$\Delta G_m^{\ominus} = \Delta G_f^{\ominus}(SiO_2, g) + 2\Delta G_f^{\ominus}(H_2O, l) - \Delta G_f^{\ominus}(SiH_4, g) - 2\Delta G_f^{\ominus}(O_2, g)$

$= \{(-856.6) + (2 \times -237.1) - 56.9 - 0\}\, kJ\, mol^{-1}$

$= -1\,387.7\, kJ\, mol^{-1}$

(d) $Si_2H_6(g) + (7/2)O_2(g) = 2SiO_2(s) + 3H_2O(l)$

$\Delta G_m^{\ominus} = 2\Delta G_f^{\ominus}(SiO_2, g) + 3\Delta G_f^{\ominus}(H_2O, l) - \Delta G_f^{\ominus}(Si_2H_6, g) - \frac{7}{2}\Delta G_f^{\ominus}(O_2, g)$

$= \{(2 \times -856.6) + (3 \times -237.1) - 127.3 - 0\}\, kJ\, mol^{-1}$

$= -2\,551.8\, kJ\, mol^{-1}$

QUESTION 10.10 (Learning Outcome 19)

For carbon, no hexahalo species are expected because the maximum coordination number is four. F^- is the smallest halogen ion; the ionic radii increase through Cl^- and Br^- to reach a maximum at I^-. The maximum coordination number for silicon is six, but the *size* of the Group IV atom increases from silicon(IV) through germanium(IV) to tin(IV). Thus, we expect it to become increasingly easy to pack the larger halogens around the central atom down the Group. This is demonstrated by the occurrence of SiF_6^{2-}, $GeCl_6^{2-}$, $SnBr_6^{2-}$ and SnI_6^{2-}.

QUESTION 10.11 (Learning Outcome 21)

Because the Si—Si single bond is strong and the 2p–2p π overlap is not good, it is preferable for Si to form single bonds rather than double. The only way found so far of preventing Si=Si bonds from polymerizing is to prevent their close approach using bulky ligands; hydrogen atoms would not be able to do this.

QUESTION 10.12 (Learning Outcome 19)

Using the values $\chi_{Si} = 1.8$ and $\chi_{Cl} = 3.0$, the ionic resonance energy calculated by Pauling's equation is

$$C(\chi_{Cl} - \chi_{Si})^2 = 96.5(3.0 - 1.8)^2 = 139\, kJ\, mol^{-1}$$

The bond enthalpy terms for Si—Si, Cl—Cl and Si—Cl are 226, 243 and 400 kJ mol^{-1}, respectively. The extra bond strength for Si—Cl is therefore measured as $400 - (226 + 243)/2 = 165.5\, kJ\, mol^{-1}$. The calculated value of the ionic resonance energy is therefore in moderate agreement with the experimental value.

QUESTION 11.1 (Learning Outcome 19)

The Lewis structure and structural formula of one resonance form of NO_3^- are shown in Structures **Q.16** and **Q.17**. The ion has three repulsion axes and is planar with a C_3 axis. There are three C_2 axes perpendicular to the C_3 axis, and so the nitrate ion has $\mathbf{D}_{3h}$ symmetry.

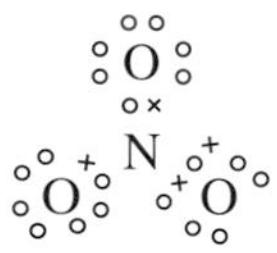

Q.16

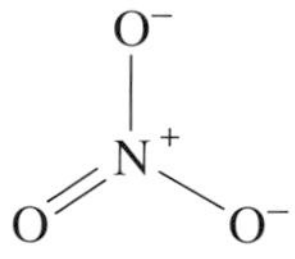

Q.17

QUESTION 11.2 (Learning Outcome 22)

Assuming three-coordination for nitrogen(V), an oxoanion with stoichiometry NO_3 is formed, which bears a charge of $(+ 5 - 6) = -1$. The corresponding oxoacid thus requires only one proton for neutrality and the formula is HNO_3 (nitric acid).

For boron, the three-coordinate oxoanion has stoichiometry BO_3, which carries a resultant charge of $(+ 3 - 6) = -3$. The neutral acid needs three protons, so the formula of the ortho acid is H_3BO_3 (orthoboric acid).

QUESTION 11.3 (Learning Outcomes 15 and 23)

Acids with $t = 0$ are for example, H_4SiO_4, H_3BO_3 and $HOCl$; acids with $t = 1$ are, for example, H_3PO_4 and $HClO_2$; acids with $t = 2$ are, for example, H_2SO_4, HNO_3 and $HClO_3$. Finally an acid with $t = 3$ is $HClO_4$.

QUESTION 11.4 (Learning Outcomes 15, 22 and 24)

The chlorine oxoacids all have one $-OH$ group, and zero, one, two and three terminal oxygens, respectively. The molecules are shown in Structures **Q.18–Q.21**.

Q.18

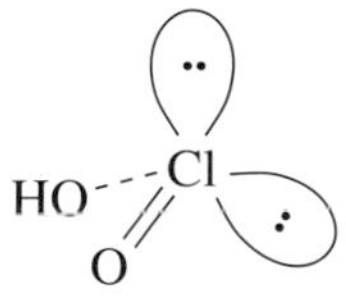

Q.19

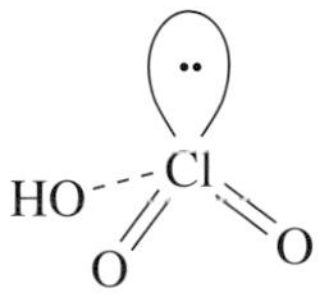

Q.20

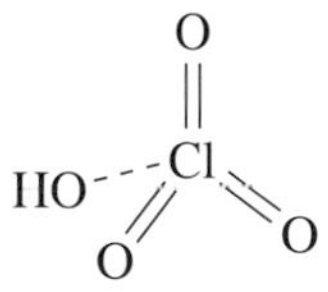

Q.21

Chlorous acid, $HClO_2$, has four repulsion axes, two of which are non-bonded pairs; it therefore adopts a bent shape. Chloric acid, $HClO_3$, has four repulsion axes, of which one is a non-bonded pair; its shape is pyramidal. Perchloric acid, $HClO_4$, has four repulsion axes, with no non-bonded pairs, and so we expect a structure based on the tetrahedral shape.

According to Pauling's second rule, the acid strength is expected to increase in the direction

$$HClO < HClO_2 < HClO_3 < HClO_4$$

QUESTION 11.5 (Learning Outcomes 15, 24 and 25)

The formulae of the fourth-Period oxoacids are estimated from their oxoanion charges as follows:

GeO_4 charge $(+ 4 - 8) = -4$. Therefore the acid is H_4GeO_4, with $t = 0$, a weak acid. In fact, the measured first dissociation constant, K_1, for germanic acid is 2.6×10^{-9} mol litre^{-1}.

AsO_4 charge $(+ 5 - 8) = 3$. Therefore the acid is H_3AsO_4, with $t = 1$, a moderately weak acid (K_1 is measured as 5.6×10^{-2} mol litre^{-1}).

SeO_4 charge $(+ 6 - 8) = -2$. Therefore the acid is H_2SeO_4, with $t = 2$, a strong acid. (K_1 is not reported. K_2 is given as 1.2×10^{-2} mol litre^{-1}; as Pauling's rules suggest, K_2 for this $t = 2$ acid is of the same order as K_1 for a $t = 1$ acid).

BrO_4 charge $(+ 7 - 8) = -1$. Therefore the acid is predicted to be $HBrO_4$, with $t = 3$, a very strong acid. In fact, it is so strong that no reliable value for K_1 has been obtained.

QUESTION 11.6 (Learning Outcomes 15 and 24)

Using Pauling's second rule,

$t = 1$ for H_5IO_6

$t = 3$ for $HReO_4$

$t = 0$ for H_3AsO_3

$t = 2$ for H_2CrO_4

so the order of acid strength should be

$HReO_4 > H_2CrO_4 > H_5IO_6 > H_3AsO_3$

$T=3 \qquad T=2 \qquad T=1 \qquad T=0$

QUESTION 11.7 (Learning Outcomes 15, 24 and 26)

The acid of stoichiometry H_3PO_3 (phosphorous acid) has a first dissociation constant typical of a moderately weak acid; that is $t = 1$. According to Pauling's second rule, the dissociation constant for an acid with $t = 0$ would be expected to be around 10^{-8} mol litre^{-1}, which suggests that H_3PO_3 is *not* the hydroxy acid $P(OH)_3$, but contains one P=O group. The correct structure for this compound **Q.22**.

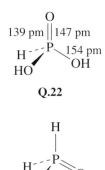

Q.22

QUESTION 11.8 (Learning Outcomes 15, 24 and 26)

A K_1 value of 1×10^{-2} mol litre^{-1} for H_3PO_2 (hypophosphorous acid) suggests that this is an acid with $t = 1$; that is, it probably contains one P=O group. The acid must accordingly contain two P—H groups and one P—OH. The likely structure is therefore the four-coordinate tetrahedral arrangement shown in Structure **Q.23**.

Q.23

QUESTION 11.9 (Learning Outcome 23)

There are three different types of fluorine atom in liquid SbF_5 (Figure Q.11). The bridging fluorines are labelled b, the fluorines *cis* to both bridges are labelled c, and those *trans* to one bridge and *cis* to another are labelled t.

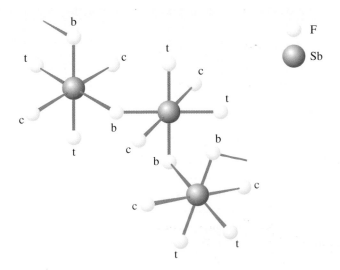

Figure Q.11 The types of fluorine atom in SbF_5.

QUESTION 12.1 (*Learning Outcomes 22 and 27*)

Sulfur has six valence electrons and so can form six covalent bonds. We also know that elements of the third Period prefer four coordination in their oxoanions. Sulfur(VI) coordinates four O^{2-} ions to form a complex of stoichiometry SO_4, with resultant charge $(+6-8)=-2$, so the oxoacid is sulfuric acid, H_2SO_4 (**Q.24**).

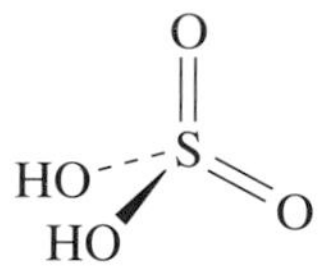

Q.24

QUESTION 12.2 (*Learning Outcomes 24 and 25*)

From Pauling's second rule we can say that, as H_2SO_4 has two terminal oxygen atoms ($t=2$), it should be a strong acid.

By Pauling's first rule $K_1/K_2 \sim 10^5$, so

$$K_2 = 1 \times 10^3 \div 10^5 = 1 \times 10^{-2}\,\text{mol litre}^{-1}$$

The measured value is $1.2 \times 10^{-2}\,\text{mol litre}^{-1}$

QUESTION 12.3 (*Learning Outcome 27*)

S_4N_4 decomposes into N_2 and S_8 as follows:

$$2S_4N_4(s) = S_8(s) + 4N_2(g);\ \Delta H_m^\ominus\ -460\,\text{kJ mol}^{-1}$$

The stability of the S_8 and N_2 molecules provides the driving force for this reaction.

QUESTION 12.4 (*Learning Outcomes 19 and 27*)

Possible resonance structures which describe the bonding in S_2N_2 are shown in Figure **Q.12**.

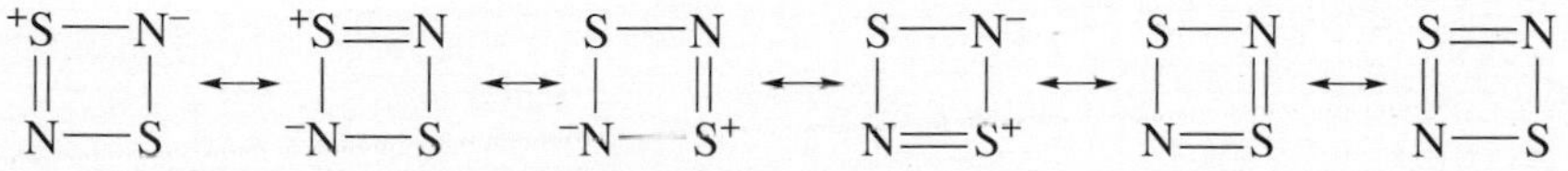

Figure Q.12

QUESTION 12.5 (*Learning Outcomes 19 and 27*)

SO_2 is a non-linear gaseous molecule with double bonds to oxygen and a coordination number of two (Structure **Q.25**). We find that selenium dioxide, SeO_2, is a chain polymer containing the linkage shown in Structure **Q.26**.

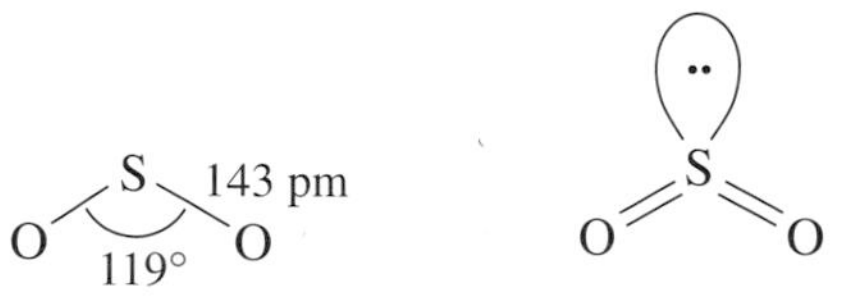

Q.25 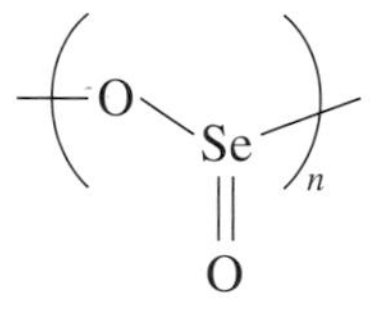   **Q.26**

There is less double bonding and the coordination number has increased to three. Tellurium dioxide contains four-coordinate TeO_4 units singly bonded into layer and three-dimensional structures. In the oxide of the metal polonium we find that PoO_2 adopts the ionic fluorite structure, and the coordination number increases to eight. Summarizing, we find a move away from covalent bonding towards ionic from the top to the bottom of the Group, in line with the increasing metallic character of the elements and at the same time we find an increase in coordination number.

QUESTION 13.1 (*Learning Outcome 18*)

(i) Aluminium precedes sulfur in Period 3 (Figure 13.1). Consequently, sulfur has the higher first ionization energy and electronegativity because these quantities show an overall increase across a Period; aluminium has the larger covalent radius because this quantity shows an overall decrease across a Period. (ii) Oxygen lies above tellurium in Group VI (Figure 13.1). Consequently, oxygen has the higher first ionization energy and electronegativity, but tellurium has the larger covalent radius. (iii) Germanium is in a later Period than chlorine, but chlorine is in a later Group (Figure 13.1). Consequently, the two effects involved in parts (i) and (ii) reinforce each other: chlorine has the higher first ionization energy and electronegativity; germanium has the larger covalent radius.

QUESTION 13.2 (*Learning Outcome 18*)

(i) $CaCl_2$; (ii) IBr; (iii) $CaMg_2$. The properties listed are those of (i) an ionic substance, (ii) a covalent substance, and (iii) a metallic substance. $CaCl_2$ is a combination of elements from Groups II and VII, so the electronegativity difference will be large, and the compound will therefore be ionic. Solid IBr contains discrete molecules. It will be covalent because it is a combination of two elements of high electronegativity from Group VII. $CaMg_2$ will be a metallic alloy because it is a combination of two metallic elements of low electronegativity from Group II.

QUESTION 13.3 (*Learning Outcome 18*)

A is tellurium and its hydride is H_2Te. The existence of the compounds AO_3 and AF_6 suggests that the highest valency and oxidation number of A is six. This implies a Group VI element. According to Figure 13.1, the semi-metal in Group VI is tellurium. The Group VI elements form hydrides of formula H_2Te.

QUESTION 13.4 (*Learning Outcome 18*)

SeO_3 is acidic, PbO is amphoteric and BaO is basic. The acidity of oxides tends to increase across a Period and decrease down a Group. In Figure 13.1 selenium lies above and to the right of both barium and lead, so on both counts its oxide should be the most acidic. As lead is later than barium in Period 6, its oxide will be the less basic. Thus, BaO will be the basic oxide and PbO will be the amphoteric one.

QUESTION 13.5 (*Learning Outcomes 18 and 21*)

PbO_2 is the oxide that decomposes. All three elements fall in the region of Figure 13.1 where the inert pair effect is strongly apparent in chemical behaviour. The effect causes the lower oxidation state to increase in stability from the top to the bottom of a Group. Thus, the oxidation state of +4 will decrease in stability from germanium to lead. PbO_2 is therefore the least stable dioxide, and on heating will be the first to decompose to an oxide in which lead has an oxidation number of +2:

$$2PbO_2(s) = 2PbO(s) + O_2(g)$$

FURTHER READING

1   E. A. Moore (ed.), *Molecular Modelling and Bonding*, The Open University and the Royal Society of Chemistry (2002).

2   L. E. Smart and J. M. F. Gagan (eds), *The Third Dimension*, The Open University and the Royal Society of Chemistry (2002).

3   D. A. Johnson (ed.), *Metals and Chemical Change*, The Open University and the Royal Society of Chemistry (2002).

4   M. Mortimer and P. G. Taylor (eds), *Chemical Kinetics and Mechanism*, The Open University and the Royal Society of Chemistry (2002).

5   P. G. Taylor and J. M. F. Gagan (eds), *Alkenes and Aromatics*, The Open University and the Royal Society of Chemistry (2002).

6   L. E. Smart (ed.), *Separation, Purification and Identification*, The Open University and the Royal Society of Chemistry (2002).

ACKNOWLEDGEMENTS

Grateful acknowledgement is made to the following sources for permission to reproduce material in this Book:

Figure 4.15: Dr Jeremy Burgess/Science Photo Library; *Figure 5.3b*: The OVONIC™ storage cell courtesy of Energy Conversion Devices, Inc. Michigan, USA; *Figure 5.3c*: Courtesy of Solectria Corporation; *Figures 6.2 and 7.1*: The Nobel Foundation; *Figure 7.8*: Courtesy of The Lawrence Berkley National Laboratory; *Figure 9.24a*: Mr Richard Whittington-Egan; *Figure 9.24b*: Topham Picturepoint – Ref: M00718696; *Figure 9.7*: Courtesy of Gesellschaft Deutscher Chemiker; *Figure 10.10*: Science Photo Library/Eye of Science; *Figure 10.11*: *Nature*, Vol. 354, p. 56; *Figure 10.15*: Courtesy of Royal Society, London; *Figure 10.18*: Adrienne Hart-Davis/Science Photo Library; *Figure 10.20*: Natural History Museum; *Figure 10.21*: Dr Jeremy Burgess/Science Photo Library; *Figure 10.22*: Natural History Museum; *Figure 10.23*: Tony Waltham, Nottingham Trent University; *Figure 11.6*: Courtesy of Scott Hughston; *Figure 11.8*: Photo Researchers Inc.; *Figure 11.14*: Courtesy of Terra Industries; *Figure 11.22*: Natural History Museum; *Figure 12.2*: Reproduced courtesy of the Library and Information Centre, Royal Society of Chemistry; *Figure 12.5, cover photo and title pages*: Natural History Museum; *Figure 12.9a*: Copyright © NASA; *Figure 12.9b*: Dr Bradford A. Smith, National Space Science Center, World Data Center for Rockets and Satellites, NASA Goddard Space Flight Centre, Greenbelt, Maryland; *Figure 12.13*: J. Edmund Whoi/Visuals Unlimited; *Figure 12.14*: American Museum of Natural History, NY; *Figure 12.19*: From *Photographic Imaging and Electronic Photography*, Sidney F. Ray (ed.), Focal Press;

Every effort has been made to trace all the copyright owners, but if any has been inadvertently overlooked, the publishers will be pleased to make the necessary arrangements at the first opportunity.

Case Study

Acid Rain: sulfur and power generation

Andrew Galwey

based on 'Sulphur, Acid Deposition and Lake Acidification'
by Tim Allott and Lesley Smart (1996)

POLLUTION AND THE ENVIRONMENT

Human activities have resulted in diverse and often extensive changes to the environment within which we live. For many of us these changes remain largely unappreciated, even unnoticed, through familiarity. Other changes to the environment that are more obviously harmful and/or offensive are referred to as *pollution*. Pollution is defined as 'contamination with harmful or poisonous substances' (*The Concise Oxford Dictionary*, tenth edition). Some industries, for example 'dirty' factories, have, in the past, devastated areas. Large-scale manufacturing processes, such as electricity generation and metal-ore refining, and transport and agriculture, have all demonstrated great potential to pollute the environment. These activities, which underlie much of our material prosperity, are pursued at a cost to the environment.

It is widely accepted that our industries cumulatively generate carbon dioxide (a '*greenhouse gas*') in sufficient quantities to cause climate change through *global warming*. Emissions of carbon dioxide have increased with the progressive expansion in the use of fossil fuels due to industrialization. The use of fossil fuels has grown more or less continually since the early 19th century. Initially, coal combustion provided the major primary source of energy, but during the 20th century the amount of petroleum burned in the internal combustion engine, and elsewhere, has become comparable with, or even greater than, the amount of coal combustion. It is generally accepted that the total quantity of CO_2 already released has resulted in detectable consequences on global climate and it has been predicted that these will persist for (at least) several decades to come.

Sulfur oxides and nitrogen oxides emitted during most types of energy production are responsible for *acid rain*, the subject of this Case Study, and also for *photochemical smog*. (Sulfur oxides are often referred to as SO_x, representing both SO_2 and SO_3. Nitrogen oxides are often referred to as NO_x. They are pronounced, respectively, as **sox** and **nox**.) Legislative controls often associated with, or prompted by, international agreements are now increasingly being introduced to limit the most widespread types of environmental damage: pollution readily traverses national boundaries. However, restrictions imposed on industrial activities to limit the generation of by-products that are regarded as unacceptable are usually opposed because they increase manufacturing costs.

It is generally recognized that the diverse impacts and stresses that we impose on the environment are concurrent, cumulative, often complex, and sometimes interrelated. The overall consequences may not be appreciated at the time. This is because the effects can be indirect and often take time to develop. The changes only become evident after the damage has been inflicted. Once these undesirable changes to the environment have been identified, they are usually followed by research to detect and measure environmental deterioration, to determine the causes and, subsequently, to identify the most effective methods for remedial action. Chemistry plays an essential part in all of these stages. As a general principle, it is usually cheaper and more efficient to remove pollutants within the factory *before* their release as, or in, effluent gases and liquid or solid wastes, than to clean up the often-widespread consequences after their dispersal into the environment. **Clean technology** develops, exploits and encourages the use of alternative chemical techniques that are inherently non-polluting.

ACID RAIN AND SULFUR RELEASE INTO THE ATMOSPHERE

2

This Case Study is specifically concerned with the origin, dispersal and probable consequences of **acid rain**. Acid rain is caused by sulfur release, mainly as SO_2, from the combustion of coal in thermal power stations used to generate electricity (Figure 2.1). The SO_2 is later converted to sulfuric acid in the atmosphere (see Section 5). Other industries, for example the smelting of sulfide ores in metal extraction, also contribute to SO_x in the atmosphere. Power stations also release nitrogen oxides, and a proportion of these appear as nitric acid in acid rain, though the significance of this contribution will not be explored here. However, depending on climatic conditions, oxidized nitrogen, NO_x, participates in various other reactions, during which significant concentrations of ozone and other unpleasant, or even dangerous, chemicals may be formed. NO_x emitted from the internal combustion engine was identified as a principal cause of the photochemical fog that polluted Los Angeles between 1940 and 1950.

Sulfur dioxide release is an important type of pollution that contributes, along with other types of emissions, to overall environmental stress. The effects of acidic gases in urban environments have been the cause of large-scale damage to historic buildings (Figure 2.2a and b). However, for health reasons, this has been much reduced by the Clean Air Act, first introduced in 1956, which allowed councils to establish Smoke Control Areas to improve air quality by the burning of cleaner fuels in these areas. Electricity generation by power stations has increased over the past few decades. These were built with high chimneys to disperse the emissions. However, these emissions have caused problems. Some decades ago it was realized that the unrestricted release of SO_x into the atmosphere was a principal source of acid rain. This realization led to concern about the possible

Figure 2.1 Residents wait for a bus as black smoke from a power plant spews sulfur dioxide into the air, 25 Dec 1995, over the Bosnian city of Tuzla.

adverse effects of acid rain on human health. It was also thought that unrestricted SO_x release was the probable cause of extensive damage to flora and fauna in central Europe and other heavily industrialized areas. SO_x emissions have resulted in long-range transnational environmental damage. Scandinavia has been blighted by acid rain from Western Europe, including the UK; Canada by pollution from industrial areas in the USA. An example of a damaged forest, attributable to acid rain, is shown in Figure 2.3. One principal source of acid rain is described in this Case Study, together with discussions of its adverse environmental effects and methods for their reduction.

The total amount of sulfur precipitated annually on land has been estimated at more than 1.0×10^8 tonnes per year. About half of this is from biological and geological processes and half is ascribed to **anthropogenic** (human-generated) activities. A high proportion of the sulfur generated by humankind has been in the gases released from coal combustion. This tends to be concentrated locally in areas downwind of the heavy industry. Other sources of sulfur oxides are mainly oil refineries, the petrochemical industry and vehicle emissions.

The following background information explains the main reasons for the choice of coal as a primary and preferred fuel in the context of the other sources of energy available.

(a)

(b)

Figure 2.2
(a) Eroded stonework, Wells Cathedral.
(b) Limestone blocks at Wells Cathedral, covered with a calcium sulfate crust, which flakes away exposing a friable surface that weathers rapidly.

Figure 2.3 Forest damage attributed to the effects of acid rain.

2.1 Coal: the most abundant fossil fuel

Deposits of coal are dispersed across all large land-masses. Estimates of total accessible reserves indicate that the amounts are sufficient to meet projected needs during the next three to five centuries. The uncertainty in this estimate arises in forecasting the ease of extraction because many seams are thin, necessitating the removal of large amounts of rock. The economic value of coal also depends on location, calorific value and concentrations of undesirable impurities (including sulfur).

In 1991, the total coal reserves were estimated to be about five times greater than the equivalent petroleum reserves and six times greater than the reserves of natural gas (based on calorific value). Petroleum and natural gas reserves combined have been projected to be capable of supplying energy demands for about 50 years.

The values for the *relative rates of use* of world energy primary sources in 1973 and 1997 (International Energy Agency) are shown in Figure 2.4.

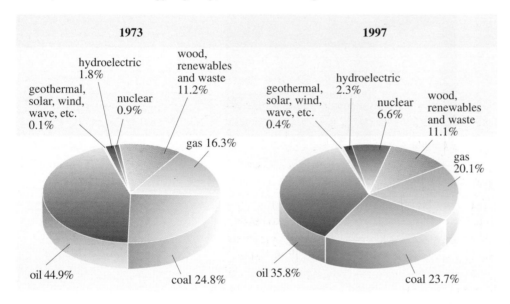

Figure 2.4
World energy production, as measured by rates of use, in 1973 and 1997.

You can see that the percentage changes in the relative amounts of the most important fuels used were comparatively small between 1973 and 1997. What the Figure doesn't show, but which is very relevant to our use of energy, is that the *total* consumption rose by 57% between 1973 and 1997.

2.2 Sulfur in oil can be extracted before use

In contrast to coal, oil is a *mixture* of molecular substances, a property that allows the sulfur to be removed by chemical treatment *before* combustion.

Hydrocracking removes sulfur from petroleum products. It is a catalytic process that involves both *hydrogenation* and *cracking* (redistribution of C—C bonds) of hydrocarbons and other components of the crude oil feedstock. Both types of reaction take place at pressures of 100–180 atmospheres and temperatures of 340–400 °C during relatively long contact times (about 1 hour) over solid catalysts.

Sulfide catalysts, such as nickel, tungsten or molybdenum sulfides, promote hydrogenation across the C—S (and C—N) links in the organic constituents of the oil, so that hydrogen sulfide (and also ammonia) are eliminated by reactions of the type:

$$R^1—S—R^2 + 2H_2 \longrightarrow R^1—H + R^2—H + H_2S \qquad (2.1)$$

where R^n are alkyl groups. Sulfur in ring structures is similarly eliminated; disulfides $R^3—S—S—R^4$ are converted into the hydrocarbons $R^3—H$, $R^4—H$, and hydrogen sulfide. The acidic sites on the catalyst surfaces facilitate the removal of nitrogen as ammonia. A variety of methods has been developed for absorbing the H_2S, including the use of solutions of amines, ammonia or alkaline salt, as well as solid oxides, especially of iron or zinc. Alternatively, aerial oxidation yields SO_2, which reacts catalytically on alumina with H_2S to generate elemental sulfur:

$$SO_2 + 2H_2S \xrightarrow{\text{catalyst}} 2H_2O + 3S \qquad (2.2)$$

Such treatment enables the reduction of concentrations of sulfur and nitrogen. Governments are currently (2002) implementing incentives to encourage greater use of low-sulfur fuels.

2.3 Other energy sources

The combustion of oil and coal leads to pollution, including acid rain, and an increase in atmospheric carbon dioxide, which may lead to global warming.

At present, and for the foreseeable future, hydroelectric, wave and wind power contribute relatively small amounts to world energy production (Figure 2.4). The equipment used in producing energy from these sources releases no significant effluent gases, but has been regarded as being unsightly and noisy.

Apart from catastrophic releases through accidents, nuclear fuels do *not* contribute directly to atmospheric pollution. However, the safe disposal of radioactive products and the decommissioning of old nuclear plants are problems that have not yet been acceptably solved. Currently it seems improbable that nuclear energy production will expand beyond the current level of about 6% unless there are significant changes in political and public attitudes. There is also the possibility that the use of solar energy may expand. Photovoltaic cells convert solar radiation directly into other forms of energy.

In assessing the overall environmental impact of these alternative methods of electricity generation, the energy consumption, materials required and transportation costs of manufacturing the equipment used — and their ultimate disposal cost or recycle value — must be included in the balance.

Coal is, by a large margin, the most abundant known fossil energy resource that is capable of meeting energy demands for some hundreds of years. However, the mining and burning of coal result in pollution. These processes must be controlled in order to prevent environmental damage.

The literature concerned with energy resources must be read with particular care if meaningful comparisons are to be made between different sets of reported data. First, the numbers are so large as to be generally of unfamiliar magnitudes. Secondly, the units can be expressed as amounts of energy obtained on combustion (British thermal units (BTU), J, etc.), or masses (tons, tonnes, lbs of coal) or volumes (barrels of oil, cubic ft of gas). Thirdly, the range of magnitudes of energy use differs considerably from country to country, depending on their relative economic prosperity.

Energy equivalent values are as follows:

1 tonne of coal $\equiv 2.88 \times 10^{10}$ J

$\equiv 0.77$ tonne oil ($\equiv 7.3$ barrels of oil)

$\equiv 750$ m^3 natural gas

These figures are based on approximate energy equivalents. The calorific values of all fuels will vary with composition and, therefore, with source.

2.4 Acid rain

Natural rainwater, in the absence of pollution, is on the acidic side of neutral. It has a pH of about 5.6, which results from the presence of dissolved atmospheric CO_2 as carbonic acid:

$$CO_2(g) + H_2O(l) \rightleftharpoons H_2CO_3(aq) \rightleftharpoons H^+(aq) + HCO_3^-(aq) \qquad (2.3)$$

Sulfur dioxide released by burning coal (and from other sources) is oxidized in the atmosphere to SO_3, which dissolves in raindrops as sulfuric acid ($H^+(aq)$ and $SO_4^{2-}(aq)$). Oxides of nitrogen released from burning coal dissolve to give some nitric acid. The acidity of water droplets in the atmosphere is increased, which has often resulted in rainwater in the pH range 4.0–4.5, depending on the local level of pollution. Exceptionally, pH values down to about 2 have been recorded (which is more acidic than lemon juice). Precipitation of acidic oxides from the atmosphere as rain means that the predominant effects appear most strongly in the prevailing downwind direction (Figure 2.5). These effects are often detectable for hundreds of kilometres. The acids that have been deposited enter the groundwater, and later streams and lakes, where they can modify the concentrations of the dissolved ions already present and thus affect the availability of nutrient ions to both flora and fauna.

The term acid rain should be understood to include **acid fog** (suspended fine water droplets of low pH) and **acid snow**. Spring thaws can release temporary flows of water containing large amounts of acid, capable of causing considerable damage to freshwater biological communities. These thaws often coincide with the onset of the growth and reproduction season.

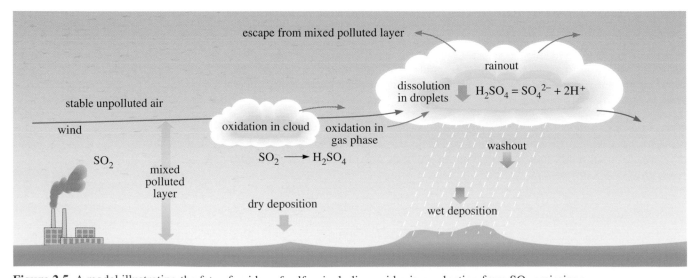

Figure 2.5 A model illustrating the fate of oxides of sulfur, including acid rain production from SO_2 emissions.

CHEMICAL CHARACTERISTICS OF COAL

Coal is the accumulated remains of plant breakdown and bacterial degradation under various reducing environments that exist in shallow and wet burial conditions. Plant material itself is chemically diverse, the two principal components being cellulose (Figure 3.1a, overleaf) and lignin (Figure 3.1b). Every coal deposit is unique, each retaining some characteristics of the types of vegetation from which it was formed, together with the specific modifications that have resulted from all the changing conditions that have occurred during its prolonged maturation. Over geological periods of time stratified layers of various thicknesses, composed predominantly of carbonaceous material, undergo slow but progressive modifications through the following sequence:

peat (formed from buried bogs) $\longrightarrow$ lignite (brown coal)
$\longrightarrow$ bituminous coal $\longrightarrow$ anthracite (derived from buried forests).

Peat can contain between 50% and 80% volatile constituents, of which a high proportion is H_2O. Bituminous coals contain 30–40% volatiles. Anthracite, which is the final form of coal, contains 5% volatiles. It is composed mainly of carbon. The compositions of coal vary significantly with location. They are dependent on the plant materials from which they were derived and the time, temperature and pressure variations that occurred during maturation of each particular deposit. Much of the coal now commercially exploited was laid down 354–290 million years ago. Geologists call this period the Carboniferous Period.

The progressive changes in composition, continued over millions of years, are accelerated by increased temperature and pressure. On deep burial, up to 10 km depth, temperatures can rise to 250 °C or even 300 °C; pressure can be up to 1 000 atmospheres. Coals are composed of the elements present in the plants from which they were derived. Initially these were predominantly C, O and H. During prolonged maturation oxygen and hydrogen are slowly expelled, predominantly as water, from the constituents of the original plant materials, together with some sulfur, nitrogen and a variety of other elements found in plants (see also Section 6.1).

Deposits also contain non-volatile residuals that constitute the ash. These include alkali and alkaline earth metals, together with the constituents of silts and clays that commonly occur in sedimentary rocks (Si, Al and Fe present as oxides, hydroxides, carbonates).

The structure of the carbonaceous phase in a mature coal consists of extended, three-dimensional condensed ring systems. These are composed predominantly of carbon in six-membered rings, although larger and smaller rings are also present. A representative structure of coal is shown in Figure 3.1c. A high proportion of the cyclic structures are unsaturated (benzene, naphthalene and other aromatic ring systems) but they do include some saturated carbons. There are also links between rings, such as CH_2 groups, and some hydroxyl groups are also present. Other elements, most notably S and N, are incorporated into these large molecules (irregular, interlinked macromolecules), for example in heterocyclic rings. Thus, a proportion of the constituent sulfur is accommodated (covalently linked) within the extended organic skeleton of coal (Figure 3.1c).

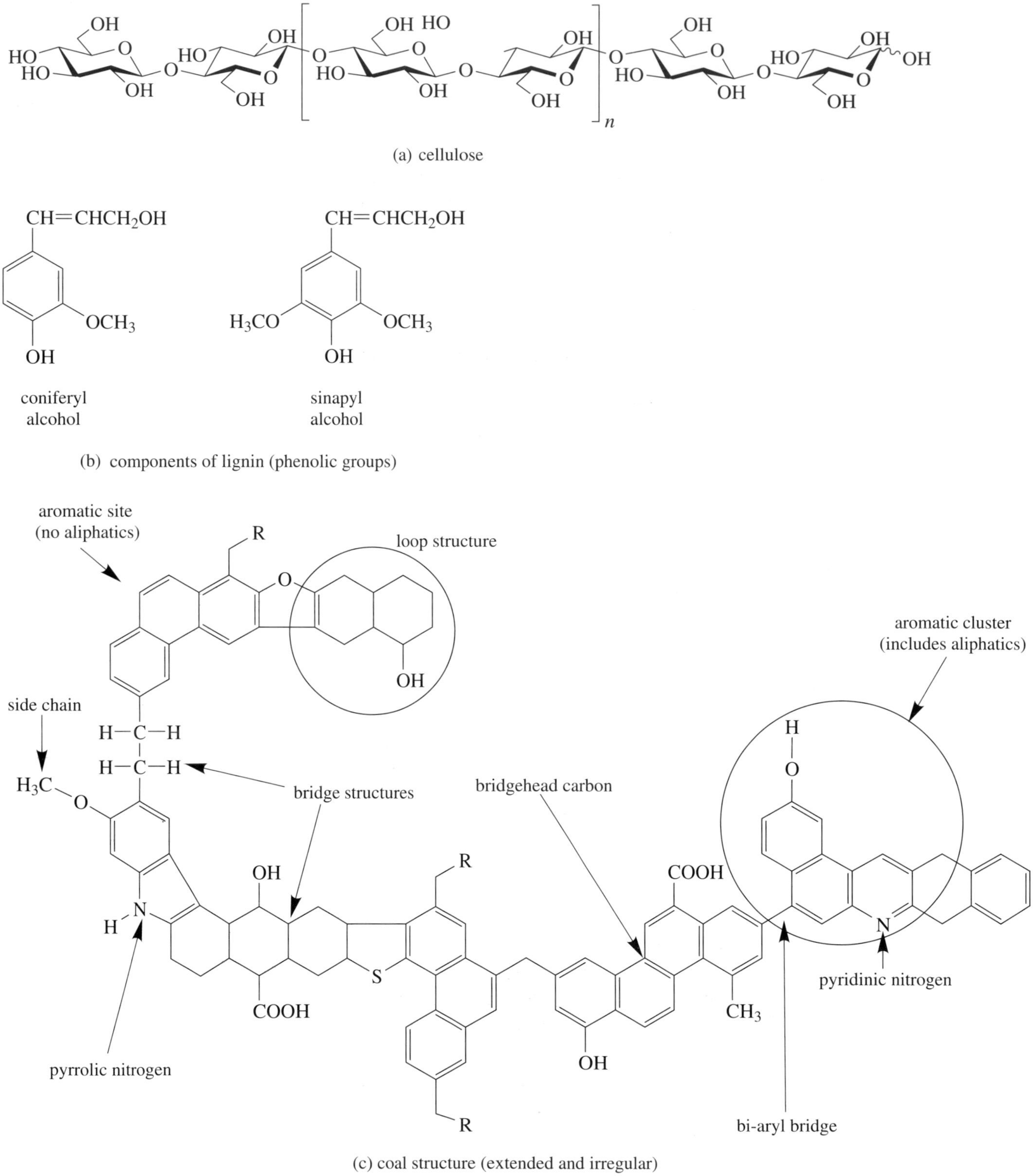

(a) cellulose

(b) components of lignin (phenolic groups)

coniferyl
alcohol

sinapyl
alcohol

(c) coal structure (extended and irregular)

Figure 3.1 Two polymeric substances, cellulose and lignin, are important constituents of plant material. The extended structure of cellulose (a) is composed of β–D–glucopyranose units, with 1,4 links between rings. The extended structure of lignin (b) is disorganized and appears random, being composed largely of coniferyl and sinapyl alcohols. (c) Coal is formed by the prolonged maturation of cellulose and lignin and other plant components during which a high proportion of the water is expelled. This gives extended but irregular structures — a proportion of a coal structure, believed to be representative, is shown.

SULFUR IN COAL

4

Sulfur is a constituent of some proteins and is present in all plants, though in small proportions (values of 5% in ash and 0.05% of dry mass have been reported). Some fraction of this will be retained as organic sulfur in the carbonaceous phase during maturation to coal. However, during the early stages of plant decay, bacterial reduction produces H_2S that precipitates FeS. Subsequent reactions yield FeS_2, which is pyrite. This is the most common sulfidic mineral. Under some conditions there may be oxidation of sulfidic minerals to sulfates of calcium (gypsum) and of iron etc., and in other circumstances the degradation of pyrite may give elemental sulfur. The relative amounts of sulfur in these forms vary, some commercial coals contain up to 3% organic sulfur and elemental sulfur can be absent. In other coals much of the sulfur is inorganic (sulfide, sulfate): there is considerable variability in the type of sulfur present.

4.1 Sulfur content of coals

The total sulfur content of coals varies widely between different deposits. Low-sulfur coals can contain 1% sulfur or less while others can exceed 10% sulfur. Many coals of economic importance contain 2–3% sulfur. The combustion of large amounts of coal results in the release of sufficient SO_2 into the atmosphere to cause significant, but often localized, environmental damage. The combustion of 1 tonne of coal yields 20 kg of sulfur dioxide for each 1% of sulfur in the fuel. The rate of present coal consumption is estimated to correspond to over 10^8 tonnes SO_2 formed through coal combustion each year. This is not all released into the atmosphere, as there are ways of removing sulfur from the coal before combustion. Also, increasing total amounts of SO_2 are now being removed by chemical methods of flue gas cleaning.

4.2 Sulfur removal before combustion

Coal requires initial pulverization, enabling the sulfur to be removed before combustion. This pre-treatment takes place prior to the fuel entering the combustion chamber. Pulverization fractures the carbonaceous material of the coal. The small crystals of inorganic impurities (including sulfides) are detached from the fuel. These can then be separated by the following physical methods.

Magnetic separation. Pyrite and other iron minerals are sufficiently magnetic to enable high field gradients (generated by a strong magnet) to remove these crystallites from the coal, which is not magnetic. Up to 90% separation efficiency has been reported.

Oxydesulfurization. The pulverized fuel is agitated with water in the presence of oxygen. This can be effective in removing inorganic oxidized sulfur as soluble salts, but the process reduces the calorific value of the fuel.

Flotation. The coal is finely ground down to release the inorganic pyrite particles. These may then be separated physically through density differences (density of pyrite: about $5.0\,g\,cm^{-3}$, density of coal: $1.2–1.5\,g\,cm^{-3}$). This method is not feasible, however, on the large scale required.

Bacterial methods for sulfur oxidation have also been considered as a means of sulfur removal.

All these methods are specific to the precombustion removal of *inorganic* sulfur and are expensive when applied to the large quantities of coal burned in power stations. The *organic* sulfur retained within the carbonaceous structure represents an appreciable contribution to acid rain and, therefore, these methods may not provide effective desulfurization.

4.3 Sulfur removal after combustion

It is more efficient, and usually essential if emission controls are to be met, to remove the SO_2 formed from the gaseous products of combustion.

Low-sulfur coals have become economically viable because they are cheaper to use, provided treating the combustion products to remove SO_2 *can be avoided.* Legislation tends to impose increasingly stringent controls on emissions, for example the European Directive (EU/98/0225) for new large plants limits the concentration of SO_2 to $200\,mg\,m^{-3}$ in the effluent flue gases. In Britain the change of fuel from coal to natural gas (largely CH_4) has resulted in an overall decrease in the amount of SO_2 released. Natural gas is low in sulfur and has a higher calorific value than coal (based on the amount of CO_2 released).

The objective of these treatments is to reduce as far as is practicable the release of sulfur (as SO_2) into the atmosphere. Considerable progress in this direction has been made through the development of chemical methods of sulfur removal from the gaseous products of coal combustion (see Section 8).

REACTIONS OF SULFUR DIOXIDE IN THE ATMOSPHERE

5

The rate at which sulfur dioxide, released into the atmosphere, becomes oxidized to the +6 oxidation state to form sulfuric acid depends on the following conditions: humidity, the amounts of dust particles, and the concentrations of transient radicals present, for example OH•. Sulfur dioxide does not react directly with molecular oxygen and/or the water vapour in the air. Some of the principal mechanisms identified as contributing to its oxidation are outlined in this section and depicted in Figure 5.1. These reactions proceed slowly, the half-lives may be several hours or days, during which the sulfur may travel in moving air masses (wind) for hundreds of kilometres from the point of release. The sulfuric acid, mainly dissolved in raindrops, may then be precipitated to cause pollution at areas remote from the point of generation and predominantly downwind.

5.1 Homogeneous oxidation in the gas phase

The principal oxidant in the atmosphere is the hydroxyl radical. This is formed by photodissociation of ozone to give atomic oxygen, which reacts with water. The hydroxyl radical participates in the oxidation of hydrocarbons (e.g. CH_4) and SO_2:

$$SO_2 + OH• \longrightarrow HSO_3• \tag{5.1}$$

Subsequent reactions of this radical, $HSO_3•$ (*not* the hydrogen sulfite ion), perhaps via $HSO_3• + O_2 \longrightarrow SO_3 + HO_2•$, lead to H_2SO_4. Alternative, perhaps concurrent, but less important routes to oxidation include the reactions of SO_2 with atomic oxygen and with $HO_2•$.

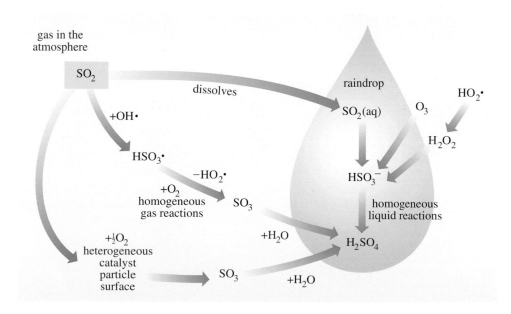

Figure 5.1
Reactions of sulfur dioxide in the atmosphere.

5.2 Homogeneous oxidation in liquid water

Oxidation of SO_2 in liquid water (raindrops) occurs through the intervention of O_3 or H_2O_2, for example:

$$SO_2 + H_2O_2 \longrightarrow H_2SO_4 \tag{5.2}$$

Small concentrations of hydrogen peroxide are present in the atmosphere. These are formed by reactions of radicals in combination such as:

$$2HO_2\bullet \longrightarrow H_2O_2 + O_2 \tag{5.3}$$

5.3 Heterogeneous oxidation catalysed by dust particles

Oxides and other solid compounds of Mn, V, Ni and Fe undergo redox reactions, and these particles catalyse oxidation through surface-adsorbed species:

$$SO_2 + \tfrac{1}{2}O_2 \text{ (+ catalyst)} \longrightarrow SO_3 \text{ (+ catalyst)} \tag{5.4}$$

The transition metals mentioned are present in particulates emitted from the combustion plant, and are active in promoting the subsequent SO_2 oxidation (by electron transfer).

5.4 Sulfate present in the atmosphere

It is believed that sulfate-containing fine particles, arising from anthropogenic activities and temporarily suspended in the atmosphere, contribute to some scattering of solar radiation. This slightly reduces the total amount of sunlight entering the energy balance in the atmosphere and thus acts in the opposite direction to the greenhouse effect. One result of efficient flue gas desulfurization (FGD), on a global scale, may be to contribute to an increase in global warming by removal or reduction of the effect from the radiation-reflective sulfate layer.

EFFECTS OF ACID RAIN IN SOILS AND GROUNDWATER

6

The sulfur dioxide released by anthropogenic activities is predominantly oxidized and converted into sulfuric acid. This falls as acid rain, which enhances the sulfate, and more significantly the hydrogen ion concentrations, in groundwater. These pH reductions, particularly to values of 5 or less, influence the distribution and availability of ions in the soil to the flora and fauna. The run-off, as streams, rivers and lakes, also undergoes an acidity increase, again affecting the microbial, fish and plant communities, which are dependent on these environments. The overall consequences of acid rain have probably not yet been completely characterized and, because of the time lag before the damage is perceived, are not fully recognized. The changes caused in one species may influence the health and success of others to which it relates through the food chain. Elucidation of the specific effects of acid rain is, therefore, complicated. In this Case Study we examine the deterioration of natural environments by reference to two of the changes most extensively investigated: the influences on tree growth and on freshwater fish populations, mainly trout. We will also consider the less obviously severe effects of acid rain in the human environment. Remember that these selected pollution effects often occur in conjunction with other consequent environmental stresses. As discussed below, the overall effects of their synergic influences may be greater than the sum of the individual contributions. Many of the changes to these complex natural systems are, as yet, imperfectly understood. Some of the adverse consequences of acid rain in the ecosystem (together with other consequences, for example the input of nutrients to soil, that may be beneficial) are indicated in Figure 6.1.

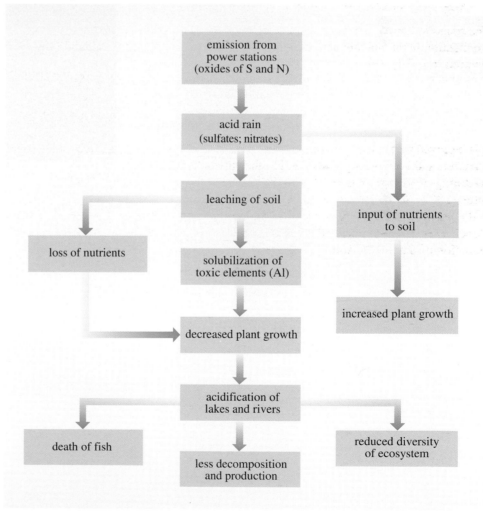

Figure 6.1 Pathways and effects of acid rain on some components of the ecosystem; some adverse and some possibly beneficial consequences are shown.

6.1 Soils: structure and composition

Soil is the porous layer of weathered rock and plant breakdown products, including air and water, which covers most of the land surface of the planet from depths of a few centimetres to several metres. It is the zone from which the roots of many forms of plant life draw water together with the elements required for their growth (including N, P, S, Mg, Na, K, Ca and trace elements, mainly metals, usually in the form of nutrient ions). The composition and structure of soils vary widely with the minerals and elements present in the underlying rocks from which they are often derived. Other contributory factors include the effects of temperature, rainfall and vegetation. Sections of most soils typically reveal a layered structure, each layer being called a *horizon*. The uppermost layer, called the O horizon, is largely made up of undecomposed plant remains. The lowest layer is largely unchanged and is composed of, or contains a high proportion of, the underlying bedrock.

Figure 6.2 shows the horizons of a typical soil structure of a *podzol*. This soil type is commonly found in cool, humid climates beneath forest cover and under moorlands. Table 6.1 and Figure 6.3 give more detailed information on soil horizons.

Figure 6.2
A typical soil profile developed under moorland vegetation (for scale the knife is about 25 cm long).

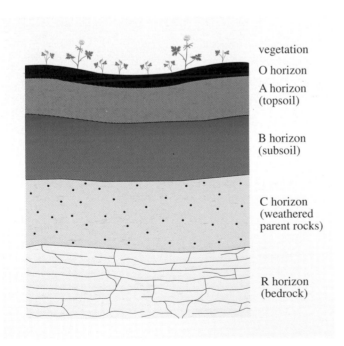

Figure 6.3
Schematic diagram of a soil profile
showing the O, A, B, C and R horizons.

Table 6.1 Soil horizons in podzols

Horizon	Colour	pH	Composition
O	black	–	*litter*: relatively undecomposed plant materials.
A	grey–black	<4	*topsoil*: contains decaying plants and breakdown products (humus), leaves, fibres, roots, etc.
B	orange–brown	4–5	*subsoil*: minerals which have been extensively weathered and leached to fine clay and silt particles. These phases exhibit ion exchange properties (see below).
C	various	5–6	*less comprehensively weathered rock*. also contains breakdown products of minerals with ion exchange capacity.
R	various	–	*unchanged bedrock*: may be igneous (e.g. granite), metamorphic (e.g. gneiss or schist) or sedimentary (e.g. limestone or shale).

6.2 Groundwater

Plant roots obtain essential elements from the groundwater in the soil. Groundwater participates in, and promotes the mobility of, the (mainly ionic) products of the slow breakdown of mineral particles, referred to as *mineral weathering*. This process is illustrated below by relatively simple (and idealized) examples. The ions released are ultimately washed from the soil horizons by *leaching*, but they may be retained temporarily by *ion exchange*. (These topics are explained later in the text.)

Mineral weathering

Several examples of the (slow) reactions that participate in the breakdown of rock minerals are given below. These illustrate how the *base cations* (Na^+, K^+, Mg^{2+}, Ca^{2+}) are released and the Al^{3+} and Fe^{2+} ions are formed under slightly acid conditions in groundwater. The minerals in most rocks are of variable compositions, depending on the elements present at their times of formation; the examples cited below represent somewhat idealized formulae. Many minerals vary over wide composition ranges but their crystal structure is constant. However, their unit cell dimensions change systematically depending on the elements present and their ratios.

The common mineral *feldspar* and the less abundant mineral *olivine*, selected as examples here, occur widely in many different rock types, and both are broken down during weathering by overall processes (physical and chemical disintegration and degradation). The typical reaction for the weathering of feldspar is shown in Equation 6.1:

$$4KAlSi_3O_8 + 4CO_2 + 22H_2O \longrightarrow 4K^+ + 4HCO_3^- + Al_4Si_4O_{10}(OH)_8 + 8Si(OH)_4 \quad (6.1)$$

feldspar $\longrightarrow$ ions + kaolinite (clay) + silicic acid

Kaolinite can react with water to give aluminium hydroxide (*gibbsite*) and hydrated silicic acid. The aluminium hydroxide dissolves in the presence of hydrogen ions, as shown in Equation 6.2:

$$Al(OH)_3 + 3H^+ \longrightarrow Al^{3+} + 3H_2O \quad (6.2)$$

The typical reaction for the weathering of olivine is shown in Equation 6.3.

$$(Fe,Mg)_2SiO_4 + 4CO_2 + 4H_2O \longrightarrow 2(Fe^{2+},Mg^{2+}) + 4HCO_3^- + Si(OH)_4 \quad (6.3)$$

olivine $\longrightarrow$ dissolved ions + silicic acid

Calcium carbonate, which is very widely distributed in nature, dissolves (to a small extent) in carbonic acid:

$$CaCO_3 + CO_2 + H_2O \longrightarrow Ca^{2+} + 2HCO_3^- \qquad (6.4)$$

The ions released during weathering participate in ion exchange equilibria between groundwater solution and adsorption at the surface sites of the mineral phases present.

Leaching

Weathering of rocks in the naturally occurring, relatively acidic, upper horizons releases ions that are retained by the ion exchange process (cations replace protons, H^+ ions, at the acidic surface sites on clays and degraded mineral particles). Aluminium and iron ions (Al^{3+} and Fe^{3+}) are carried downwards by the rainwater that percolates through the soil and may enter streams. However, an increase in pH results in the precipitation of $Al(OH)_3$ and $Fe(OH)_3$ in the lower horizons. In some soils, the formation of a dense deposit of mainly iron hydroxide between the rock particles clogs the drainage channels forming an *iron pan*. This results in waterlogging and the retention of sufficient water to form a boggy or marshy area.

Ion exchange

Ion exchange is an important property of clays and silt particles. Clays are very finely divided, recrystallized products of rock weathering, which are less than $4\,\mu m$; silt particles are $4–60\,\mu m$. The acidic sites on the outer faces of these high-surface-area alumino-silicate phases attract and establish equilibrium between retained surface-adsorbed ions and those dissolved in groundwater (Na^+, K^+, Mg^{2+}, Ca^{2+}, etc.). This equilibrium effectively provides a reservoir of nutrients for plants because, when ions are withdrawn from groundwater into the plant root system, the equilibrium restores the availability of the ions required for plant growth.

EFFECTS OF ACID RAIN ON NATURAL ENVIRONMENTAL SYSTEMS

7

Identification of the specific causal effects responsible for environmental damage is exceptionally difficult. This is because natural systems have complex interrelationships, involving numerous variables, and are often incompletely understood. And it can be challenging, sometimes apparently impossible, to design experiments intended to differentiate between possible specific effects and to measure their individual contributions to environmental damage while mimicking field conditions. In some situations, such as the weathering of rocks or the growth of trees, experiments may require time-scales that are impractically long. Pollution effects are multiple, synergic, and take place in a context of changing land use and slow climate change. Characterization of the damage specifically caused by acid rain must be distinguished from those effects that arise both naturally and through increased levels of ozone, nitrogen oxides and other anthropogenic emissions.

We will consider two effects of rising acid levels in groundwater to illustrate the complexity of research directed towards the specific effects on the ecosystem of the increase of H_2SO_4 in rain. These are the effects on tree growth and the influence of pH on fish in streams and lakes. At first it was considered possible that the observed low pH values in some soils, streams and lakes in northern Europe could be the continuing consequences of natural changes following the last ice age, which occurred comparatively recently on a geological time-scale. Useful information about the changes of acidity of lakes in Scotland and in Scandinavia was obtained by examination of sediments. The study of lake sediments and the history of lakes is called *paleolimnology*.

7.1 History of pH changes in lakes in Scotland

Sediments deposited in lakes contain the skeletal remains of diatoms (Figure 7.1, overleaf). These are microscopic, single-celled organisms. The texture of the cell wall is highly specific to each species. These silica structures are well preserved and provide a record of the diatoms living in a lake at different stages of the lake's history. It has been established conclusively that different species of diatoms live in different pH ranges, so that the history of acidity within a lake can be determined from the pattern of skeletons detected in the sediments buried at different depths (see Figure 7.2, overleaf). Ages can be determined through the use of radiometric techniques, based on ^{14}C and ^{210}Pb, which can be used to date remains found at different levels below the surface.

Sediment cores were taken from Lough Round in Galloway, southern Scotland, in 1989. These cores show a progressive, systematic reduction in the populations of those diatoms that are sensitive to acid conditions, starting about 150 years ago (Figure 7.2). The inferred average pH in 1850 was about 5.6. This was followed by a progressive diminution to around pH 4.7 in about 1930, a value that has persisted to the present day. The only realistic explanation for this change is the effect of

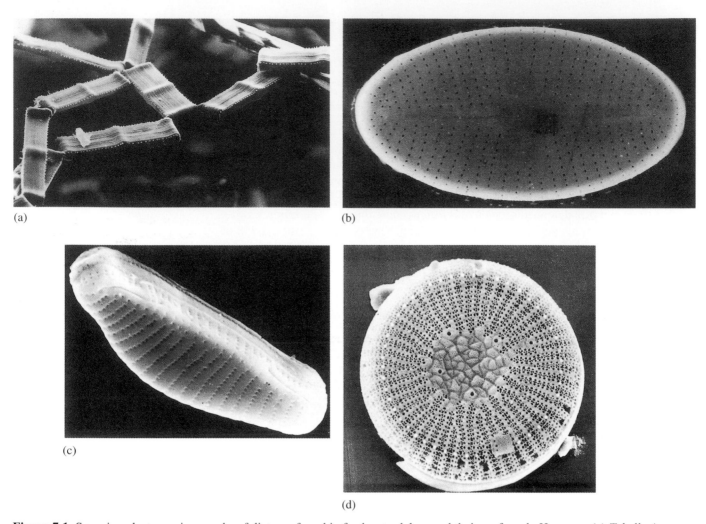

Figure 7.1 Scanning electron micrographs of diatoms found in freshwater lakes, and their preferred pH ranges. (a) *Tabellaria quadriseptata* (pH < 5; magnification ×650); (b) *Achnanthes minutissima* (pH 5.0–5.5; magnification ×12 000); (c) *Eunotia* sp. (pH 5.0–5.5; magnification ×6 000; (d) *Cyclotella kützingiana* (pH ≈ 7; magnification ×13 000).

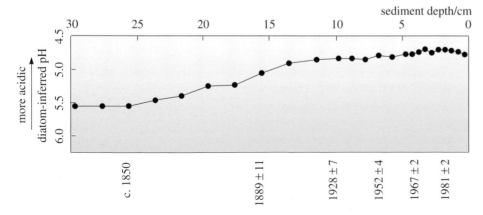

Figure 7.2 Graph of pH (inferred from diatom populations) against year (inferred from sediment depth). Data from Lough Round in Scotland.

acid emissions from anthropogenic (industrial) activities — it cannot be ascribed to any continuing natural process. Support for this conclusion is provided by ash and soot particles with highly characteristic shapes that can reliably be identified as particulate emissions released during coal burning. The detection of these particles in the sediments provides evidence of increased acidity (Figure 7.3). The relative abundances of these markers at different depths of burial correspond well with the historical levels of activity by coal-burning plants. Many similar studies in different localities have concluded that the recent rise in acidification of forests and heathlands is not a continuing natural process but is directly attributable to industrial pollution, and acid rain in particular, during the last 150 years or so.

7.2 Acid rain and forest damage

Sulfur dioxide is toxic to plants, inhibiting growth at concentrations of < 0.1 p.p.m. This *direct* effect of sulfur combustion occurs only in the vicinity of the emission source. Most of the SO_2 released is eventually oxidized in the atmosphere where the *indirect* effects of acid rain, through pH changes, influence the availability of nutrients in the soil.

It has been concluded that damage to forests (Figure 2.3) is frequently associated with disturbances to the supply of plant nutrients, particularly Mg and K, but Ca, P, S and N also appear to be involved. Thus, the main effects of acid rain are believed to be indirect. Serious forest decline has occurred during the last few decades in central Europe, notably in Germany, and in parts of North America. These declines can usually be mitigated by appropriate corrective changes in the supply of the appropriate elements essential to the plants concerned. Many of the forests now surviving tend to be concentrated at sites that are generally poorly supplied with nutrients or located in climatically less-favoured regions. This is the current situation because agriculture has thrived for millennia on the most fertile land available, from which the trees have been removed.

It is particularly difficult to measure experimentally the concentrations and total amounts of ions in the vicinity of the roots that are available to, and can be taken up by, trees. Micro-site variations of a particular species can even be significant between individual, neighbouring plants. An alternative method that avoids the difficulties that arise in making meaningful analyses of groundwater is to measure the quantities of various ions taken into a growing tree. This is done by determining the increase in the amounts of ions present in the leaves. This may be correlated with soil chemistry, to some extent, to identify nutrient deficiencies.

Magnesium is an essential constituent of the green pigment, chlorophyll, which is used by plants to manufacture sugars from carbon dioxide, water and sunlight. When the supply of magnesium is insufficient the leaves turn yellow, and the vital activity of the plant is restricted. Some internal redistribution can maintain the greatest activity of chlorophyll production at the most effective sites but the plant cannot thrive and extreme deprivation of magnesium results in death. There is evidence to show that, in some species, aluminium in the groundwater opposes the uptake of magnesium and calcium. Acid rain appears to increase the concentrations of aluminium ions in solution, by reactions of the types mentioned under mineral weathering above. Acidification can also result in the diminution of the concentrations available of other required nutrients, such as Na, K and Ca, by

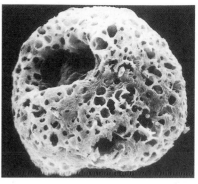

Figure 7.3
An electron micrograph of a carbonaceous particle ($\times$ 1 700).

facilitating elution. In the context of other stresses, and for trees growing on marginal sites with generally poor soil, it is difficult to distinguish quantitatively the specific contribution made by acid rain. However, it appears that there is enhanced mobility of soluble cations, resulting in their loss from the soil. There may also be opposition to Mg uptake as a result of an increase in Al in the groundwater.

7.3 Lake acidification and freshwater fish

The concentrations of dissolved ions in freshwater streams and lakes are relatively small so the introduction of a strong acid causes a large pH change. The *buffering capacity* of such water, that is its ability to sustain a constant hydrogen ion concentration, is low and is often controlled by the amount of hydrogen carbonate ions in solution (see Equation 2.3). The pK_a values for the first and second dissociation steps of carbonic acid are 6.37 and 10.25, respectively. In the presence of dissolved CO_2, while HCO_3^- ions remain present, the rates of pH change are relatively reduced. However, the decrease becomes more rapid after the CO_2 has been displaced. The dissolution of carbonate minerals, $CaCO_3$ and $MgCO_3$, together with CO_2, increases the buffering capacity of water (see Equation 6.4).

The dispersal of crushed limestone into acidified lakes diminishes the effects of acid rain. This process has been widely undertaken in Sweden where the acidification is ascribed to industrial pollution and the catchment areas contain few carbonate rocks. Large-scale liming projects have resulted in the return of wildlife to some areas. Reapplication of crushed lime is required at intervals to maintain this improvement.

Acidification of streams and lakes results in the loss of various types of fauna and flora that can only thrive in species-specific pH ranges. Crustaceans and snails require almost neutral waters. Salmon and trout fail to survive at pH ranges below 5.5; pike and perch are slightly more acid tolerant while eels can live at acidities to about pH 5.0. At high latitudes the thaw of acid-containing snow in spring results in a sudden pH drop (short periods below pH 5.0) in rivers. This is a problem for fish survival as it occurs at about the time that the fish are spawning — a critical coincidence. The loss of microscopic plant and animal life results in clear, apparently pure, water that is almost sterile and is certainly not characteristic of a healthy environment. Only those few plants that are capable of tolerating conditions of high acidity may colonize this environment.

Acidification has other indirect effects on species living in lakes and rivers. At low pH the rock-weathering processes result in the release and mobilization of relatively increased amounts of elements toxic to life, including Pb^{2+}, Cd^{2+} and Hg^{2+}. Acid rain may vary all equilibria within the groundwater: some elements become more mobile and are more rapidly washed to the streams, while the availabilities of others increase. The concentrations of aluminium ions (present as several hydrated species, including $Al^{3+}(aq)$, $AlOH^{2+}$, $Al(OH)_2^+$ and $Al(OH)_4^-$; the proportions vary with pH) also rise with acidity. This increase is believed to result in the precipitation of a gelatinous material within the gills of fish. This precipitation is ascribed to a discontinuous rise in pH between the low pH river water and the body fluids of the fish. The deposit causes death through oxygen deprivation and suffocation by blocking the oxygen-exchange surfaces. While aspects of these explanations of the loss of wildlife from aquatic systems remain incomplete, there is little doubt that acidity changes modify the balances in nature. The impacts vary with species, but overall contribute to the decline in species diversity and ultimately the health of our environment.

7.4 Acid rain in the human environment

Low concentrations (e.g. 1 p.p.m.) of SO_2 are not immediately toxic to humans. However, the effects of prolonged exposure, together with other factors such as asthma and old age, may be more significant, particularly when particulate material is also being breathed. In general, plants are more susceptible than animals to damage through exposure to SO_2.

One obvious result of acid rain is the corrosive degradation of the external surfaces of buildings. The outer surfaces of limestone are particularly susceptible to attack by acids. The roughened surfaces generated in this way retain black particles of soot that can darken and change the appearance of many older structures (see Figure 2.2b). Some buildings in our town centres have been subjected to expensive cleaning processes to restore their original appearances. Statues, containing fine detail, are readily destroyed by low-pH rain, and restoration programmes for cathedrals may require modern replacements for those figures that have suffered greatest weathering and irreversible degradation.

CLEANING PROCESSES IN POWER STATIONS

8

In thermal power stations used for electricity generation, the coal fuel is carried by an air stream into the furnace combustion zone in the form of a finely crushed powder. The heat released by carbon oxidation converts the water into steam, which then drives turbines that are used to generate the electricity. Various processes take place to remove pollutants from the flue gases before they enter the atmosphere.

Staged combustion

Air is admitted to the oxidation zone in controlled amounts. This maintains a maximum temperature *below* the range where the principal constituents of air combine to form nitrogen oxides at successive stages during the coal combustion. This is an effective way of minimizing NO_x production, though some NO_x may be formed through oxidation of the constituent nitrogen present in the carbonaceous phase of the coal.

Slagging

Air admission to the combustion zone is controlled so that sulfur is oxidized to SO_2 only, thereby minimizing SO_3 formation. This is because the sulfates that arise from reactions with metals present (such as Mg, Ca and Fe) may melt to form *slag*. Slag is an accumulation of non-volatile materials that can adhere to the furnace walls, thereby opposing the transfer of heat to the water beyond. The presence of SO_3 can also result in boiler corrosion.

Note: The admission of excess air, above that required for combustion, is undesirable because the heating of this air would be a waste of thermal energy in hot, but unused, expelled gases.

Treatment of combustion products

The flow of gases leaving the hot combustion zone is principally composed of nitrogen (from the air), a small amount of residual oxygen, carbon dioxide, sulfur dioxide and much of the non-volatile particulate matter that is referred to as **fly ash**. To meet emission standards, now mandatory in many countries, these gases are subjected to two cleaning processes:

- **electrostatic precipitation** (ESP), which removes the suspended particles (fly ash), and
- **flue gas desulfurization** (FGD), which is expected to remove up to about 95% of the SO_2 formed.

The sulfur removed during coal combustion generates gypsum (see below) and slag material. Slag has no commercial value. By contrast, the sulfur removed from oil and gas can often be recovered in the form of useful materials, including elemental sulfur. NO_x is not usually removed, but a small proportion may be retained during FGD.

Figure 8.1 shows a partial flow diagram for a coal-fired combustion plant. In this example, limestone is injected into a sequence of combustion zones, of which one is shown. Any sulfate formed is removed either as unburned residuals from the furnace or at the particulate collector stage (fly ash) which is not shown.

$$SO_2 + \tfrac{1}{2}O_2 + CaCO_3 \longrightarrow CaSO_4 + CO_2 \qquad\qquad (8.1)$$

There is a subsequent sulfur removal stage that may operate through any of the chemical FGD methods described in Section 8.2.

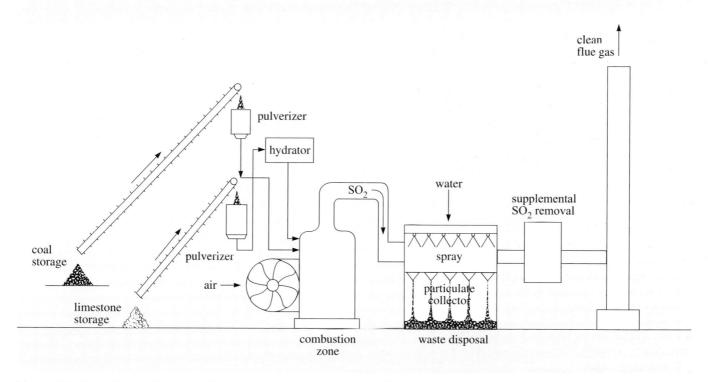

Figure 8.1 Flow diagram for a coal-fired combustion plant in which sulfur oxides are removed by reaction with limestone within the furnace. A subsequent FGD stage is also included, leading to the emission of relatively clean flue gas.

8.1 Electrostatic precipitation (ESP)

The effluent gases from the combustion furnace include suspended particles of ash that have become electrically charged within the flame and during energetic (collisional) encounters; those bearing like charges will *not* coalesce. This suspension is passed down tubes containing a central metal rod maintained at up to 100 000 volts with respect to the outer wall. Particles are attracted and discharged; the finely divided material readily coalesces. These larger grains of material settle and are collected for disposal, often as landfill. The European Directive (EC No 52/2000), applicable to new plants only, requires emissions of particulates in the gases expelled into the atmosphere to be maintained below 50 mg m^{-3} for small plants and below 30 mg m^{-3} for large plants.

8.2 Flue gas desulfurization (FGD)

The wet method of flue gas desulfurization brings the flue gases containing finely crushed limetstone into contact with water. Both these materials are required in large quantities — both are cheap and usually readily available. The overall sequence of chemical steps can be summarized by Equation 8.2:

$$SO_2 + \tfrac{1}{2}O_2 + 2H_2O + CaCO_3 \longrightarrow CaSO_4 \cdot 2H_2O + CO_2 \qquad (8.2)$$

The particulate solid product, gypsum ($CaSO_4 \cdot 2H_2O$), is then collected. This is commercially useful because it can be manufactured into plasterboard for use by the building industry. Alternatively, it can be disposed of safely as waste in landfill.

Four steps have been identified in controlling the rate of the overall chemical changes.

1 *Absorption of SO_2.* Gases are brought into contact with water droplets to form sulfurous acid, which then dissociates:

$$SO_2 + H_2O \rightleftharpoons HSO_3^- + H^+ \rightleftharpoons SO_3^{2-} + 2H^+ \qquad (8.3)$$

2 *Oxidation of HSO_3^-.* The kinetics of the oxidation of the HSO_3^- ion have not been completely agreed, but the reaction is believed to be catalysed by the small amounts of transition metal ions (catalytic, redox reaction), which are always present, at least in trace amounts, in the aqueous phase. This step is:

$$HSO_3^- + \tfrac{1}{2}O_2 \longrightarrow H^+ + SO_4^{2-} \qquad (8.4)$$

3 *Dissolution of the limestone.* Limestone is not very soluble, but the equilibrium is:

$$CaCO_3 + 2H_2O \rightleftharpoons Ca^{2+} + H^+ + HCO_3^- + 2OH^- \rightleftharpoons Ca^{2+} + 2H^+ + CO_3^{2-} + 2OH^- \qquad (8.5)$$

4 *The calcium and sulfate ions react to form gypsum.* Gypsum is relatively insoluble and from reactions (8.4) and (8.5) above:

$$Ca^{2+} + SO_4^{2-} + 2H_2O \rightleftharpoons CaSO_4 \cdot 2H_2O \qquad (8.6)$$

This proposed mechanism has been used to account for the several steps that contribute to the overall reaction. Plant equipment must be designed to maximize contact between the large volume of gas leaving the furnace and the water solvent/limestone mixture and, subsequently, to ensure rapid interaction between the limestone and the acid. Purity, particle size and the amount of retained water in the gypsum manufactured must all be controlled to yield a product that is acceptable for use as a building material. This method is relatively simple and is based on cheap raw materials (air, water and chalk) being widely and readily available in the large quantities required. The product, gypsum, is non-polluting and has a commercial value.

8.3 Other methods of sulfur removal

The following alternative chemical approaches to FGD have been proposed.

1 *Sodium hydroxide* can be used to absorb SO_2 from the flue gases. HSO_3^- is oxidized. $CaCO_3$ is then added to precipitate calcium sulfate and regenerate (most of) the alkali.

$$SO_2(aq) + OH^-(aq) \longrightarrow HSO_3^-\,(aq) \xrightarrow{\;O_2\;} SO_4^{2-}(aq) \qquad (8.7)$$

$$Ca^{2+}(aq) + SO_4^{2-}(aq) \longrightarrow CaSO_4 \cdot 2H_2O \qquad (8.8)$$

$$\text{gypsum}$$

Variations of this wet regenerative technique that have been proposed and investigated include the absorption of SO_2 by Na_2SO_3, Na_2CO_3, K_2CO_3 or $Mg(OH)_2$, or by buffered citric acid or FeS slurry, from which SO_2 and/or sulfur are recovered.

2   *Dry regenerative techniques* have been suggested, including the adsorption of SO_2 either on an activated carbon bed, followed by oxidation to sulfuric acid, or on CuO ($\longrightarrow CuSO_4$), which is later reduced by hydrogen, releasing SO_2.

3   *Dry calcium carbonate or calcium hydroxide* is injected into the combustion flame and the oxidized sulfur is retained by the solid particles. This process is, however, not very efficient as excess solid must be used. $Ca(OH)_2$ is more expensive to prepare but is more efficient in sulfur removal.

4   *Seawater scrubbing.* The flue gases can be passed through a tower where they make contact with droplets of seawater. This method is capable of removing around 90% of the SO_2 present (see Equations 8.3 and 8.4 above). Seawater is slightly alkaline (pH $\approx$ 8) and contains oxygen which converts dissolved sulfite into sulfate, a normal component of the ocean. Once the initial capital investment has been made, and the equipment has been installed on a coastal site that has a suitable sea circulation pattern, natural (and abundant) saltwater is admitted and the effluent containing the slightly enhanced sulfate concentration is harmlessly dispersed back to the ocean. No chemicals are added and there is no disposal problem.

ACID RAIN IN CONTEXT

9

This Case Study has been concerned with acid rain, which is just one limited aspect of pollution. However, the lessons learned have important consequences for environmental protection more generally. Widespread appreciation of the damage that can be inflicted by lack of restrictions on waste disposal, and sulfur dioxide in flue gases in particular, have strongly influenced industrialists and politicians in alerting us all to the necessity for maintained vigilance, if we are not to suffer from the unexpected, unwelcome and often expensive consequences of our activities. The strong warning given by the acid rain experience has led to remedial action, but the effects, though now less severe and obvious, are nonetheless still present. Emissions continue to inflict damage in places where plant, animal and insect communities are especially sensitive. The maintained large-scale emissions of CO_2 to the atmosphere continue to contribute to the greenhouse effect, with unknown long-term consequences.

In this Case Study we have considered emissions of SO_x, produced through coal combustion, which have resulted in acid rain. This is not the only possible source of anthropogenic sulfuric acid. It may also (and additionally) be released into the atmosphere during ore smelting of commercially important metals that occur as sulfides, such as lead, zinc, nickel, copper and mercury. Efforts are continuing to reduce emissions from these sources; one approach has been to develop alternative (non-combustion) methods of extracting metals from their ores (clean technology). Furthermore, while SO_x remains a significant pollutant in many places, other types of acid rain are known. NO_x, mainly generated in internal combustion engines, may produce nitric acid, which is another cause of acid rain. This, in some atmospheric conditions (often associated with the west coast of North America), leads to photochemical fog, a brown haze capable of causing discomfort (and worse) to humans. It can also cause economic damage to crops. However, NO_x emissions can be reduced by changing the fuel : oxygen ratio within car engines and by fitting catalytic converters to treat the exhaust gases.

Sulfur is a less severe problem for fuels based on petroleum because it can be extensively removed by catalytic hydrodesulfurization during the refining process. Unfortunately this method is not applicable to coal. Nevertheless, debates continue about what upper limit of sulfur can be regarded as 'acceptable' in the large amounts of petroleum burned. All the SO_x present is released untreated into the atmosphere, much of it in urban areas and at low levels.

Finally, it should also be remembered that there are natural sources that discharge acidic compounds into the atmosphere, such as SO_2 and HCl from volcanoes, and sulfur compounds that are formed in marine biological cycles. These represent a balance in nature that, as seen here, is readily and unacceptably disturbed by anthropogenic activities.

In some countries, including the UK, coal consumption is decreasing. The high cost of sulfur removal is one significant economic motivation towards replacing the exploitation of natural carbon deposits with natural gas (CH_4) as a primary energy source. Methane fuel contains less sulfur and yields proportionately less CO_2 for each unit of electricity generated (thus making a smaller contribution to the greenhouse effect, another reason for the preferred use of this fuel). It may well be that, during the foreseeable future, coal usage will continue to decline as a large-scale provider of electrical energy through preferred use of alternative generation methods. These include nuclear power (possibly, eventually, by fusion), wind, tide and wave energy, fuel cells and photocells converting sunlight directly into electrical power. Nevertheless, coal continues to represent a massive, and long-term, accessible (if inconvenient and sometimes dirty) reserve of energy based on carbon derived from plants and so represents fossilized sunlight energy. If we can reduce the levels of pollution that have historically been associated with the use of this fuel to acceptable levels — although it is difficult to obtain an agreed definition as levels are becoming progressively more stringent — and increase the efficiency of energy release, it seems unlikely that this prolific primary source of power will be abandoned in the immediate future.

ACKNOWLEDGEMENTS

Grateful acknowledgement is made to the following sources for permission to reproduce material in this book:

Figure 2.1: AP Photo/Peter Dejong; *Figures 2.2a and b*: Building Research Establishment; *Figure 2.3*: Still Pictures/Mark Edwards; *Figure 6.2*: Centre for Ecology and Hydrology; *Figure 7.1*: Environmental Change Research Centre, UCL; *Figure 7.3*: Neil Rose, Environmental Change Research Centre, UCL.

Every effort has been made to trace the copyright owners, but if any has been inadvertently overlooked, the publishers will be pleased to make the necessary arrangements at the first opportunity.

Case *Study*

Industrial inorganic chemistry

Alan Heaton

Liverpool John Moores University

and Rob Janes

INTRODUCTION

1

The main aim of this Case Study is to give you an overview of the industrial production of inorganic chemicals — one of the most successful sectors of the UK manufacturing industry. Table 1.1 shows world production levels of some major industrial chemicals. It is no accident that the main economic powers in the world host substantial chemical industries, and inorganic chemicals dominate the upper half of the chemicals 'top twenty', providing no less than eight of the top ten! It is important as well to distinguish between the two categories of inorganic chemical: those manufactured on the very large scale — **heavy inorganic chemicals** — and those made on a much smaller scale for specialist applications — **speciality inorganic chemicals**.

Table 1.1 Production statistics of inorganic chemicals

Chemical	World production/ million tonnes per year
ammonia	114.22
chlorine	40
fertilizers	138
hydrogen peroxide	1.5
phosphoric acid	35.85
sodium hydroxide	39
sulfuric acid	135.67
titanium dioxide	3.7

Adapted from Royal Society of Chemistry data, 1997.

You have already met the reactions underlying a number of industrial inorganic processes. Although these will not be repeated here, we will review a number of the practical and economic factors that need to be considered when producing these chemicals on a large scale.

HEAVY INORGANIC CHEMICALS

2

2.1 Sources of industrial inorganic chemicals

One of the major differences between the organic chemicals industry and the manufacture of inorganic chemicals on the large scale lies in the source of raw materials. Most industrial organic chemicals can be derived from a single source: oil and natural gas, coal, or carbohydrates (biomass, and animal and vegetable oil and fats). In contrast, industrial inorganic chemicals are produced from a wide variety of sources.

2.1.1 Ores

The most important raw materials are metal ores, for example metal oxides, sulfides, carbonates and halides. Some specific examples are bauxite (aluminium oxide), pyrite (iron (II) sulfide), limestone (calcium carbonate) and salt (sodium chloride). Metals are extracted from their ores, for example aluminium from bauxite, and can then be converted into their salts. In some cases processing does not yield the metal but another compound — limestone on heating above 1 170 °C is converted to quicklime (CaO). This reacts with water to give slaked lime ($Ca(OH)_2$). A number of non-metals are also extracted from minerals. One example is phosphorus, which is produced by the reduction of phosphates with coke and sand in an electric furnace.

2.1.2 Air

The air is another raw material for the chemical industry. Ignoring water vapour, the amount of which varies from region to region, its principal components are nitrogen (78.03%), oxygen (20.95%), argon (0.93%) and carbon dioxide (0.04%), together with tiny amounts of the noble gases and hydrogen. For the large-scale production of industrial gases from air, *cryogenic distillation* is the most economical method — air is liquefied and separated into nitrogen, oxygen and argon. The first step involves getting the air so cold that it forms a liquid. This process can be traced back to the work of Linde and Hampson at the beginning of the 20th century, who used the cooling effect resulting from *adiabatic expansion of gases*. Air is typically compressed to about 6 bar, cooled below ambient temperature (at which point impurities are removed) and, following further cooling, is expanded into a distillation column. As the constituents of air have different boiling points and densities, the liquid air is then distilled in much the same way as mixed liquids, for example petroleum, are fractionated.

Customers requiring large quantities of gases have them transported and stored as liquids (Figure 2.1).

Figure 2.1
A tanker for the transport of liquid nitrogen

2.1.3 Water

Seawater is a valuable source of inorganic chemicals, specifically calcium and magnesium salts and bromine. Although their concentration is low when compared with, say, that of a metal in a metal ore, this is offset by seawater being free and easily pumped ashore to the processing plant. For example, the world's most important source of bromine is the Dead Sea in Israel. Another example of water as a valuable source of an inorganic chemical is described in Box 2.1.

BOX 2.1 Sodium carbonate and the soda lakes of Kenya

Sodium carbonate (commonly known as soda ash) has been used for thousands of years, particularly in glass manufacture and for washing clothes. Today it is also used as an alkali in place of sodium hydroxide in many applications. Other uses are in water purification, polymer production, tanning, effluent neutralization and sugar extraction. It is manufactured industrially by the Solvay (or ammonia–soda) process, which relies on the precipitation of sodium bicarbonate (sodium hydrogen carbonate) from the reaction of ammonia and carbon dioxide (produced by the action of heat on limestone) with a solution of sodium chloride. The bicarbonate is then heated to produce soda ash. This may appear to be a rather convoluted process for a reaction which may be represented as:

$$2NaCl(aq) + CaCO_3(s) = Na_2CO_3(aq) + CaCl_2(aq) \qquad (2.1)$$

However, the reverse reaction is thermodynamically more favourable. If calcium chloride and sodium carbonate are mixed, sodium chloride and calcium carbonate are immediately formed — this is fortunate as otherwise the White Cliffs of Dover would have dissolved into the salt water of the English Channel long ago!

Sodium carbonate also occurs naturally as the minerals *natron* (Na_2CO_3), *thermonatrite* ($Na_2CO_3.H_2O$) and *trona* ($Na_2CO_3.NaHCO_3.2H_2O$). These ores can either be used directly for a number of applications, or processed to give a purer product. Sodium carbonate is also found in lakes which contain alkali brines. One of the most important examples is Lake Magadi (Figure 2.2) in Kenya, located southwest of Nairobi, near the Tanzanian border. It is one of a string of alkaline lakes and hot-springs situated in the Rift Valley, which separates east and west Kenya. These lakes are unique because their water is highly concentrated sodium carbonate solution. Poor drainage, and a high rate of evaporation of surface lake water, results in high levels of soda ash, which may be isolated. Indeed, Lake Magadi has been a commercial source of soda ash since the early 1900s, and still contributes to the local economy, with over 300 000 tonnes being extracted annually. The main export markets are in South East Asia and South Africa. Furthermore, these highly alkaline conditions are an ideal breeding ground for algae and certain species of fish which, in turn, are enjoyed by birds, particularly flamingos, which flock to the lake in their millions, adding to the attraction of the region to tourists.

Figure 2.2 Vast quantities of naturally occurring trona deposits are found in Lake Magadi, Kenya.

2.1.4 Elements

As the above example demonstrates, most elements are found in a combined state as chemical compounds: processing is required to extract them. In contrast, just a few elements occur in elemental form. Apart from nitrogen and oxygen, the most important of these is sulfur, which is found in underground deposits, notably in Poland and the USA. It was not until the beginning of the 20th century, when the elegant Frasch recovery process (devised by a Canadian scientist, Hermann Frasch), was introduced, that the deposits became commercially important. In the Frasch process (Figure 2.3), three coaxial tubes are inserted into the sulfur beds in a shaft of about 30 cm diameter. High pressure steam at 165 °C is forced down the outer pipe to melt the elemental sulfur, which is then blown up inside the second pipe with the

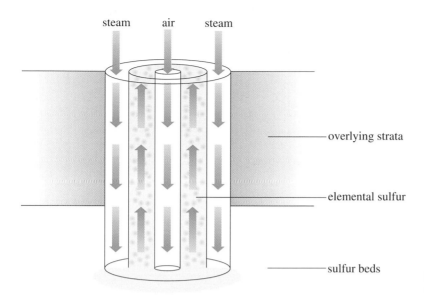

steam   air   steam

— overlying strata

— elemental sulfur

— sulfur beds

Figure 2.3
The Frasch sulfur extraction process.

assistance of compressed air passed down the centre pipe. The beauty of this process is that it encompasses in a single stage both mining and purification. The high quality product (99% pure sulfur) is thus obtained relatively cheaply.

Interestingly, the major source of sulfur nowadays is via the desulfurization of crude oil and natural gas — testimony to the enormous scale on which these materials are refined, since they only contain a few per cent of sulfur-containing compounds. These compounds have to be removed because they tend to poison the catalysts used in later (downstream) reactions and pollute the environment.

2.2 The production of selected heavy inorganic chemicals

This category encompasses many of the traditional inorganic chemicals used commercially, which tend to be fairly simple compounds. All are made on a *large scale* with individual plants usually having the capacity to produce at least 100 000 tonnes of product per annum. This means that they have to operate in a **continuous mode** in order to achieve these enormous throughputs of materials. Reactants are constantly being fed into the plant at one end whilst products are continually removed from the other. This goes on for 24 hours a day, 365 days per year, except for shutdown for maintenance, or in an emergency.

As shown in Table 1.1, the number of chemicals manufactured as bulk commodities is large. Consequently, in this account we have focused on just a few representative examples that embody the major considerations which underpin the operation of a viable manufacturing unit. These are: sulfuric acid; phosphoric acid; ammonia, nitric acid and nitrates; the chlor-alkali industry; and the inorganic fluorine industry.

Continuous-mode operation has important implications for the location of a production plant. In order to minimize the transportation of large quantities of chemicals, the plant is ideally located near to both sources of raw materials and also the users of the product(s). It is often difficult to achieve both these requirements, so the emphasis tends to be on the former. Another important consideration for these giant plants is proximity to large amounts of water. This is needed both for cooling, and for diluting treated waste or effluent before discharge or disposal. The siting of plants near to the mouths of river estuaries fulfils these requirements, and also enables large amounts of chemicals to be brought in and sent out by ship. Hence the UK chemical industry has major concentrations at Runcorn and Widnes (Figures 2.4 and 2.5) on the River Mersey, Teesside on the River Tees, Grangemouth on the River Forth, and Baglan Bay in South Wales.

These geographical considerations are exemplified by looking at one aspect of the growth of the chemical industry in the north east of England. The work of Fritz Haber in Germany in 1908 showed that at high pressure and temperature, and using an iron catalyst, ammonia could be made from the reaction of nitrogen and hydrogen. Following further developmental work by Carl Bosch, the chemical company BASF established the first production plant in Ludwigshafen in 1912.

Figure 2.4
The Winnington alkali chemical production plant, near Runcorn, Cheshire, operated by the Brunner-Mond chemical company.

Figure 2.5
A postcard from Widnes, depicting early chemical production in the area. The wording on the reverse is *Widnes with the lid off makes the sun hide its face.*

The ammonia was used to manufacture fertilizers and explosives for use in the first World War. The British government of the day realized there was an urgent need for the UK to produce its own synthetic ammonia and the chemical manufacturer Brunner-Mond was charged with setting up a plant. After obtaining BASF secrets from two engineers in Alsace (following its return to French rule after the war), Brunner-Mond was able to build their first production plant. Billingham on Teesside (Figure 2.6) was chosen as the site for a number of reasons. The River Tees provided a port and there is a ready supply of water. A nearby power station was under construction at the time, and the proximity to the Durham coalfields meant coal could be easily and cheaply transported to the plant. Coal was required to provide the high temperatures required for the Haber process, and also as a source of hydrogen. The first ammonia was produced on Christmas Eve, 1923.

Subsequently, inorganic chemical production expanded and diversified into other areas. Anhydrite (calcium sulfate) was mined for use in sulfuric acid production, nitric acid and fertilizer production was introduced, and later organic chemicals and polymer production began, expanding to nearby Wilton. Ammonia and mineral acids are still manufactured to this day on the Billingham site.

Figure 2.6 Billingham on Teeside, the site of the first ammonia production plant in the UK.

2.2.1 Sulfuric acid

Sulfuric acid is the strong acid most widely used by the chemical industry, due to its low cost of synthesis. It is manufactured by the **contact process**, which is covered in detail earlier in this Book. Here we will concentrate only on one stage — which illustrates some of the principles important in plant design — the combination of sulfur dioxide with oxygen to give sulfur trioxide in the presence of a catalyst (usually V_2O_5):

$$2SO_2(g) + O_2(g) = 2SO_3(g) \quad \Delta H_m^{\ominus} \sim -100 \, kJ \, mol^{-1} \tag{2.2}$$

A closer look at this step will give you some idea of the practical aspects of bulk chemical production, which as you will see, draw upon some fundamental chemistry. This is an equilibrium reaction, and since it is exothermic, Le Chatelier's principle tells us a low temperature will favour SO_3 production. However, at low temperatures the reaction rate will be slow. Hence a compromise must be reached.

Owing to the reaction being highly exothermic it is difficult to achieve a good yield using a single reactor, since as it warms up it will throw the reaction into reverse. The solution is to use a series of reactors — up to four — with cooling of the gases in between. In Reactor 1 about 60% of the sulfur dioxide is converted into sulfur trioxide at 600 °C. At the exit from Reactor 4, about 95% of reactants have been converted into products.

Sulfuric acid plant managers are almost unique in being delighted whenever the price of oil, and hence energy, increases! This is because they are net 'exporters' of energy, since as we have seen, the reactions in the plant are highly exothermic, and the sale of this energy reduces the manufacturing cost of the sulfuric acid.

2.2.2 Phosphoric acid

The most important application of sulfuric acid is in the preparation of phosphoric acid, by the so-called **wet acid process**. This produces phosphoric acid from the reaction of dilute sulfuric acid with mineral phosphate.

$$Ca_3(PO_4)_2(s) + 3H_2SO_4(aq) = 3CaSO_4(s) + 2H_3PO_4(aq) \tag{2.3}$$

The product is of lower purity than the phosphoric acid produced by the thermal process described below, but this costs about three times as much to produce. The gypsum (calcium sulfate), is a by-product that is widely used in the building industry as a low cost, fire-retardant wall board. After filtration, the dilute phosphoric acid is concentrated by evaporation to a rather dark impure brownish liquid, which is suitable for use in fertilizer manufacture.

The **thermal process** forms an alternative route to produce phosphoric acid. It involves production of elemental phosphorus as an intermediate, and may be used when food quality chemicals are required (Figure 2.7). The first step is the reduction of phosphate by carbon in the presence of silica, which forms a removable silicate slag with the calcium. The reduction is carried out at 1 600 °C in an electric furnace.

$$2Ca_3(PO_4)_2(s) + 10C(s) + 6SiO_2(s) = P_4(g) + 6CaSiO_3(l) + 10CO(g) \tag{2.4}$$

Phosphorus is removed as a vapour and condensed under water. Phosphoric acid can then be made by the oxidation of phosphorus and subsequent treatment of the phosphoric oxide with water.

$$P_4(s) + 5O_2(g) = P_4O_{10}(s) \tag{2.5}$$

$$P_4O_{10}(s) + 6H_2O(l) = 4H_3PO_4(l) \tag{2.6}$$

Figure 2.7
Food grade phosphoric acid is found in cola.

The industrial-scale availability of phosphoric acid, as well as elemental phosphorus whose extraction was discussed above, opens up an enormous number of wide-ranging applications for these products. They are now of huge commercial importance (Figure 2.8 overleaf).

2.2.3 Ammonia, nitric acid and nitrates

Ammonia plants utilizing the Haber process are amongst the largest in the chemical industry, with some having capacities of 500 000 tonnes per annum! These plants are so large in order to take advantage of the economy of scale. Increasing the scale significantly reduces the cost per tonne, mainly due to the capital costs of building the plant, where a tripling of capacity results in only a doubling of the cost.

Looking more closely at the Haber process in practical terms, we note that, like sulfur trioxide formation in the contact process, it is an equilibrium reaction and also exothermic.

$$N_2(g) + 3H_2(g) = 2NH_3(g) \quad \Delta H_m^{\ominus} = -90\,kJ\,mol^{-1} \tag{2.7}$$

Again we have a compromise between equilibrium yield and an acceptable reaction rate, since the former is favoured at low temperatures and the latter at higher ones. This dilemma is overcome by operating at $400\,°C$ in the presence of an iron oxide-based catalyst. Since the reaction involves four molecules of reactants forming two of products, Le Chatelier's principle predicts that increasing the pressure will move the equilibrium in favour of the products. Thus the reaction is carried out under high pressure, although process improvements have allowed a reduction in pressure from > 250 atm to about 80 atm. In the not too distant future further reductions can be expected due to a new catalyst being introduced — ruthenium metal impregnated on graphitic carbon. Since this will be about 50 times more expensive than the old catalyst, it must deliver very significant improvements to be economically viable.

It is interesting to note that the hydrogen required for the Haber process comes from an *organic source* — North Sea gas (methane).

Two stages are needed, with the first known as **steam reforming**:

$$CH_4(g) + H_2O(g) = CO(g) + 3H_2(g) \tag{2.8}$$
$$CH_4(g) + 2H_2O(g) = CO_2(g) + 4H_2(g) \tag{2.9}$$

These reactions take place at $750\,°C$, and are carried out over nickel-based catalysts. The **water gas shift reaction** is then employed to remove carbon monoxide and produce more hydrogen:

$$CO(g) + H_2O(g) = CO_2(g) + H_2(g) \tag{2.10}$$

In practice this is carried out in two stages: at $400\,°C$ over an iron catalyst, and then at $200\,°C$ over a copper one. Finally the carbon dioxide is removed by bubbling the gaseous mixture through either potassium hydroxide or diethanolamine solution. It is important that the oxides of carbon are reduced to a very low level otherwise they may poison the catalyst in the Haber process.

In addition to fertilizer production, one of the most important uses for ammonia is in the manufacture of nitric acid by the Ostwald process, as discussed earlier in this Book. In essence, ammonia is oxidized over a catalyst to oxides of nitrogen, which are dissolved in water to form a concentrated solution of the acid. From a purely practical standpoint, the initial oxidation step is worth a closer look.

$$NH_3(g) + \tfrac{5}{4}O_2(g) = NO(g) + \tfrac{3}{2}H_2O \quad \Delta H_m^{\ominus} = -950\,kJ\,mol^{-1} \tag{2.11}$$

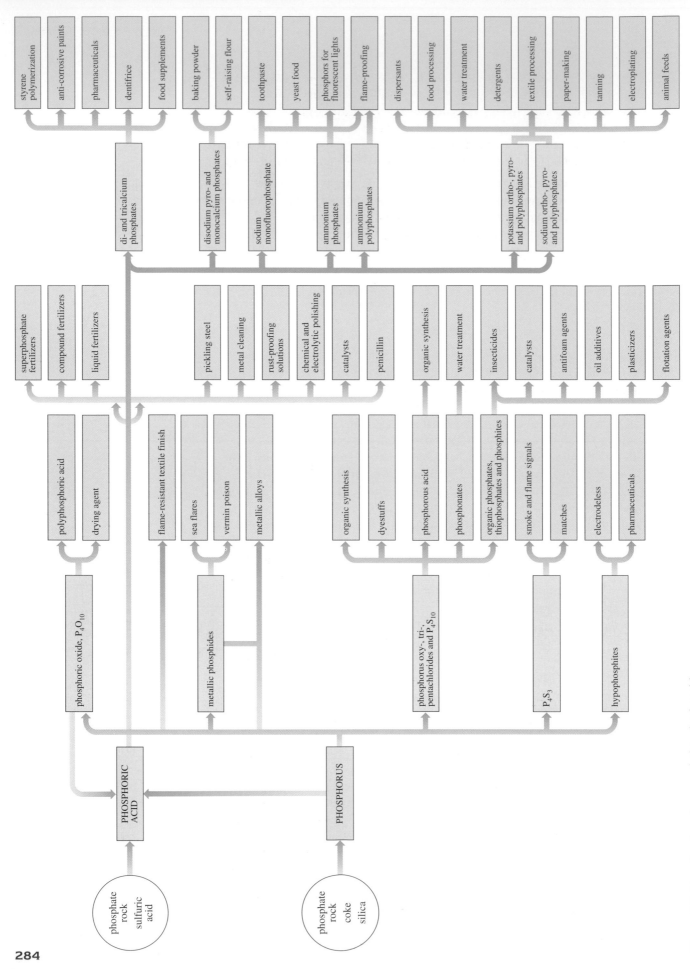

Figure 2.8 Uses for phosphorus and phosphoric acid.

This process is very tricky to control in practice, because it is accompanied by a number of side reactions, which are also very fast; for example, formation of nitrogen and water. These are minimized by careful reactor design and very fine temperature and flow rate control. Incredibly, the actual contact time in the reactor is a mere 10^{-4} s — a very fast reaction indeed! The gases enter at 300 °C and leave at 900 °C due to the highly exothermic nature of the reaction. The catalyst for this step is a platinum–rhodium gauze.[*]

Nitrates, which find extensive application as fertilizers, are generally obtained from mineral deposits, but such sources have now been supplemented by manufacture from nitric acid. The industrial production of ammonium nitrate (one of the most important fertilizers, due to its high nitrogen content) is beset with one major problem. You may remember that heating ammonium nitrate above about 200 °C results in an explosive decomposition, and in a large-scale synthesis this could be catastrophic! Ammonium nitrate is prepared industrially by the neutralization of aqueous nitric acid by gaseous ammonia, followed by evaporation of water. Molten ammonium nitrate at about 170 °C cascades down a tall tower, up which cool air is flowing. Tiny spheres of ammonium nitrate are then removed from the base of the tower. The following precautions are routinely taken to minimize the explosion risk:

(i)   gaseous ammonia is injected at various stages to maintain the pH of molten NH_4NO_3 above 4.5, below which sensitivity to decomposition becomes a problem;

(ii)   pressure build up during neutralization is avoided; and

(iii) various substances, which can catalyse the decomposition (e.g. oil, chlorides and metals such as cobalt) are excluded. This is also an important precaution when storing the compound.

2.2.4 The chlor-alkali industry

An important sector of heavy inorganic chemical manufacturing is the production of chlorine and sodium hydroxide — the chlor-alkali industry. The manufacture of these chemicals has a long history. Today they are produced simultaneously by the electrolysis of sodium chloride solutions, but this was not always the case. The two chemicals were originally manufactured by different routes. In the 19th century chlorine was made by the oxidation of hydrogen chloride (itself made by reaction of salt with sulfuric acid) using the *Deacon process*. Sodium hydroxide was prepared by the reaction of calcium hydroxide with sodium carbonate — the *lime-soda process*.

In the UK, the first electrolytic plant operated by the Castner-Kellner Company in Runcorn started production in 1897. The electrolysis of brine (aqueous sodium chloride) using the so-called membrane cell, is now the method of choice (see *Metals and Chemical Change*[1]). The earlier mercury cells were the cause of much environmental concern, and their energy consumption was enormous. ICI's plant at Runcorn, Cheshire had cell rooms the size of two football pitches and used as much electricity as the whole of the nearby city of Liverpool! The energy usage of the membrane cells is significantly lower — note that even a small saving of, say, 50p per tonne, really adds up for a 100 000-tonnes-per-year plant. Membrane cells are small and easy to assemble. For example, the ICI FM21 cell has electrodes of only 0.21 m^2, and 50 such cells would provide a plant with an output of 50 000 tonnes per

[*]This reaction is demonstrated at a laboratory scale in 'The p-Block elements in action' on one of the CD-ROMs accompanying this Book. ⌨

annum. Hence the trend now is to erect these on the sites of chorine users, for example in vinyl chloride production which is used to make PVC, rather than undertake the difficult task of transportation (see Box 2.2). This is a major departure from the original requirement to site chlor-alkali plants near to the source of brine, which was the reason for the growth of the industry at Runcorn, close to a rocksalt mine inland in Cheshire.

There is, however, a drawback with membrane technology, due to the hostile environment in which the cells operate. The highly alkaline environment means that the expensive membranes must be replaced every 2–3 years.

BOX 2.2 Chemicals on the move: cracking the Hazchem code

Tankers containing chemicals are often seen on our roads and motorways. The chemicals they contain range from highly toxic to harmless. When transporting chemicals, a haulier must be aware of the nature of the load, and the vehicle must carry the appropriate hazard warning panel to advise on actions to be taken in the event of spillage. An example of such a panel is shown in Figure 2.9 for substance 1017 — chlorine, labelled 2XE. You can see what this means by consulting the Hazchem scale (Figure 2.10).

The number 2 means the fire brigade can use water, fog or spray equipment on this chemical. The X is in the 'contain' section, indicates the spillage must be contained, and not allowed to enter drains or watercourses. If it was in the 'dilute' section, the substance could be flushed away. The X also indicates that full body protective clothing must be warn if tackling a spillage; if there were a V next to the letter (not the case here) then the chemical is deemed to be highly reactive and potentially explosive. Finally, the letter E advises evacuation of the area should be considered.

Figure 2.9
Hazchem transportation details for chlorine.

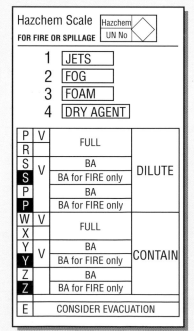

Figure 2.10
Interpretation of the Hazchem notation.

Since the production of chlorine and sodium hydroxide is inextricably linked, a surge in demand for one creates problems with sales of the other. For example, increased demand for chlorine for vinyl chloride production meant that there was an abundance of sodium hydroxide. As a result it replaced sodium carbonate in some of its applications.

The uses of sodium hydroxide are many and varied. A large proportion finds application as a reagent in organic chemical plants, and in the synthesis of other inorganic chemicals. It is also used in household cleaning products, including oven and drain cleaners, and finds widespread application in the food industry, principally for breaking down proteins. One lesser known application of sodium hydroxide is to stick salt crystals to the surface of pretzels. The dough is coated with sodium hydroxide which acts as a glue for the salt, and in the subsequent baking process it is converted to harmless sodium carbonate (Equation 2.12).

$$2NaOH(s) + CO_2(g) = Na_2CO_3(s) + H_2O(g) \qquad (2.12)$$

Over one-third of chlorine produced is used to make PVC via ethylene dichloride (1,2-dichloroethane).

$$H_2C{=}CH_2 \longrightarrow ClH_2CCH_2Cl \longrightarrow H_2C{=}CHCl \longrightarrow {-}{\left(\!{-}CH_2{-}\underset{Cl}{CH}{-}\!\right)}_n^{-} \qquad (2.13)$$

ethene   1,2-dichloroethane   chloroethene   PVC
(ethylene dichloride)   (vinyl chloride)

The next major use is the production of other inorganic chemicals. These include sodium and calcium hypochlorites, which are useful oxidizing agents and find application as bleaches, disinfectants and in sewage treatment. Interestingly, these are made by the reaction of the two main products of sodium chloride electrolysis:

$$Cl_2(aq) + 2OH^-(aq) = Cl^-(aq) + ClO^-(aq) + H_2O(l) \qquad (2.14)$$

Other significant uses are in the production of chlorinated solvents and chlorofluorocarbons (CFCs) and their replacements.

2.2.5 The inorganic fluorine industry

Given the extreme nature of fluorine, you might be slightly surprised to learn that fluorine, fluorides and hydrofluoric acid form the basis of a large and diverse sector of the inorganic chemicals industry. Indeed they are the common link between aluminium production, fire-fighting agents, Gore-Tex™ fabric, glass etching, non-stick frying pans, petrochemicals and refrigeration.

Fluorine is only found in nature in the combined state in minerals, two of the most important being *cryolite* ($3NaF.AlF_3$) and *fluorspar* (CaF_2). Cryolite is found only in Greenland, but little is mined. Instead large quantities of synthetic cryolite is made for use in aluminium production (see below). Fluorspar is widely distributed, the main UK deposits being in the Derbyshire Peak District. UK production of fluorspar exceeds 200 000 tonnes per annum. Over half of this is used for the manufacture of anhydrous hydrogen fluoride, from which most other fluorine-containing compounds are made.

The reaction is straightforward, although heating is required because it is endothermic.

$$CaF_2(s) + H_2SO_4(l) = 2HF(g) + CaSO_4(s) \qquad (2.15)$$

The product is either liquefied by refrigeration or dissolved in water to give hydrofluoric acid. Inevitable impurities in this process are calcium carbonate and silica from the raw materials. These also react:

$$CaCO_3(s) + H_2SO_4(l) = CaSO_4(s) + H_2O(l) + CO_2(g) \qquad (2.16)$$

$$SiO_2(s) + 4HF(l) = SiF_4(g) + 2H_2O(l) \qquad (2.17)$$

Silicon tetrafluoride itself reacts with hydrofluoric acid to give fluorosilicic acid (H_2SiF_6). Since these impurities consume sulfuric and hydrofluoric acids it is essential to use fluorspar which is very pure. Incidentally, there are vast quantities of the by-product calcium sulfate stored in the old salt mines in Cheshire, and a fortune awaits anyone who can find a use for it!

Hydrofluoric acid has to be handled with great caution since it is extremely corrosive to skin, eyes, mucous membranes and the lungs. It causes deep-seated burns if spilt on the skin and even after treatment these are very slow to heal.

Even more care has to be exercised with fluorine gas due its great reactivity; for example, if the neat gas contacts any organic material, the latter bursts into flames. However, the element has now been tamed, being produced by the electrolysis of potassium hydrogen difluoride (KHF_2) containing hydrofluoric acid, using carbon anodes and steel cathodes in a steel cell at 90 °C. The fluorine gas contains about 4% HF which is removed by reaction with sodium fluoride to give sodium hydrogen difluoride. The fluorine is either used directly or else liquefied for storage. Such cells are operated by British Nuclear Fuels (BNFL) at its site near Preston.

So why is the inorganic fluorine industry so important?

The most important use of hydrofluoric acid is the manufacture of cryolite and aluminium fluoride, which are used as molten electrolytes in the electrolysis of *bauxite* to produce aluminium. The second major use, which was on a par with that just mentioned but has now declined somewhat, is in the production of organofluorine compounds. These include the chlorofluorocarbons (CFCs), which are now being phased out due their adverse environmental effects, although interestingly their replacements, the HFCs (hydrofluorocarbons) and HFAs (hydrofluoroalkanes) also require HF for their manufacture. However, in their major use as aerosol propellants the CFCs have been replaced by non-fluorine-containing compounds — the so called 'ozone friendly' compounds. Other related fluorine-containing compounds of importance are the fire extinguishant BCF™ (bromochlorodifluoromethane, CF_2BrCl), and Halothane or Fluothane™ ($CF_3CHBrCl$), once the most widely used general anaesthetic.

Fluoropolymers are ubiquitous and have special properties. The best known is PTFE (polytetrafluoroethylene), which is widely used as a coating for non-stick cooking ware, under its trade names Teflon™ or Fluon. Gore-Tex™ is made from expanded PTFE, and is used in outdoor clothes as it allows the fabric to breathe, whilst still retaining excellent water repellency. Similar polymers are used to give water and stain resistance to textiles, for example, Scotchguard™.

In addition to cryolite and aluminium fluoride, there is a wide range of inorganic fluorides of commercial importance; these include sulfur hexafluoride, used as an electrical insulating gas. Sodium fluoride has been used in the fluoridation of water, although it is being replaced by fluorosilicic acid, and its salts. Tin (IV) fluoride (SnF_4) is used in toothpaste to prevent dental decay, and boron trifluoride (BF_3) is a widely used catalyst for reactions in the petrochemical industry.

The first major use of fluorine was in the separation of uranium ^{235}U from its 238 isotope (^{238}U) via the formation of uranium hexafluoride (UF_6):

$$UO_2(s) + 4HF(g) = UF_4(s) + 2H_2O(g) \tag{2.18}$$

$$UF_4(s) + F_2(g) = UF_6(g) \tag{2.19}$$

The uranium hexafluoride is then fed to a gas centrifuge plant in which $^{235}UF_6$ is separated from the $^{238}UF_6$. The enriched $^{235}UF_6$ is used to prepare the fuel for the nuclear electric power plant. In the UK, a gas centrifuge plant is situated at Capenhurst near Chester.

Finally, several metal fluorides are used in place of fluorine gas to fully fluorinate organic molecules — the reaction is a lot less exothermic and therefore more controllable. A frequently used reagent for this purpose is cobalt (III) fluoride (CoF_3). The best known example of its use is the conversion of naphthalene into perfluorodecalin ($C_{10}F_{18}$). This compound is used as an electrical insulator, and has been tested as artificial blood because of its ability to dissolve large amounts of oxygen, coupled with its non-toxicity and chemical inertness.

SPECIALITY INORGANIC CHEMICALS

3

As the name implies, these chemicals are manufactured for more specialist uses, and as a consequence are produced on a much smaller scale, but command a much higher price than those produced in bulk. In general much research and development has occurred, often involving collaboration between university and industrial researchers. These chemicals tend to be made in **batch mode**; that is, in the same way in which we normally carry out reactions in the laboratory, where reactants are loaded into a reaction vessel and heated for a prescribed length of time. The products are isolated, and the process repeated to produce more.

As with the heavy inorganic sector, the list of speciality chemicals is vast. In this section we only consider a small selection of some common speciality inorganic chemicals to illustrate both their wide variety and diversity of applications.

In fact it is difficult to place many chemicals squarely into either the heavy or speciality categories. A chemical which is typical of this 'grey area' is titanium dioxide, which is produced on a vast scale, and may be thought of as straddling both markets. Its uses are extremely varied (Figure 3.1), being found as a pigment in paints, inks and varnishes, as a whitener in paper, and as a filler in rubber and plastic products. It is also an important ingredient in sunscreen lotions. More specialized uses are as catalysts and catalyst supports and in humidity sensors.

3.1 Inorganic chemicals in electronics

Most people associate electronic devices with silicon chips, and indeed the key component of digital electronics is the silicon integrated circuit. It is interesting to note that in a similar manner to the way the heavy inorganic sector is associated with certain geographical locations, the microelectronics industry also has certain focal points. We frequently hear about 'Silicon Valley' in California and 'Silicon Fen' near Cambridge, UK. One reason for the growth of microelectronics in these areas is the proximity to universities and academic input into research and development.

The silicon integrated circuit utilizes interconnected transistors at the surface of the silicon chip which process electronic signals. However, it is the subtle interactions between silicon and other elements that produce the useful effects, and which are central to the functioning of these devices. The Group IV/14 element, silicon, is the pre-eminent elemental semiconductor (as discussed earlier in this Book), whose conductivity is increased by doping at the parts per million level (see *Molecular Modelling and Bonding*[2]). Impurities from Group V/15, such as phosphorus or arsenic, have five valence electrons, and the extra electron is only weakly bound to the phosphorus, playing no part in bonding in the lattice. These extra electrons are responsible for increasing the conductivity of the semiconductor. Such dopants are known as *donors*. Alternatively, impurities from Group III/13, such as boron or aluminium, with three valence electrons, need one extra electron to satisfy their bonding requirements. A bonding electron from an adjacent silicon can move to the

Figure 3.1
Titanium dioxide is the most commonly used white pigment in paints. It is also used as a sunblock, preventing harmful UV radiation from reaching the skin and reflecting visible light.

impurity, generating vacancies, or positive 'holes'. Again there is a net movement of electrons and conductivity rises. These impurities are known as acceptors.

Silicon that contains donor impurities is known as n-type silicon, while silicon doped with acceptor impurities is known as p-type. Many semiconducting devices consist of a layer of p-type silicon on top of a layer of n-type silicon — the so-called p–n junction. An electric current can flow in only one direction across this junction. The extra electrons in the n-type silicon can move across into the p-type material, but electrons will not flow the other way, as the n-type material already has an excess of electrons.

Doping at these minute levels means that unwanted impurities in the host semiconductor must be reduced to very low levels: 99.999 9 to 99.999 999% pure materials are often required.

Silicon is obtained by the reduction of silicon dioxide with coke in an electric arc furnace:

$$SiO_2(s) + 2C(s) = Si(l) + 2CO(g) \tag{3.1}$$

Purification methods have been developed which usually involve conversion of silicon into volatile silanes which are then subjected to fractional distillation, and converted back to silicon. The Siemens process, Equations 3.2 and 3.3, involves the production of $SiHCl_3$ by reacting silicon with HCl. Following fractional distillation, the halide is reduced with hydrogen at about 1 000 °C to produce rods of ultra-high purity silicon (Figure 3.2).

$$Si(s) + 3HCl(g) = SiHCl_3(g) + H_2(g) + \text{impurities} \tag{3.2}$$

$$SiHCl_3(g) + H_2(g) = Si(s) + 3HCl(g) \tag{3.3}$$

Figure 3.2
Ultra-pure silicon used for making computer chips.

Semiconductors are also used in many optical devices. These tend to be so-called compound semiconductors, such as gallium arsenide (GaAs) or indium phosphide (InP) which can both emit light (as lasers and light-emitting diodes) or sense it (as photodetectors).

Metal oxides also feature prominently in the electronics industry. These include the *perovskite*, barium titanate ($BaTiO_3$), and various rare-earth oxides, which are used as capacitors: zirconia (ZrO_2) and tin (IV) oxide (SnO_2) are used in gas sensors, and alumina (Al_2O_3) and beryllium oxide (BeO) are used as substrates for integrated circuits due to their good insulating properties.

3.2 Inorganic chemicals in medicine

Inorganic chemicals have been used in medicine since ancient times, although perhaps today we might question the efficacy of these remedies. One of the earliest examples was the use of mercury as a drug by Greek physician Hippocrates; later Paracelsus used mercurous chloride as a diuretic. In the 19th century, mixtures of gold compounds with sodium chloride were used to treat syphilis — unsuccessfully!

However, metal compounds do play a vital role in modern medicine. One of the most important breakthroughs was the discovery of the anti-tumour properties of platinum complexes, notably *cis-platin* (Structure **3.1**), by Barnett Rosenberg and his colleagues in the 1960s. Platinum drugs are now widely used in cancer chemotherapy, with particular success in the treatment of testicular and ovarian tumours. Other examples include sodium bicarbonate and magnesium and

3.1

aluminium oxides, which are used in antacid formulations, and lithium compounds, which are used in the treatment of manic depression. Gold compounds such as aurothioglucose have been found to be effective in the treatment of rheumatoid arthritis.

3.3 Inorganic chemicals as colours

Inorganic chemicals have been used for many centuries as pigments for decoration. Figure 3.3 shows a painting which dates back 25 000 years, found in the deep recesses of caves in Altamira, northern Spain. The Magdalenian artists used mixtures of coloured oxides from the earth, which they applied to the cave walls, probably mixed with water and fat, with their fingers. The reason why these early paintings survived the centuries is that they were not exposed to the destructive forces of wind, rain and light, problems which still beset the paint manufacturer of today.

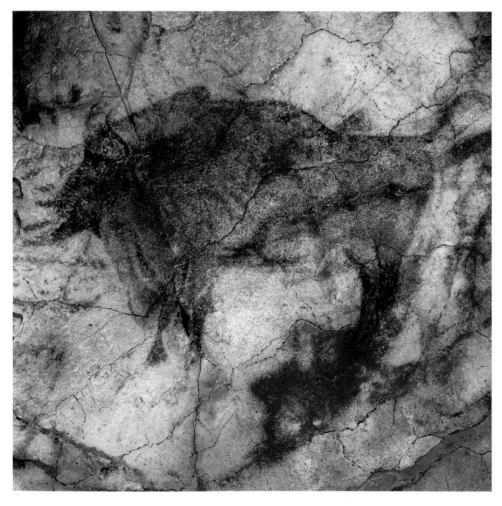

Figure 3.3
An example of a cave painting — a bison — found at Altamira, northern Spain.

Iron oxides, which were undoubtedly components of these prehistoric paints, still find application today as low-cost, low-toxicity pigments in paint, plastics, rubbers and cosmetics. The three main colours available are red (*haematite*; Fe_2O_3), yellow (*goethite*; FeO.OH) and black (*magnetite*; Fe_3O_4), and other colours can be produced by blending these three. The oxides are generally prepared by precipitation from aqueous solution:

$$4FeSO_4(aq) + O_2(g) + 8NaOH(aq) = 2Fe_2O_3(s) + 4Na_2SO_4(aq) + 4H_2O(l) \qquad (3.4)$$

The iron (II) sulfate is often obtained as a by-product from steel pickling or titanium dioxide manufacture.

A much wider range of colours is obtained from the cadmium sulfoselenide pigments. Cadmium sulfide (CdS) has a bright yellow colour, and by careful substitution of Zn or Hg for Cd, or Se for S, hues ranging from green through orange to deep maroon may be produced. Cadmium pigments are used for colouring plastics and in specialist paints; they are also combined with ground glass to form enamels. However, these compounds are highly toxic and it is likely the trend in the pigment industry will be to move away from these materials. For ceramic and glass applications, 'purple of Cassius' enamels based on colloidal gold[3] are less toxic, and can give colours in the range light pink to deep purple.

Lead compounds have also been widely used as pigments. Lead chromate, $PbCrO_4$, is the yellow pigment used in the double yellow lines on roads. *White lead*, $PbCO_3.Pb(OH)_2$ was for many years the pigment of choice for white paint. However, toxicity concerns meant that it was replaced by titanium dioxide (TiO_2), the most important white pigment in current use. TiO_2 is extremely efficient at scattering light due to its high refractive index,[3] and blends with other pigments to produce paler colours. To produce matt or semi-gloss effects the paint surface is rendered less reflective by incorporating large particles, such as calcium carbonate, barium sulfate or silica.

SUMMARY

This account of industrial inorganic chemical production is certainly not comprehensive. However, from this 'snapshot' you should be able to appreciate something of the scale and diversity of the industry and some of the commercial considerations underlying the production of these chemicals, which are so important to our everyday lives.

FURTHER READING

1   D. A. Johnson (ed.), *Metals and Chemical Change*, The Open University and the Royal Society of Chemistry (2002).

2   E. A. Moore (ed.), *Molecular Modelling and Bonding*, The Open University and the Royal Society of Chemistry (2002).

3   L. E. Smart and J. M. F. Gagan (eds), *The Third Dimension*, The Open University and the Royal Society of Chemistry (2002).

ACKNOWLEDGEMENTS

Grateful acknowledgement is made to the following sources for permission to reproduce material in this book:

Figure 2.1: Courtesy of BOC; *Figure 2.2*: David Keith Jones/Images of Africa Photobank; *Figure 2.4*: Courtesy of Brunnermond Ltd; *Figure 2.5*: Courtesy of Barry Miller; *Figures 2.9 and 2.10*: Courtesy of Chemfreight Dangerous Goods Training Ltd; *Figure 3.2*: Courtesy of Intel Corporation; *Figure 3.3*: Spanish Tourist Office.

Every effort has been made to trace all the copyright owners, but if any has been inadvertently overlooked, the publishers will be pleased to make the necessary arrangements at the first opportunity.

INDEX

Note: Principal references are given in bold type; picture and table references are shown in italics.

M

N

CD-ROM INFORMATION

Computer specification

The CD-ROMs are designed for use on a PC running Windows 95, 98, ME, 2000 or XP. We recommend the following as the minimum hardware specification:

Processor	Pentium 400MHz or compatible
Memory (RAM)	32MB
Hard disk free space	100MB
Video resolution	800 × 600 pixels at High Colour (16 bit)
CD-ROM speed	8 × CD-ROM
Sound card and speakers	Windows compatible

Computers with higher specification components will provide a smoother presentation of the multimedia materials.

Installing the CD-ROMs

Software must be installed onto your computer before you can access the applications. Please run INSTALL.EXE from either of the CD-ROMs.

This program may direct you to install other, third party, software applications. You will find the installation programs for these applications in the INSTALL folder on the CD-ROMs. To access all the software on the CD-ROMs you must install QuickTime and Adobe Acrobat™ Reader.

Running the applications on the CD-ROM

You can access the *Elements of the p Block* CD-ROM applications through a CD-ROM Guide (Figure C.1) which is created as part of the installation process. You may open this from the **Start** menu, by selecting **Programs** followed by **The Molecular World**. The CD-ROM Guide has the same title as this book.

The Data Book is accessed directly from the **Start | Programs | The Molecular World** menu (Figure C.2) and is supplied as an Adobe Acrobat document.

Problem solving

The contents of this CD-ROM have been through many quality control checks at the Open University and we do not anticipate that you will encounter difficulties in installing and running the software. However a website will be maintained at
    http://the-molecular-world.open.ac.uk
that details solutions to any faults that are reported to us.

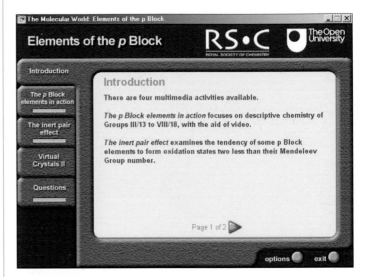

Figure C.1 The CD-ROM Guide.

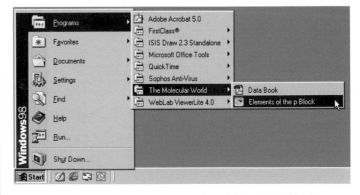

Figure C.2 Accessing the Data Book and CD-ROM Guide.